NATO Science for Peace and Security Series B: Physics and Biophysics

NATO Science for Peace and Security Series

Sub-Series B. Physics and Biophysics

Volumes in the Physics and Biophysics Sub-Series deal with scientific solutions to security threats arising from nuclear and conventional agents and weapons. Topics such as detectors, sensors, multisensor processing, explosion detection, biosensors and biotechnology, and nuclear safety and protection are covered in this sub-series.

Dorin Radu • Mirsada Hukić • Ashok Vaseashta
Editors

Countering Hybrid Threats Against Critical Infrastructures

Springer

Editors
Dorin Radu
Faculty of Civil Engineering
Transylvania University of Braşov
Braşov, Romania

Mirsada Hukić
Department of Medical Sciences
Academy of Sciences and Arts of Bosnia
Sarajevo, Bosnia and Herzegovina

Ashok Vaseashta
International Clean Water Institute
Manassas, VA, USA

ISSN 1874-6500 ISSN 1874-6535 (electronic)
NATO Science for Peace and Security Series B: Physics and Biophysics
ISBN 978-94-024-2303-7 ISBN 978-94-024-2304-4 (eBook)
https://doi.org/10.1007/978-94-024-2304-4

This Springer imprint is published by the registered company Springer Nature B.V.
The registered company address is: Van Godewijckstraat 30, 3311 GX Dordrecht, The Netherlands

Foreword

In general, critical infrastructure (CIS) refers to the systems, facilities, and assets that are vital for the functioning of society and the economy. These are considered vital because their disruption would impact public safety, security, and health or economic stability, including physical and virtual components that are interconnected and/or interdependent. There are 16 broad classifications of CIS sectors that are part of a complex, interconnected ecosystem, and such systems, facilities, and assets are vital to modern functional societies. Yet the CIS faces various threats, and any disruption to these sectors could have potentially debilitating national security, economic, and public health or safety consequences. The threats include cyber and physical attacks, natural disasters, supply chain disruption, public health emergencies, technological dependencies, and hybrid threats—which operate in conjunction with two or more threat vectors. The protection of CIS infrastructure is fundamental and strategic for national security, public safety, economic stability, and societal resilience. It is of vital importance to study and mitigate threat vectors and find ways to enhance the resiliency of such systems. An understanding of such threats, finding strategic solutions, and enhancing resiliency will render CIS safe, secure, and sustainable.

I am very pleased that NATO funded an Advanced Research Workshop (ARW), the first of its kind at the Transilvania University of Brasov, to bring several experts from NATO and partner countries to assess the scope and find strategic solutions to counter the threat to CIS. In total, 17 experts presented various aspects of threats to existing CIS and ways to mitigate such challenges. In addition, the ARW hosted three working groups dealing with hybrid threats, risks, and vulnerabilities; technology innovations to counter threat vectors; and risk assessment. The objectives of each working group were to identify problem sets and create an informational document consisting of solutions, recommendations, challenges, and improvements in current approaches. These are documented in the ARW proceedings, which consist of 16 chapters, including 3 working group recommendations. The chapters are authored by scientists and researchers who have made significant contributions to this emerging field. The chapters are well-written and are addressed to a broad readership. NATO administrators, CIS practitioners, experts, and the academic community, including students, will find these topics exciting and timely. I am certain that the readers will find the NATO ARW proceeding as a valuable tool for developing research projects.

I congratulate the editors and authors for this excellent work!

Dean, Faculty of Civil Engineering Ioan Tuns
Transilvania University of Braşov, Braşov, Romania
Member of Romanian Academy of Technical Sciences
Bucharest, Romania

Preface

In an increasingly interconnected and technologically advanced world, critical infrastructures—such as transportation, energy, healthcare, and communication systems—form the backbone of modern society. However, these vital systems are not immune to emerging threats, particularly hybrid threats that blend conventional and unconventional tactics to exploit vulnerabilities.

Critical infrastructure refers to the essential systems and assets vital to a nation's security, economy, public health, and safety. It encompasses various sectors, including energy, water, transportation, healthcare, information technology, food, finance, and emergency services. Each of these infrastructures plays a crucial role in society's functioning, as it provides the necessary services and support for daily life. However, these critical infrastructures face aperiodic threats from hybrid tactics that combine conventional and unconventional methods to challenge national and international security. Hybrid threats include cyber-attacks that target critical infrastructure, such as ransomware attacks on hospitals, disinformation campaigns aimed at manipulating public perception, terrorism that employs traditional military tactics and cyber means, economic coercion used to destabilize nations, and proxy warfare through third parties alongside cybersecurity operations.

The interplay between critical infrastructure and hybrid threats reveals significant vulnerabilities, as these systems are often interconnected. An attack on one sector, like energy, can have cascading effects on others, such as healthcare, leading to widespread disruption. The goal of hybrid threats is to create fear and disrupt normal operations, which can result in financial losses and diminish public confidence.

To combat these challenges, it is essential to implement a range of mitigation strategies that focus on enhancing the resilience of critical infrastructure. Regular risk assessments can help identify vulnerabilities, while advanced cybersecurity measures protect IT systems. Additionally, public–private partnerships can strengthen security frameworks, and crisis management plans can be developed and rehearsed for potential hybrid threat scenarios. Understanding and addressing the interconnections between critical infrastructure and hybrid threats is crucial for ensuring national security and public safety.

This proceedings volume is the result of the Advanced Research Workshop (ARW), *Countering Hybrid Threats Against Critical Infrastructures*, an event which was organized in Braşov from 21 to 23 May 2024, by the Transilvania University of Braşov, Faculty of Civil Engineering together with the Academy of Sciences and Arts from Bosnia and Herzegovina. The activity was supported by the NATO Science for Peace and Security (SPS) program to promote dialogue and practical cooperation between NATO member states and partner countries based on scientific research, technological innovation, and knowledge exchange to support civil security-relevant activities that respond to NATO's strategic objectives. The goal was to raise awareness about the significance of hybrid threats to infrastructure and to propose solutions for countering and preventing these threats through collaborative efforts among participating countries. The event brought together academics, policymakers, and end-users to evaluate requirements and perceptions and to consider ways that the scientific community can provide the solution pathways to meet such challenges. Various academics, including experts with international reputations and expertise in structural integrity, risk management applied in structures, vulnerability assessment, and artificial intelligence fields, contributed to this volume, to review the state-of-the-art and evaluate the latest developments and technological innovations through the exchange of information and best practice.

The book consists of 16 chapters that evaluate solutions for developing more effective governance in support of critical infrastructure through awareness of the importance of infrastructure hybrid threats, presenting solutions by countering and preventing various types of threats through collaborations between countries, and assessing an overview of existing defense tools and resources. The contents of this volume result from the collaboration between scientists and experts from NATO and partner countries in sharing good practices and evaluating strengths and weaknesses in counteracting terrorist threats in critical infrastructures.

The first three chapters present the results of the three ARW Working Groups (WGs)—first related to critical infrastructures and modern societies, second related to hybrid threats, risks, and vulnerabilities, and the third deals with technological innovations to counter threat vectors. A chapter by Gor et al. describes an important aspect of water and wastewater critical infrastructure and ways to protect them from hybrid threats. A chapter by Bolgen et al. discusses technological innovations for detection, decontamination, and protection against chemical and biological threats. A review by Bancila et al. provides an innovative technological solution using temporary fiber-polymer-filled composite bridges for disaster recovery and military operations. Radu et al. provide an assessment of existing steel bridges which pose a great challenge to critical infrastructure. Rocha et al. discuss cybersecurity maturity models and the challenges and opportunities associated with their implementation. The rest of the book is complemented by various other aspects of critical infrastructure protection that were discussed during the advanced research workshop.

Threats against critical infrastructures are diverse and evolving, encompassing a range of risks from cyber-attacks and physical sabotage to natural disasters and emerging technology vulnerabilities. The interconnected nature of these infrastructures means that disruptions in one sector can have far-reaching impacts on others, posing significant challenges to national security and public safety. To effectively mitigate these threats, a comprehensive approach is essential, involving risk assessment, enhanced cybersecurity measures, regular personnel training, and strong academic–stakeholder partnerships. Through the result of the authors' input, this volume attempts to succeed in focusing on enhancing the security and resilience of critical infrastructures by addressing the diverse aspects of threats.

Brașov, Romania Dorin Radu
Sarajevo, Bosnia and Herzegovina Mirsada Hukić
Manassas, VA, USA Ashok Vaseashta

Contents

About the Authors

Mirsada Hukić, MD, PhD, is a Professor Emeritus at the University of Sarajevo, specializing in microbiology and immunology, with a subspecialty in virology. She is a member of the Academy of Sciences and Arts of Bosnia and Herzegovina (Department of Medical Sciences), the European Academy of Sciences (Medicine and Life Science Division), the Bosnian-Herzegovinian American Academy of Arts and Sciences, and the International Bioethics Committee of UNESCO. Additionally, she is the Director and Founder of the Institute for Biomedical Diagnostic and Research Analysis and the Director of the Center for Disease Control and Geohealth Studies at the Academy of Sciences and Arts of Bosnia and Herzegovina. She serves on the Board of Directors at the Bosniak Institute Adil Zulfikarpašic Foundation in Sarajevo and the Bosniak Institute Stiftung Adil Zulfikarpasic in Zurich. She is also a former President of the Special Olympics of Bosnia and Herzegovina. Professor Hukic has over 50 years of extensive experience in clinical diagnostics of infections and research on the ecology, etiology, and kinetics of microorganism relationships among various living organisms. Her work explores the role of micro-organisms in processes that can be both beneficial and harmful to life on Earth. Emerging and re-emerging pathogens, synthetic microorganisms, and new diseases hold a special place in her research.

Ashok Vaseashta, PhD, D.Sc, Acad**,** serves as the CEO/CTO and Executive Director of Research at the International Clean Water Institute in Virginia, USA. He also holds several honorary positions, including Professor at Transylvania University of Brasov; Chaired Professor of Nanotechnology at the Ghitu Institute of Electrical Engineering and Nanotechnologies, Academy of Sciences of Moldova; Academician at the Euro-Mediterranean Academy of Arts and Sciences; and Senior Strategic Research Advisor to several organizations. Inspired by nature and guided by societal necessities, he strives for technological innovations to address the global challenges of the twenty-first century. He is a scholar, visionary, strategist, and dedicated futurist providing strategic leadership to promote and advance research initiatives and priorities using data-driven decisions. He received his PhD from Virginia Polytechnic Institute and State University, Blacksburg, VA, in 1990, followed by Kobe's postdoctoral fellowship. Following his PhD, he served as a Professor and Researcher at Virginia Tech and Marshall University. He also served as the Director of Research at the Institute for Advanced Sciences Convergence and International Clean Water Institute for Norwich University's Applied Research Institutes, Vice-Provost (Rector) for Research in South Carolina, and Executive Director and Chair of the Institutional Review Board at a State University in New Jersey. His earlier honorary positions included a Visiting Professor at the Riga Technical University, Latvia, 3 Nano-SAE Research Centre, University of Bucharest, Romania, and Transilvania University of Brasov, Romania, and a Visiting Scientist at the Weizmann Institute of Science, Israel. He served two rotations at the U.S. Department of State, first as a strategic S&T advisor in the Bureau of International Security and Nonproliferation, Office of Weapons of Mass Destruction and Terrorism, and subsequently as a U.S. diplomat. His research interests span foresight, nanotechnology, environmental and ecological science, green chemistry, the environmental impact of micro-/nano-plastics, and critical infrastructure safety and security—all using the nexus of advanced technological solution platforms. He is the author and editor of 19 books and has published over 350 articles in scientific journals, book chapters, and at conferences.

He serves on the editorial boards of several international journals and is an active member of various professional organizations.

Dorin Radu, PhD, is Associate Professor at the Faculty of Civil Engineering, Transilvania University of Braşov, where he serves as Vice Dean for Research. He received his PhD in civil engineering from the Polytechnical University of Timişoara, having as cotutors Prof. Radu Băncilă and Prof. Aleksandar Sedmak from the University of Belgrade. Since 2003, he teaches courses at the Faculty of Civil Engineering in Braşov, covering various fields, with focus on static and stability of structures, theory of elasticity and plasticity, steel structures, and building information modeling. He is a specialist in structural integrity and has many contributions in research directions dealing with structural assessment, neural networks applied in civil engineering, new materials for sustainable constructions, environmental protection, and energy conservation. He is the author of more than 60 scientific papers, 30 of which have been published in renowned scientific journals—Q1 and Q2 quartiles. He serves as director or member for several research projects focused on structure assessment, material behavior, and sustainability. He participated in a large number of conferences as a keynote speaker, where he presented his research with the results being chosen for publication in renowned scientific journals. He is a member of the European Structural Integrity Society and has been invited to hold lectures at several universities in Romania, Serbia and Croatia. He is a member of Romanian's national technical committees for norms and standards. He is an organizer, editor, and reviewer for international conferences and reputable international journals.

Chapter 1
Critical Infrastructures and Modern Societies

Konstantinos Papatheodorou, Carmine Baruffo, Nimet Bölgen, Marijana Hadzima-Nyarko, Ashok Vaseashta, Alina Bărbulescu, Dorin Radu, Konstantinos Ntouros, Elena Tzanou, and Mirko Rocca

Abstract As our world becomes increasingly interconnected, the concept of critical infrastructure has become paramount in understanding the resilience of modern societies. Critical infrastructures, such as power grids, communication networks, and transportation systems, are the backbones of societal sustainability, safety, and security. These infrastructures are not only essential for the day-to-day functioning of communities, but they also play a crucial role in enabling economic competitiveness, public health, and the ability to cope with crises. However, the growing complexity and interdependence of these critical systems have also introduced greater vulnerabilities. The disruption of one critical infrastructure can have cascading effects on others, leading to widespread societal disruption. This phenomenon is exacerbated by the reliance on automated control systems and the increased connectivity of these systems through the internet, which further exposes them to various threats, including malevolent attacks, natural disasters, and random failures. To address these challenges, it is essential to understand the operational environment of

---

K. Papatheodorou · K. Ntouros · E. Tzanou
International Hellenic University, Nea Moudania, Greece

C. Baruffo
NATO Centre of Excellence Defence Against Terrorism, Norfolk, VA, USA

N. Bölgen
Mersin University, Mersin, Türkiye

M. Hadzima-Nyarko
Josip Strossmayer University of Osijek, Osijek, Croatia

A. Vaseashta
International Clean Water Institute, Applied Research, Manassas, USA

A. Bărbulescu · D. Radu (✉)
Transilvania University of Braşov, Braşov, Romania
e-mail: dorin.radu@unitbv.ro

M. Rocca
Riga Technical University, Rīga, Latvia

D. Radu et al. (eds.), *Countering Hybrid Threats Against Critical Infrastructures*, NATO Science for Peace and Security Series B: Physics and Biophysics, https://doi.org/10.1007/978-94-024-2304-4_1

critical infrastructures and the potential threats they face. The paper aims to present a review of critical infrastructure threats through modern society's countermeasures tools.

Keywords Critical infrastructure · Resilience · Threats · Risk assessment

1.1 Introduction

Critical infrastructure can be defined as indispensable assets and facilities that play a vital role in a nation's societal, economic, and security well-being, covering areas such as energy, telecommunication, water, healthcare, transportation systems, and more [1]. These critical infrastructures are essential for ensuring the uninterrupted operation of society and are vulnerable to a range of risks including technical malfunctions, natural calamities, and cyber incursions [2, 3].

Safeguarding and fortifying critical infrastructure is of utmost importance for national security, given that any disturbance could directly impact the stability and safety of a country, thereby necessitating comprehensive strategies for crisis management [4, 5]. The interconnection of different sectors of critical infrastructure, makes resilience assessment a quite complex and difficult task, underscoring the need for a human-centric perspective in evaluating and improving the resilience of these systems.

The first formal federation definition of Critical Infrastructure was developed in 1996 when President Clinton signed Executive Order EO-13010 which designated eight government critical information sectors [6]. In 2013 the White House released Presidential Policy Directive PPD-21 [7], which defined the designated critical infrastructure sectors and the sector-specific-agencies. PPD-21 identifies 16 critical infrastructure sectors built on the extensive work done to date to protect critical infrastructure and describes a national effort to share threat information, reduce vulnerabilities, minimize consequences, and hasten response and recovery efforts related to critical infrastructure (Fig. 1.1).

Nowadays, Critical Infrastructure is recognized by the majority of governments as the body of systems and networks—physical or virtual assets that are essential such that their continued operation is required to ensure the security of the state, nation, its economy, and the public's health and/or safety. Critical infrastructures are facilities and services vital to the basic operations of society. Sectors that are considered "critical infrastructure" vary among different countries, but most would comprise: energy, water, food, transport, telecommunications, health care, as well as banking and finance.

Any physical or virtual information system that controls, processes, transmits, receives, or stores electronic information in any form that is vital to the functioning of a society's economy, public health, safety, security, or any combination of the above, including the food and agriculture sectors, transportation systems, energy and water supply, financial services, telecommunications & broadband internet, defense, and more.

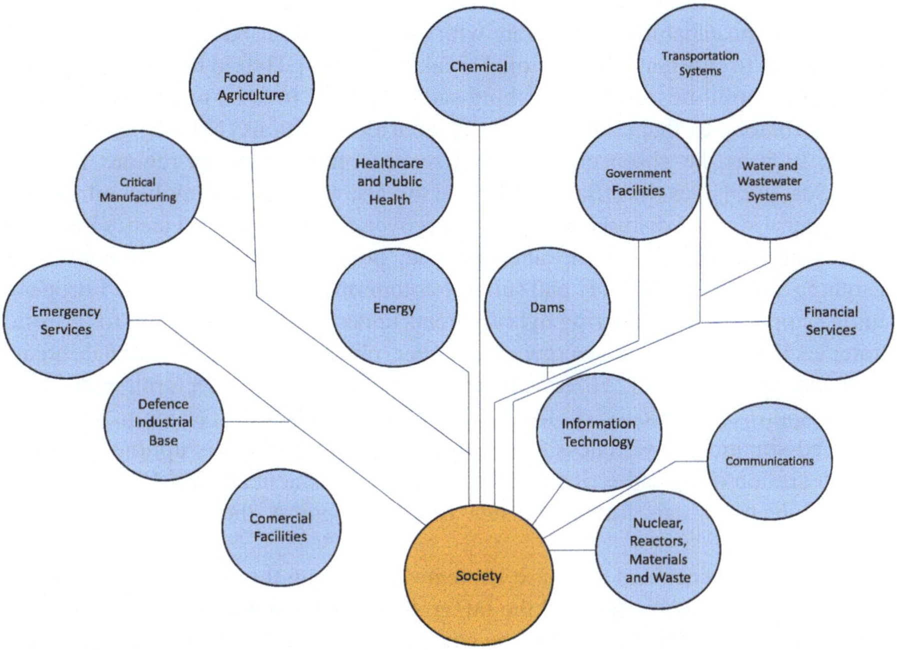

Fig. 1.1 Critical infrastructure sectors in the US according to PPD-21

1.2 Hybrid Threats and Critical Infrastructures

The landscape of global security has undergone a significant transformation in recent decades, with the emergence of hybrid threats posing new challenges to the resilience of critical infrastructures. The increasing reliance on interconnected digital systems and the growing integration of cyber and physical domains have amplified the vulnerability of critical systems, rendering them susceptible to a diverse array of malicious actors and disruptive events [8–11].

One of the defining characteristics of hybrid threats is their ability to leverage a combination of conventional and unconventional tactics, blending military, political, economic, and technological means to undermine the stability and functioning of targeted systems [8, 11].

Hybrid threats can significantly affect critical infrastructure by exploiting the interconnectedness and dependencies of various systems, often leading to cascading failures and widespread disruptions. These threats encompass a variety of strategies, including cyber-attacks, physical sabotage, and information warfare, that can aim at crucial societal functions like energy, finance, transportation, and communication systems [12, 13]. The growing intricacy and digital transformation of critical infrastructures, such as the incorporation of Industrial Control Systems (ICS) and the utilization of emerging technologies like IoT and cloud computing, render them more susceptible to such multifaceted assaults [14, 15]. For example, cyber-attacks on ICS have the capability to interfere with the supervision and regulation of distant sites, leading to substantial operational breakdowns [16]. Hybrid threats also exploit human susceptibilities, such as phishing and/or bribery, in order to penetrate critical systems, thereby further complicating the security conditions [17]. The geopolitical setting, such as the absence of regional security frameworks in regions like the Black Sea, intensifies the efficacy of hybrid threats, which aim at civilian infrastructure to cause societal disturbances [18]. Moreover, the use of hybrid tactics blurs the distinction between peacetime and wartime, posing a challenge for nations in designing and applying timely and suitable countermeasures [19, 20]. The European Union's approach to countering hybrid threats underscores the necessity for holistic strategies that tackle the convergence of cyber, physical, and informational threats to critical infrastructure [21]. To counter hybrid threats, a multisectoral, multidisciplinary approach is necessary, focusing on resilience-based decision-making and enhanced situational awareness to manage and recover from disruptions [14, 19]. The integration of AI and serious gaming in risk assessment and the design of resilient architectures can further bolster defenses against these complex hybrid threats [22].

Efficient resilience and response strategies, including the development of adaptive decision-making models and the integration of situational awareness tools, are extremely important for mitigating the consequences of these threats and for ensuring the protection of critical infrastructure.

1.2.1 Critical infrastructure Protection Against Hybrid Threats

Safeguarding vital infrastructure from blended risks demands a versatile strategy that merges diverse tactics and approaches. A holistic decision-making scheme, like the fusion of the OODA loop (Fig. 1.2) and PDCR (Plan for the Dissemination and Communication of the action results) frameworks, can boost flexibility and swift reactions to cyber threats, thereby ensuring cyber resilience [15]. Employing

Fig. 1.2 OODA loop

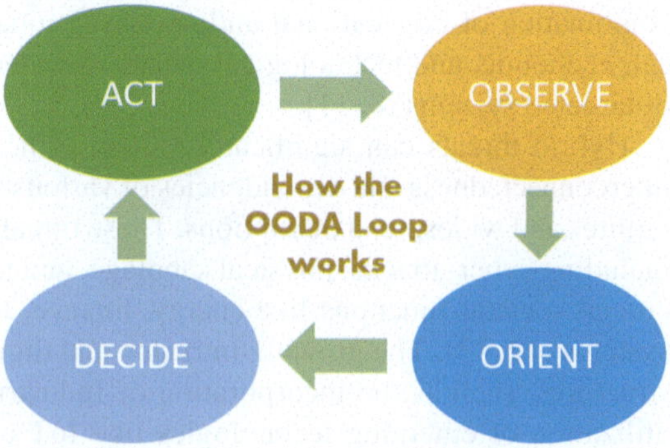

distributive identification and dynamic programming can elevate the dependability of managerial choices concerning the security status of information systems, enabling prompt and efficient countermeasures against external cyber-attacks [23].

Decision-making based on resiliency, taking into account the dependencies and social resilience of digital systems, plays a pivotal role in tackling the far-reaching consequences of hybrid threats across various sectors [12]. Proactively utilizing non-lethal agents and upholding international humanitarian law, can additionally fortify the safeguarding of critical infrastructure components during hybrid conflicts [14]. The European Union's systematic strategy to combatting hybrid threats, encompassing economic, legislative, political, and informational instruments, underscores the significance of a comprehensive comprehension and united response to such threats [21]. Ensuring awareness, resilience, and response capabilities is crucial, given that hybrid warfare blurs the distinction between peacetime and wartime, posing challenges in devising suitable countermeasures [24]. The Black Sea region's encounter with hybrid warfare emphasizes the necessity for a Critical Infrastructure Protection (CIP) framework to heighten resilience [17]. Realistic scenarios and proactive measures before threats emerge, can assist in forecasting and reducing their impact, thus ensuring effective recovery during crises [25]. In conclusion, evaluating and enacting strong laws, such as Poland's National Critical Infrastructure Protection Program, is crucial for establishing a firm basis for safeguarding critical infrastructure [26].

1.2.2 Chemical and Biological Threats

The increased use of chemical and biological agents in the present-day era spans various sectors, including agriculture, medicine, industry, and even personal care products. While the abundance of such products is a sign of technological advancement, such agents also pose potential risks to human health, the environment, and global security. From a security standpoint, the dispersion of these agents can occur naturally, incidentally, accidentally, and intentionally. The increasing complexities in national security necessitate the continuous evolution within the sphere of chemical and biological threat detection, decontamination, and protection. Countering chemical and biological threats involves complex challenges including the type of agent used, level of exposure to warfare agents, and contamination. For chemical agents, typically, it is their concentration above LD50 (median lethal dose required to kill a member of the tested population). A lower LD_{50} is indicative of higher toxicity. However, for biological agents, the symptoms appear after a certain period, rendering attribution a difficult process. Furthermore, due to a very large variety of chemical and biological agents, reactions arising from varying concentrations, response of mixtures, and also variation in response time of chemical and biological agents, it is beyond the scope to present an exhaustive summary. While we benefit from the technological innovations and availability of a variety of products, it is necessary to develop contingencies for safeguarding, especially emergency responders, warfighters, and safety personnel. Specially designed personal protective systems that can protect their user not only passively, but actively, and can detect, adsorb, degrade, and decontaminate chemical/biological warfare agents. Realizing that biological warfare agents can be microorganisms, and chemical agents can be toxins, organometallics, toxic industrial chemicals (TICs), toxic industrial materials (TIMs), or explosives, typically used in unconventional battlefields to kill, seriously injure, or incapacitate exposed individuals, different chemical, biological, and force protective gears are designed for such specific purposes. Additional protective measures against chemical and biological agents are respirators, skin protection, detection, decontamination, and medical countermeasures. Many chemical and biological warfare agents are respiratory and dermally active. Protective fabrics provide barrier protection by adsorbing or blocking toxic agents. The development of protective materials includes catalytically active nanofibers, sensors, and selective gas permeating properties. Electrospun nanofibrous materials enable the development of effective systems that offer chemical and biological agents' detection, decontamination, and protection. Additional countermeasures include bioinformatic tools, regular training, and chemical-biological-radiological-nuclear (CBRN) reservists who specialize in CBRN protection and are available on a volunteer basis in the event of an emergency. These reservists work in close coordination with medical personnel to develop training exercises, and inventory of countermeasures fully stocked for emergencies. The CBRN reservists are trained for improved explosive devices (IEDs, also termed as CBRNE (E for explosives)) and are trained in sensors and detectors to detect the release of strong hydrocarbons. Governments and international bodies have established regulations to control the use and disposal of these agents. However, enforcement and global cooperation are essential to mitigate risks. Public awareness and education about the benefits and risks of chemical and biological agents are crucial for informed decision-making and responsible use. Hence, while CRRNE agents have brought significant advancements in various fields, their increased use necessitates careful management to avoid negative consequences for human health, environmental sustainability, and global security.

1.2.3 Countering Hybrid Threats

Countering hybrid threats requires a multidisciplinary approach for which it is crucial to define what hybrid threat and critical infrastructure are. The scientific community, as a key stakeholder, could make a valuable contribution by defining and designing a structured and prioritized approach to categorize critical infrastructure and examining its specific vulnerabilities and risks.

Not all critical Infrastructure systems are the same. While there are common categories, including communication, transportation, food and water, and energy, their criticality could differ from country to country. The risk analysis process further helps to assess the risks based on the peculiar vulnerabilities of each infrastructure while making recommendations on how to mitigate and manage the associated risks.

From this perspective, the "Risk Assessment Methodologies for Critical Infrastructure Protection" edited by the European Commission Joint Research Centre Institute for the Protection and Security of the Citizen is a valuable tool for risk assessment. Even though this offers 21 different assessment methodologies, new Emerging Disruptive Technologies (EDT) could require new methodologies and approaches.

Defending critical assets from hybrid threats is a challenging effort. A congested threat environment requires resilience as part of a wider approach in which security is applied by design. The EU and NATO share a common interest in preventing disruptions to critical infrastructure that provides essential services to citizens and supports their economy. Through the work of the NATO-EU Task Force on the resilience of critical infrastructures, they have confirmed that they have a shared view of the potential threats and risks. The conclusions and recommendations made by the task force are a good place to start.

Developing advanced research means thinking "over and above"—trying to guess how technologies, including Artificial Intelligence, Cyber, and Bio-Nano technologies, could represent a threat or an opportunity for worldwide peace and security. Being aware of the current threats and their possible evolutions remains crucial. UAVs and 3D printing pose a challenge for critical infrastructure protection because they are technology that is inexpensive and easy to access. Furthermore, the Internet of Things and the Internet of Behaviors could represent associated risks since backdoors could exploit cyber vulnerabilities.

Critical Infrastructure Protection encompasses structures, systems, and processes, and it requires an overarching approach that should not dismiss the establishment of a security organization tailored for the protection of critical infrastructure, a change in the security mindset, the training of personnel, and of course, the management of sufficient resources.

Key factors of success for Critical Infrastructure Protection remain the prevention of assessed risks. The main effort and contribution from the scientific community could come from the development of new technologies or assets capable of deterring, detecting, and denying physical or virtual threats against critical infrastructures, realizing an effective risk management approach, enhanced by the Public-Private partnership and information sharing.

1.3 Public Response to Hybrid Threats

Public reaction to hybrid challenges is shaped by various elements that differ significantly from responses to conventional challenges. Hybrid challenges, blending traditional and unconventional strategies, exploit weaknesses in cyber, physical, and human realms, rendering them complex as compared to conventional threats [27, 28]. The public's viewpoint and response are greatly influenced by the information sphere, where misinformation and propaganda play pivotal roles in shaping perspectives and provoking actions [16, 24]. Anxiety and defensive reactions are elevated in hybrid challenges due to their secretive and unpredictable nature, leading to notable fear and psychological impacts among the population [29]. The interconnectivity of modern technologies and vital infrastructures also implies that disruptions can have far-reaching and immediate consequences, further complicating and enhancing public reactions [30, 31]. Strategic communication and resilience are crucial in lessening these implications, given that standard law enforcement measures often prove to be inadequate [32, 33]. Moreover, the geopolitical background highlights the involvement of both state and non-state entities in utilizing hybrid approaches to gain strategic benefits, thus impacting public opinion and governmental strategies [34]. A unified response to hybrid challenges demands a holistic strategy encompassing enhancing cybersecurity, strategic communications, and public awareness to improve the Public's capacity to respond efficiently during emergencies. In contrast to traditional threats which may typically require direct military or law enforcement actions, hybrid threats necessitate a deeper understanding of the socio-political landscape to understand their diverse nature in order to coordinate efforts across the various sectors involved.

1.3.1 Assessing Hybrid Threat Risks—The Role of State Authorities

State authorities have a pivotal role in evaluating hybrid threat risks through designing strategic messages, boosting situational awareness, and utilizing analytical instruments for gauging and alleviating these risks. They must concentrate on developing a knowledge structure to shape public strategies in evaluating risks linked to the information sphere and misleading campaigns [35]. It is essential for authorities to foster international cooperation and regulations to effectively address hybrid threats targeting critical infrastructures [28]. Safeguarding worldwide security demands joint reactions to intricate challenges, including hybrid threats, a fact that calls for a thorough analysis of the nature of these risks [30]. Developing models for analyzing hybrid risks, like the suggested hyper-forest of multi-trees concept, aids in comprehending enemy tactics, thereby assisting in devising suitable countermeasures [36]. NATO and its member nations should adjust to legal hurdles tied to data collection, intelligence exchange, and susceptibility assessment to reinforce societal strength against hybrid threats [37]. The use of composite indicators, such as the Normandia Index, provides an analytical instrument for predicting risks and supporting policy formulation on a national scale [38]. Managing cybersecurity risks also contributes, by utilizing public repositories and classification instruments to quantify possible risks and build risk evaluation systems [39]. The emotional and psychological impacts of hybrid threats, need also to be taken into account, as they sway public responses and policy choices [29]. Democratic states confront the obstacle of harmonizing surveillance for security with safeguarding civil liberties, necessitating a thorough analysis and execution of responsibility mechanisms [40]. In general, state authorities should merge these varied strategies to effectively evaluate and mitigate the multi-dimensional risks brought by hybrid threats.

1.3.1.1 Public Behavior

To that end, analysis of the factors that affect Public behavior during emergencies is quite important. Public behavior during emergencies is influenced by a multitude of psychological factors. The way individuals perceive risks, their emotions, level of recognition affect the way they respond during crises [41]. A person's psychological state, with feelings of tension and panic, greatly influences the specific needs for psychological support, where factors like age, gender, profession, and educational background must also be considered [42]. According to the Compensatory Control Theory, individuals seek order and predictability when their sense of control is threatened a fact that drives them to behaviors like belief in conspiracy theories and irrational thinking [43]. Moreover, the amplification of risk and emotional mechanisms also shape public psychological reactions, with the disclosure of information, trust in government, and social assurance standing as crucial factors [44]. Positive personality characteristics and emotions, coupled with a supportive social setting, nurture selfless actions that can counteract adverse consequences during crises [45]. Strategies employed by the government to release information, influence the development of public opinion by impacting perceptions such as perceived risk and emotional position [46]. The degradation of the public sense of safety emerges as another important element, impacted by governmental and media responses, individual comprehension of risk and readiness for emergencies, and collective psychological behavior [47]. Social identity and standards play a central role in promoting collective support, where a shared identity fosters unity and solidarity among people, ultimately enhancing collective psychosocial strength [48]. Understanding all these conditions is imperative to comprehend in order to plan for efficient crisis management and to enhance public and community resilience during emergencies.

1.3.1.2 Public Communication

Public communication campaigns have displayed varying degrees of effectiveness in addressing hybrid challenges and shaping public behavior, predominantly dependent on their blueprint, execution, and the fusion of strategic principles. These initiatives are formulated to educate and influence behaviors on a large scale within a specified timeframe, utilizing diverse communication efforts and messages across multiple platforms [49]. Through the utilization of distinct communication strategies like guiding, showcasing, rallying, and engaging, these initiatives try to stimulate societal transformation and accomplish the campaign objectives [50]. When it comes to hybrid threats, the European Union (EU) employs strategic messaging and sanctions as central policy mechanisms to effectively tackle destabilization challenges [51]. The effectiveness of these initiatives lies in their ability to categorize audiences based on their readiness to adopt recommended behaviors, ensuring tailored and targeted communication to efficiently shape public perceptions and responses towards hybrid threats and improve Public capacity to respond [52].

1.3.2 Natural Disasters and Critical Infrastructures

Disasters are natural (geophysical and other processes in nature) and man-made (anthropogenic disasters) events that have a negative impact on people, societies, and the environment.

Natural disasters as harmful events for people, their material assets (critical infrastructures), and the environment that occur on/in different spheres of the Earth (lithosphere, hydrosphere, atmosphere, and biosphere) [53]. They can therefore be categorized as follows: geo-hazards (such as earthquakes, volcanic eruptions, and tsunamis), meteorological hazards (such as storms and extreme temperatures), hydrological hazards (such as floods and landslides), climatological hazards (such as drought and forest fires), biological hazards (such as epidemics and insect infestations) and extraterrestrial hazards (such as air eruptions and space weather) [54].

All natural disasters cause severe damage to critical infrastructure [55]. Any breach in infrastructure poses a serious threat to the proper functioning of these systems and can lead to property damage, human casualties, and significant economic losses [56].

Today's world, characterized by an ever-increasing reliance on digital technologies, has also seen an alarming increase in the frequency and severity of natural disasters. A sometimes-neglected consequence is their significant impact on cyber security.

Infrastructure vulnerability, communication breakdown, data loss, phishing and fraud, and vulnerable supply chains are some examples of how natural disasters and cybersecurity are intertwined. Natural disasters often disrupt critical infrastructure such as power lines, communication networks, and data centers. When these critical systems fail, businesses and people become vulnerable to cyberattacks. When infrastructure becomes unstable, communications can fail, leading to a lack of coordination between response teams. In such cases, cybercriminals can take advantage of the chaos and confusion by spreading disinformation or launching attacks on emergency services. Physical damage to data centers and storage facilities can lead to data loss [54].

Critical infrastructure protection is one of the most important measures to mitigate the consequences of natural disasters. The main objective of protecting critical infrastructure from the effects of natural disasters is to maintain the continuity of its operations [56]. Reducing the impact of natural disasters on people and critical infrastructure includes measures to prevent or reduce the possibility of physical threats and social disruption [57]. There are two predominant types of natural disaster impact reduction: structural and non-structural. Structural mitigation refers to the design (e.g. seismic design codes), construction, maintenance, and renovation of physical structures and infrastructures to withstand the physical forces and impacts of natural disasters. In contrast, non-structural mitigation is concerned with reducing the vulnerability of the population, physical structures, and infrastructure to hazardous conditions. Non-structural mitigation approaches include legally adopted urban planning measures that take into account the potential impact of disasters, regulating development in high-risk zones such as landslide-prone hillside terrain and coastal areas targeted by storm waves, etc. [58].

1.4 Conclusions

The modern world has become inextricably linked, with critical infrastructure playing a vital role in the function and security of our societies. This reliance on advanced digital technologies and networked systems has ushered in significant gains in service delivery and economic productivity, yet has also rendered these critical systems vulnerable to manipulation by malicious actors.

The growing dependence of communities on the seamless operation of power, water, transportation, and communication networks to sustain daily life has paradoxically increased the susceptibility of these interdependent infrastructure systems to disruption, as the compromise of one network can cascade to impact the performance of others.

The integration of cyber-physical systems across critical infrastructure has introduced a diverse array of vectors through which nefarious groups and state actors can seek to undermine the stability and resilience of modern societies.

With the increasing frequency and severity of natural disasters, the need for effective cybersecurity measures during and after them is growing. A proactive approach to cybersecurity can help people, organizations, and governments better prepare for the perfect storm when it occurs. This includes disaster recovery plans, redundancy (implementing redundant systems and data backups), employee training (training employees to recognize phishing attempts and fraud), secure remote working, and collaboration (encouraging public and private sector cooperation in disaster response).

The relationship between natural hazards such as earthquakes, and hybrid threats to critical infrastructure is inherently complex, wherein natural hazards present an immediate physical menace to infrastructure, whereas hybrid threats capitalize on vulnerabilities within both physical and digital spheres.

It is thus apparent that a multidisciplinary and multisectoral, holistic approach that integrates seismic risk management with initiatives aimed at countering hybrid threats, is necessary to effectively mitigate these combined risks.

This holistic strategy must incorporate risk assessment at preliminary stages and the seamless integration of safety and security protocols throughout all phases of infrastructure design and operation. Furthermore, the augmentation of situational awareness and preparedness is of paramount importance.

The General Public serves as a pivotal actor in the mitigation of risks associated with both earthquakes and hybrid threats through diverse mechanisms, encompassing perception, participation, and collaborative efforts with governmental and scientific entities. The engagement of the public can significantly bolster the efficacy of disaster preparedness and response strategies; thus, the public's reactions to hybrid threats and natural disasters such as earthquakes, can profoundly impact the safety and resilience of critical infrastructures.

In pursuit of this objective, it is essential to address the challenges and principal issues that the public encounters during earthquakes and other emergencies, including but not limited to panic, traffic congestion, unfamiliarity with safe/refuge areas, insufficient knowledge regarding evacuation routes, reluctance to partake in risk mitigation efforts, non-adherence to guidelines, and a deficit of education and comprehension concerning the associated risks.

There exists a pressing need to inspire the public to cultivate a greater interest in acquiring knowledge about and engaging in risk mitigation activities, which includes personal and familial preparedness planning and readiness for emergency response through organization and training.

References

1. Slavimir V, Martin B (2023) Cyber security of critical infrastructure. Pravo 40(2):77–88. https://doi.org/10.5937/ptp2302077
2. Mengling T (2022) Critical infrastructures overview: past, present and future. Sustainability 14(4):2233–2233. https://doi.org/10.3390/su14042233
3. Lehto M (2022) Cyber-attacks against critical infrastructure. In: Lehto M, Neittaanmäki P (eds) Cyber security. Computational methods in applied sciences, vol 56. Springer, Cham, pp 3–42. https://doi.org/10.1007/978-3-030-91293-2_1
4. Ewa Ż (2020) Critical infrastructure in the shaping of national security. Sec Def Q 28(1):70–81. https://doi.org/10.35467/SDQ/118585
5. Furuta K, Komiyama R, Kanno T, Fujii H, Yoshimura S, Yamada T (2022) Resilience analysis of critical infrastructure. WSEAS Trans Comput 21:58–65. https://doi.org/10.37394/23205.2022.21.8
6. Clinton William J (July 15, 1996) Executive order eo 13010 critical infrastructure protection. FAS.org. Federation of American Scientists. Retrieved April 2, 2016 (figure from Gartner Inc. 2022)
7. PPD-21 Presidential Policy Directive—Critical Infrastructure Security and Resilience, White House, 12 Feb 2013
8. Goldman H, McQuaid R, Picciotto J (2011) Cyber resilience for mission assurance. In: 2011 IEEE international conference on technologies for homeland security (HST), Waltham, MA, pp 236–241. https://doi.org/10.1109/ths.2011.6107877
9. Linkov I, Kott A (2018, January 1) Fundamental concepts of cyber resilience: introduction and overview. Cornell University. https://doi.org/10.48550/arxiv.1806.02852
10. Román R, Nájera P, López J (2011, September 1) Securing the Internet of things. IEEE Comput Soc 44(9):51–58. https://doi.org/10.1109/mc.2011.291
11. Warkentin M, Willison R (2009, April 1) Behavioral and policy issues in information systems security: the insider threat. Palgrave Macmillan 18(2):101–105. https://doi.org/10.1057/ejis.2009.12

12. Francesco M (2022) Applying resilience to hybrid threats in infrastructure, digital, and social domains using multisectoral, multidisciplinary, and whole-of-government approach. NATO Sci Peace Secur Ser. https://doi.org/10.3233/nicsp220017
13. Tikanmäki I, Ruoslahti H (2022) How are hybrid terms discussed in the recent scholarly literature? Proc 21st Euro Conf Cyber Warfare Secur 21(1):296–304. https://doi.org/10.34190/eccws.21.1.457
14. Kawka W (2022) Conditions for using nonlethal weapons of the new generation as components of critical infrastructure systems in the environment of hybrid threats. Problemy Techniki Uzbrojenia i Radiolokacji 159:7–45. https://doi.org/10.5604/01.3001.0015.8552
15. Shypovskyi V (2023) Decision-making process model for cybersecurity protection of critical infrastructure objects under the hybrid threats influence. Soc Develop Secur 13(3):43–55. https://doi.org/10.33445/sds.2023.13.3.3
16. Valenza F (2022) Next generation of hybrid threats. In EICC 2022: Proccedings of the european interdisciplinary cybersecurity conference (EICC 2022), June 15–16, 2022, Barcelona, Spain. ACM, New York, NY, 1 pages. https://doi.org/10.1145/3528580.3535333
17. Vevera A-V, Georgescu A, Cîrnu C-E (2022) A critical infrastructure perspective and systems perspective on hybrid threats in the Black Sea region. Roman Milit Think 2022(1):64–81. https://doi.org/10.55535/rmt.2022.1.04
18. König S, Schauer S (2019) Cascading threats in critical infrastructures with control systems
19. Miroslav M (2018) Critical national infrastructure in the scope of asymmetric security threats involved in hybrid warfare concept. Social Science Research Network
20. Roman D (2017) Cyber-attacks as a threat to critical infrastructure, p 259. https://doi.org/10.17234/INFUTURE.2017.26
21. Buhm-Suk B (2022) Countermeasures for hybrid threats: the experience of the European Union and its member states. Her Russ Acad Sci 92(S4):S315–S320. https://doi.org/10.1134/s1019331622100033
22. Bruzzone AG, Massei M, Ferrari R (2022) Hybrid and cyber threat in towns and critical infrastructures and industrial plants. https://doi.org/10.46354/i3m.2022.dhss.006
23. Serhiy T, Serhii B, Buchyk O, Kulinich O (2022) Protection of state management of critical infrastructure objects under the influence of cyber attacks. Ìnfokomunìkacìjnì tehnologìï ta elektronna ìnženerìâ 2(2):33–41. https://doi.org/10.23939/ictee2022.02.033
24. Tikanmäki I, Ruoslahti H (2022) How are hybrid terms discussed in the recent scholarly literature? Eur Conf Cyber Warf Sec 21(1):296–304. https://doi.org/10.34190/eccws.21.1.457
25. Figuli L, Dvorak Z (2023) Threats on the critical infrastructure elements. In: Volume 63: Monitoring and protection of critical infrastructure by unmanned systems, NATO science for peace and security series—D: information and communication security. IOS Press, Amsterdam, pp 196–205. https://doi.org/10.3233/NICSP230015
26. Zbigniew C, Sławomir Ż, Henryk W (2023) Critical infrastructure threats. Studia Administracji i Bezpieczeństwa 13(13):263–272. https://doi.org/10.5604/01.3001.0016.2902
27. Apostol AC, Nicoleta C, Nastase M (2022) Societal resilience, A key factor in combating hybrid threats. Int Conf Know Based Org 28(2):107–115. https://doi.org/10.2478/kbo-2022-0057
28. Korystin O, Svyrydiuk N (2023) Assessment of hybrid threats and capabilities to counter them in the formation of strategic communications. Uzhhorod Nat Univ Herald Ser Law 2(77):66–74. https://doi.org/10.24144/2307-3322.2023.77.2.11
29. Lupulescu G-D (2023) Hybrid threats-possible consequences in societal contexts. Eur Conf Cyber Warf Sec 22(1):616–622. https://doi.org/10.34190/eccws.22.1.1119
30. Grybko O (2021) Countering hybrid threats in the context of a changing global environment. https://doi.org/10.52363/PASSA-2021.1-7
31. Andrzej J (2020) Countering hybrid threats: resilience in the EU and NATO's strategies. J Politic Stud 2020:5–26. https://doi.org/10.12775/CJPS.2020.001
32. Miroslav M (2017) Hybrid security threats and contemporary approach to national security. Social Science Research Network

33. Korzeniewicz RP, Casullo ME (2009) Responding to disaster: two logics of demands and the politics of hybridity. Soc Mov Stud 8(2):95–114. https://doi.org/10.1080/14742830902770241
34. Blanton JF (2013) Finding the balance to combat a hybrid threat
35. László H (2020) A hibrid konfliktusok felívelési szakasza, avagy a fenyegetés észlelésének, megelőzésének és kezelésének nemzetbiztonsági aspektusai 148(4):49–64. https://doi.org/10.35926/HSZ.2020.4.4
36. Zharikova M, Pickl S(2023) A framework for hybrid risk analysis. https://doi.org/10.54941/ahfe1002850
37. Cooper CG (2022) From preparing for war to protecting the peace—legal hurdles for effectively dealing with hybrid threats in NATO1. Revue de droit militaire et de droit de la guerre 59(2):142–164. https://doi.org/10.4337/mllwr.2021.02.01
38. A Irena, CS Banacu, M Andreica, L Ivan (2023) Composite indicators used in measuring hybrid threats. Proc Int Conf Busi Excel, 17: 882–894. https://doi.org/10.2478/picbe-2023-0081
39. Banks R, Jones J, Hazzazi N, Garcia P, Zimmermann R (2021) Hybrid security risk assessment model. In: Latifi S (ed) ITNG 2021 18th international conference on information technology-new generations. Advances in intelligent systems and computing, vol 1346. Springer, Cham. https://doi.org/10.1007/978-3-030-70416-2_10
40. Atkinson C, Chiozza G (2021) Hybrid threats and the erosion of democracy from within: US surveillance and European security. Chin Polit Sci Rev 6:119–142. https://doi.org/10.1007/s41111-020-00161-2
41. Liu J, Hashim NB, Zainudin SSBS, Razi SABHM (2023) The influencing factors of public communication behavior for public health emergencies. Int J Acad Res Business Soc Sci 13(5):1020–1032. https://doi.org/10.6007/ijarbss/v13-i5/16846
42. Su F, Fan B, Song N, Dong X, Wang Y, Li J, Xue B, Qiao X (2021) Survey on public psychological intervention demand and influence factors analysis. Int J Environ Res Public Health 18(9):4808. https://doi.org/10.3390/IJERPH18094808
43. Rao T, Zhu X, Yang S-L, Bai J (2022) Compensatory control in public emergencies. Xinli Kexue Jinzhan 30(5):1119–1119. https://doi.org/10.3724/sp.j.1042.2022.01119
44. Yi C, Li S (2021) A study on the formation and evolution model of social psychology in major public health emergencies. E3S Web Conf 290:02027. https://doi.org/10.1051/E3SCONF/202129002027
45. Zhenling L, Jie L (2020) Analysis and enlightenment of altruistic behavior in public emergencies based on big data and positive psychology. In: 2020 international conference on modern education and information management (ICMEIM), Dalian, China, pp 252–256. https://doi.org/10.1109/ICMEIM51375.2020.00065
46. Zhou X, Zhao Y, Wang T, Guo L, Niu W (2019) The influencing factors of public opinion diffusion behavior in public emergencies: moderate role of government information publishing strategy. Association for Computing Machinery, New York. https://doi.org/10.1145/3371238.3371249
47. Lu Y, Qiu J, Jin C, Gu W, Dou S (2021) The influence of psychological language words contained in microblogs on dissemination behaviour in emergency situations—mediating effects of emotional responses. Behav Inform Technol 41(6):1337–1356. https://doi.org/10.1080/0144929X.2021.1886325
48. Yang J, Xu W (2018) Research on the influencing factors of public sense of security and countermeasures in major emergencies–an empirical analysis on the public sense of security in view of the explosion in Binhai new area in Tianjin. In: Proceedings of the eleventh international conference on management science and engineering management 11. Springer International Publishing, pp 1653–1666. https://doi.org/10.1007/978-3-319-59280-0_139
49. Giumelli F, Cusumano E, Besana M (2017) From strategic communication to sanctions: the European Union's approach to hybrid threats. In: A civil-military response to hybrid threats. Springer International Publishing, pp 145–167. https://doi.org/10.1007/978-3-319-60798-6_8
50. Treverton GF, Thvedt A, Chen AR, Lee K, McCue M (2018) Addressing hybrid threats. Swedish Defence University, pp 84–86. https://www.hybridcoe.fi/wp-content/uploads/2018/05/Treverton-AddressingHybridThreats.pdf

51. Bünzli F, Eppler MJ (2019) Strategizing for social change in nonprofit contexts: a typology of communication approaches in public communication campaigns. Nonprof Manag Leader 29(4):491–508. https://doi.org/10.1002/NML.21346
52. Rice RE, Atkin CK (2012) Public communication campaigns. SAGE Publications, Thousand Oaks
53. Mijalković S, Cvetković V (2013) Vulnerability of critical infrastructure by natural disasters. Belgrade: In: Proceedings "National Critical Infrastructure Protection, Regional Perspective", pp 91–102
54. https://www.2-sec.com/the-perfect-storm-how-natural-disasters-impact-cybersecurity/
55. Degg M (1992) Natural disasters: recent trends and future prospects. Geography 77(3):198–209
56. Hromada M, Lukas L (2012) Critical infrastructure protection and the evaluation process. Int J Disast Recov Business Continuity 3(November):1–10
57. Mokhtari M, Faridi P, Masoodi M, Mehran AS (2023) Perspective chapter: a global view of natural hazards related disasters [Internet]. In: Natural hazards—new insights. IntechOpen, Rijeka. https://doi.org/10.5772/intechopen.111582
58. Murray T (2012) Critical infrastructure protection: the vulnerability conundrum. Telematics Inform 29(1):56–65. https://doi.org/10.1016/j.tele.2011.05.001

Chapter 2
Hybrid Threats, Risks, and Vulnerabilities—Critical Infrastructure Resilience Solutions Toolkit for Cross-Sectoral Applications

Ashok Vaseashta, Alexandru C. Hudişteanu, Ilija Djugumanov, Marcel H. Arama, Dejan Vasović, Marija Milenković, Nimet Bolgen, Mihai Codreanu, and Carmen Maftei

Abstract Hybrid threats to critical infrastructure represent a multifaceted challenge that requires a coordinated and comprehensive approach to mitigate. By addressing the vulnerabilities and implementing robust security measures, nations can enhance the resilience of their critical infrastructure against these evolving threats. The work reported here stems from collective discussions and inputs from a working group on Critical Infrastructure: Hybrid threats, risk assessments, and vulnerabilities. The working group focused on identifying the scope of threats, risks, and vulnerabilities to develop a comprehensive solution toolkit for cross-sectoral

A. Vaseashta (✉)
International Clean Water Institute, Applied Research, Manassas, VA, USA

A. C. Hudişteanu
Maritime Security Centre of Excellence (MARSEC COE), Türkiye & Romanian Naval Forces, Istanbul, Türkiye

I. Djugumanov
Euro-Atlantic Council of North Macedonia, Member of Atlantic Treaty Association, Skopje, North Macedonia

M. H. Arama
Centre of Excellence Defence Against Terrorism (COE-DAT), Devlet Mahallesi İnönü Bulvarı Süleyman Emin Caddesi No: 65 06420 Çankaya, Ankara, Türkiye

D. Vasović
Faculty of Occupational Safety, University of Nis, Niš, Republic of Serbia

M. Milenković
University Square, Nis, Republic of Serbia

N. Bolgen
Chemical Engineering Faculty, Mersin University, Department, Mersin, Türkiye

M. Codreanu · C. Maftei
Civil Engineering Faculty, Transilvania University of Brasov, Brasov, Romania

D. Radu et al. (eds.), *Countering Hybrid Threats Against Critical Infrastructures*, NATO Science for Peace and Security Series B: Physics and Biophysics, https://doi.org/10.1007/978-94-024-2304-4_2

applications. Based on the recommendations of the working group, we believe that the solutions and best practices can be widely applied across the spectrum of threats and vulnerabilities. This report focuses on an overview of hybrid threats and how several solution pathways encompassing technical, social, and policies can be applied to bridge vulnerabilities.

Keywords Hybrid threats · Critical infrastructures · Risks · Vulnerabilities · Resilience solutions toolkit · Cross-sectoral

2.1 Introduction

Critical infrastructure (CIS) comprises several essential systems and assets that are vital to a nation's security, economy, public health, and safety. These include sectors such as energy, transportation, water supply, telecommunications, finance, and healthcare. Disruption of CIS poses a serious challenge, with far-reaching implications for national security, economic stability, public safety, and overall societal well-being. Their disruption can lead to cascading failures across multiple sectors, amplifying the impact of a single event. Critical infrastructures are the backbone of functional modern societies and any incident, natural, accidental, incidental, or intentional will have serious consequences, such as national security, economic instability, public safety and health, social unrest, loss of public trust, and cascading failures across sectors. Safeguarding CIS is a vital, strategic security concern and a challenge for all nations since malign actors and adversaries continue to look for vulnerabilities, actively targeting the Allied nations' CIS in pursuit of their hidden agenda to destabilize and undermine democratic societies.

Among many challenges, hybrid threats pose significant risks to these systems which arise by combining cyber, physical, and psychological operations to disrupt, degrade, or destroy critical infrastructure. Many efforts to define and categorize hybrid threats, key infrastructure sectors, critical infrastructure or entities, and their vulnerabilities have been made over the years and a great deal of literature already exists on the matter. However, agreeing on key definitions or framework components tends to escape policymakers and experts at times, due to several reasons beyond the scope of this publication and also how cooperative policymaking mechanisms operate in general. Negotiations on terms and definitions tend to take longer than practitioners are comfortable knowing and waiting for, as the latter's job is to prevent, respond, recover, and manage consequences from the effects of successful threat vectors. As an example, the United Nations Convention on the Law of the Sea (1982 UNCLOS) [1] required some 9 years to reach a framework that policymakers could agree on in principle, during the third UNCLOS (UNCLOS III) [2] between 1973 and 1982. It further took another 12 years for the Treaty to come into force in 1994, following the ratification of its 60th member from Guyana. Policy-making and legal definitions alignment are usually long and arduous processes, especially true where more Nations are involved, as might be the case for NATO. When developing and using new and emerging technologies, including in the field of artificial intelligence (AI) and data collection techniques, in addition to agility, it is critical to consider new opportunities to improve resilience, potential vulnerabilities, and cascading failures as countermeasures to hybrid threats to reduce overall risks across the sectors.

A comprehensive security approach across the administration and society, acting in a more strategic, synergistic, and coordinated manner across all relevant policy areas. To increase resilience and counter hybrid threat vectors, new tools should be used effectively to provide a means of identifying such threats and addressing any direct investment (foreign and domestic) that may affect security, law, and public order. The protection of national critical infrastructures, as well as of the functions and services critical to the proper functioning of the state, the economy, and society, is a key priority and requires an approach at the level of the whole of the government approach [3, 4], and society. Work in this direction must take into account the strong interdependencies between the various critical functions and services, including financial services, the key role of the private sector, the changing security environment, and emerging risks, both in the physical and cyber fields. In addition to legal requirements, agreements should be promoted with private sector infrastructure and service owners and operators to ensure the continuity of and access to critical services, including beyond force majeure, by ensuring an acceptable level of preparedness to respond. All relevant threats, as well as the flexibility to address and mitigate the effects of low-probability, high-impact events. Since the responsibility for critical infrastructure, protection is primarily a matter of international importance, and the high degree of cross-border and cross-sectoral interdependencies requires coordinated or, where appropriate, legally harmonized efforts at the international level, including the smooth functioning of the internal policies, the working group articulated a solution toolkit that can be applied cross-sector with a goal that understanding and mitigating such threats will make critical infrastructure, and society in general, safe, secure, and resilient.

2.2 Critical Infrastructure

Critical infrastructure refers to the systems, facilities, and assets that are vital for the functioning of society and the economy. A generic overview of CIS and how it relates to society is shown in Fig. 2.1.

These are considered essential because their disruption would impact public safety, security, and health or economic stability. Such CIS includes both physical and virtual components that are interconnected and interdependent. In the US, the Presidential Policy Directive 21 (PPD21) [5] advances a national unity of effort to strengthen and maintain secure, functioning, and resilient critical infrastructure consisting of (but not limited to), the following, viz.: Energy sector—Nuclear power plants, Electrical grids, Oil and natural gas production facilities, pipelines, and fuel storage; Chemical sector—Petrochemical manufacturing, Agricultural related

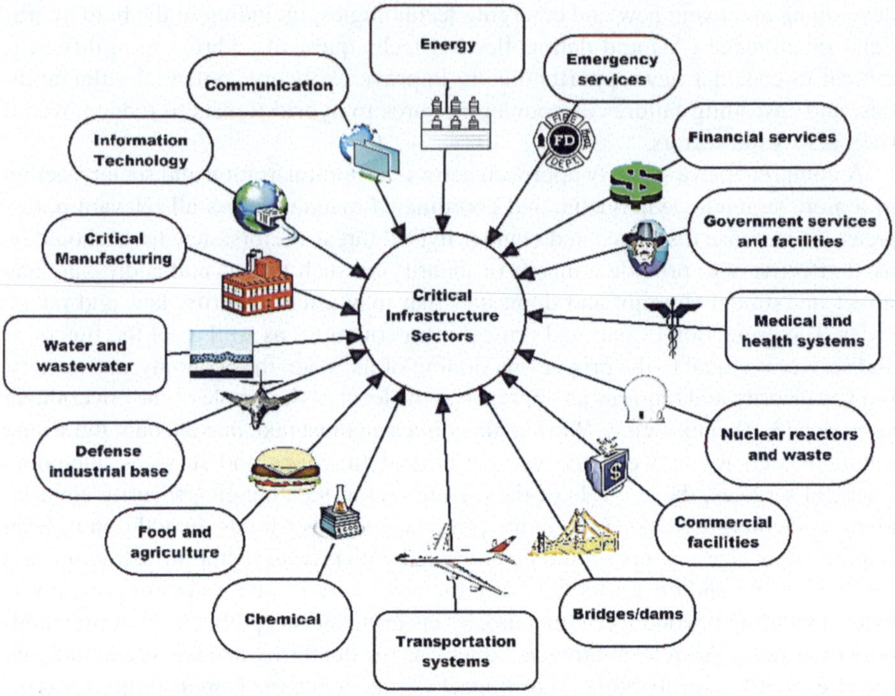

Fig. 2.1 An overview of the scope of critical infrastructure

chemical production, and general purpose chemical distribution; Transportation sector—Aviation hubs, seaports, railways, highways, especially intermodal systems, underwater cables and pipelines, bridges, and public transit systems; Water and wastewater distribution systems—water treatment plants, reservoirs, dams, pumping stations, and sewer systems; Communications sector—telecommunication networks, internet service providers, mail and package delivery services, and satellite systems; Financial services sector—banks, stock exchanges, and payment clearinghouses; Healthcare—hospitals, clinics, and medical supply chains; Emergency services—law enforcement, fire departments, social services and emergency management systems; Food and agriculture—farms, food processing facilities, food supply distribution networks, and food safety systems; Government—defense industrial base, federal government facilities, and national security systems; Information Technology—data centers, critical software and hardware, cybersecurity systems, and internet infrastructure; and Government service—social, legal, and regulatory aspects. Any disruption to these services by hybrid [3, 4] or dual-use [6] threats would present a serious challenge, that can lead to cascading breakdowns across multiple sectors. We now discuss hybrid threats and how they impact CIS, including risks and potential vulnerabilities.

2.3 Hybrid Threats

The vast literature on hybrid threats is a testament to the subject's relevance and contemporaneity. As an example, the European Centre of Excellence (CoE) for Countering Hybrid Threats in Helsinki, Finland, characterizes hybrid threats as follows [7], viz. coordinated and synchronized actions that deliberately target systemic vulnerabilities, through a wide range of means; exploit thresholds of detection and attribution and the interface spaces (war-peace, internal-external, local-state, national-international, friend-opponent). The effect ranges from influences to bypassing, to rendering ineffective or actively harming entire systems/the targets while influencing different forms of decision-making at the local (regional), State, or institutional level to obtain the agent's strategic goals. Certain examples of hybrid threats specific to CIS are:

2.3.1 Cyber-Attacks and Ransomware

Hackers may launch directed cyber-attacks against CIS to disrupt services, extract sensitive information, or cause physical damage, including, but not limited to, disrupting power grids and/or contaminating water treatment facilities [3, 8–10]. Furthermore, malicious actors may launch ransomware attacks to halt critical service systems, demanding ransom payments to re-establish functionality. Other forms of threats include distributed denial of service (DDoS) attacks, inundating the system with spam emails, introducing viral signatures through attachments, spoofing, and conducting illegitimate businesses using unused email accounts [11], just to name a few.

2.3.2 Disruption of Physical Assets

These kinds of disruptions include, (a): Coordinated Physical Attacks: Attacks on physical assets such as power plants, distribution pipelines, or transportation hubs which can cause widespread disruptions. (b): Insider Threats: Individuals with authorized access to critical infrastructure may be coerced or recruited to sabotage systems from within. Insider threats are usually hard to detect at the outset and are much more complex to mitigate.

2.3.3 Dissemination of Dis/Misinformation

In the present democracy, propagating mis/disinformation contributes to undermining public trust, which includes propagating false information through targeted campaigns to alter the public's perception of CIS, leading to panic, mistrust, or reduced cooperation during emergencies. Another major impact is social engineering, such that the attackers may use misinformation to manipulate employees or decision-makers into taking measures that compromise security [6, 12, 13]. Propagation of misinformation has recently become a major election interference issue causing mass confusion and public distrust. Using some automation tools, some groups have tried "believability experiments", i.e., to study how believable a misinformation campaign might appear to the general public.

2.3.4 Financial Instability Threats

Saboteurs, by manipulating or attacking key supply chains, can create shortages, disrupt Supply Chains, or cause financial market destabilization, thus causing disruption of essential services. This results in financial misappropriations due to economic pressures and some people, with advanced knowledge, may take undue advantage due to large market fluctuations. Furthermore, coordinated cyber-attacks on financial systems and coordination by some powerful alliances can destabilize the financing and operation of major financial critical infrastructure, causing widespread panic.

2.3.5 Political and Regulatory Manipulation

These influence operations are mostly by the special interest groups, who, by covert attempts, try to sway political or regulatory decisions related to the management and security of critical infrastructure, which may not be in the best interest of the public. Undue political pressure by adversaries may push for regulations that unintentionally weaken the resilience of critical infrastructure sectors causing regulatory sabotage.

The hybrid threats are, in general, a wider selection of means and tools than the ones conventionally employed for competing and contesting against targets or opponents, and not, an aggregation of concerted actions used in creative manners. These are, for example—information campaigns, control of narratives, social media influence campaigns, regular hard-power measures, and kinetic effect actions, as shown in conceptual Fig. 2.2.

A critical component for the instrumentalization of hybrid threats is the element of plausible deniability. Being able to attribute entities as responsible for their

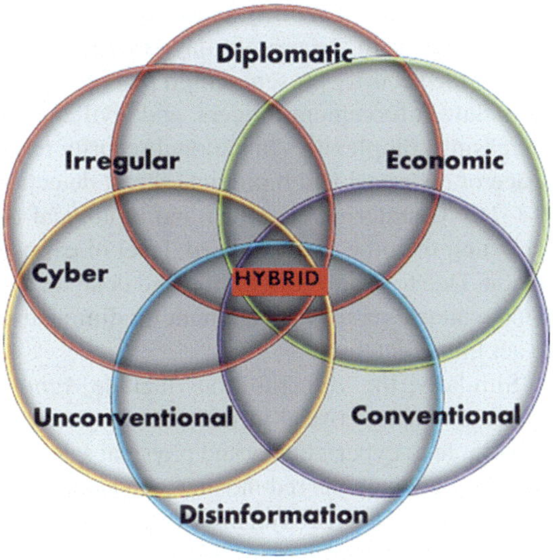

Fig. 2.2 Venn diagram shows that hybrid threats incorporate overlapping modes

actions, especially for hard measures and kinetic effect actions, defeats the purpose of utilizing hybrid threat campaigns. A would-be attacker would prefer to stay as far as possible from responsibility for the attacks, all the while reaping the benefits and building/shoring effects of other (sometimes softer) tools and actions. Hybrid threats are generally targeting critical systems and processes, at various pressure points and at various moments, making the existence of a hybrid threat campaign a daunting task under normal conditions. The target vulnerabilities are selected for their potential for negative cascading and/or build-up effects, or, at a minimum, to create the conditions for subsequent critical vulnerabilities to appear and enable further exploitation.

2.3.6 *Risks and Vulnerabilities*

The following risks and vulnerabilities to critical CIS are envisaged. The list of risks is not exhaustive and hence it is not limited only to the following risks and vulnerabilities. *Functional Disruption*: Hybrid threats can cause significant disruptions of operations in critical infrastructure, leading to the loss of essential services, such as electricity, water, or transportation. *Economic Disruption*: The impact on critical infrastructure can have severe economic consequences, including loss of productivity, increased costs, and damage to key industries. *General Public Safety*: Attacks on critical infrastructure can pose direct threats to general public safety, such as contaminated water supplies or failures in emergency response systems. *National Security*: Compromised or vulnerable critical infrastructure can weaken a nation's defense capabilities, making it more vulnerable to further attacks or coercion. *Loss of Public Trust*: Misinformation and operational failures can erode public confidence in the reliability and security of critical infrastructure. *Chem-Bio Threats Related Risks*: In the present-day era, the rising chemical and biological threats to worldwide security present a high priority for the safety of medical personnel, soldiers, law enforcement officers, and civilians, including critical infrastructure. The increasing complexities in national security require continuous evolution within the space of chemical and biological threat detection, decontamination, and protection [14–16]. Countering chemical and biological threats involves complex challenges including the type of agent used, level of exposure to warfare agents, and contamination. Due to ease of dispersion and delay in time response, biological agents complicate attack vector management by diminishing the capability of first responders to act immediately.

Similarly, the vulnerabilities include *Aging Infrastructure*, viz.: (a): Outdated Systems—many critical infrastructure systems rely on outdated technology that is vulnerable to cyber-attacks and physical sabotage. (b): Limited Redundancy—older systems may lack the redundancy needed to quickly recover from attacks or disruptions. *Interconnected Systems*: (a): Complex Dependencies—the interconnected nature of modern critical infrastructure means that disruption in one sector can have cascading effects across multiple sectors (e.g., an attack on the power grid affecting water supply and healthcare). (b): Supply Chain Vulnerabilities—global supply chains supporting critical infrastructure may be vulnerable to disruption, especially if they rely on a limited number of suppliers or geopolitical hotspots. *Human Factors*: (a): Insider Threats—employees with access to the operational details of critical systems may be exploited through coercion, bribery, or social engineering. (b): Lack of Training—inadequate training in cybersecurity and crisis response can leave personnel unprepared to process the challenges of hybrid threats. *Regulatory Gaps*: (a): Inconsistent Standards—especially varying security standards and regulations across different sectors or regions can create vulnerabilities. (b): Slow Policy Response—the rapid evolution of hybrid threats may outpace the development and implementation of relevant regulations and policies, which are inherently slow.

2.4 Cross-Disciplinary Solutions Toolkit of Best Practices

Critical infrastructure security involves protecting the systems, facilities, and services that form the foundation of modern society. This includes water supply systems, sewage and waste management, public health facilities, emergency services, telecommunications, and other essential services. The security of CIS is critical for public safety, economic stability, and the overall functioning of communities. In general, a cross-disciplinary solution toolkit of best practices is an organized collection of strategies, methodologies, and tools drawn from various fields to address complex problems. The approach leverages the strengths and insights of multiple disciplines, fostering innovation and comprehensive solutions and often integrates the best practices from various sectors. Since the WG had representations from multiple disciplines, the objective was to develop cross-disciplinary solutions that integrate the best practices from these sectors, which requires collaboration across disciplines, including science, engineering, policymaking, and community engagement, to create resilient and sustainable systems that protect critical environmental entities from various threats [17–19]. This may not be, always, the case, as hybrid events are not predictable, and due to uncertainty, the solutions may have to be formulated on-the-go or on a case-by-case basis. We list below some of the core principles of developing such a toolkit, viz.: *Identify core disciplines*: Identify the disciplines that are relevant to the problem at hand and recognize the strength of each discipline. *Develop a framework for integration*: Develop a common vocabulary to facilitate communication across disciplines, i.e. Establish collaboration protocols. *Curate best practices from each discipline*: Identify and document best practices, tools, and methodologies from each discipline. *Adapt and integrate*: Modify these practices to fit the specific context or problem, ensuring they can work together effectively.

An effective toolkit consists of problem-solving methodologies, analytical tools, creative techniques, human-centered approaches, and project management frameworks. The toolkit must undergo testing an iterative process with feedback loops and options for continuous improvements. This stage follows the implementation strategy ensuring that it meets the unique needs of each situation. To develop a community of practice, the toolkit must be documented for knowledge sharing with implement metrics to evaluate the success of the toolkit in solving problems, for scalability and adaptation.

Hence, by systematically integrating methodologies, tools, and practices from multiple fields, such a toolkit can address complex challenges more effectively than a single-discipline approach. However, the caveat is that while policymakers differ on definitions and frameworks, hybrid threats can and do capitalize on existing or newly found vulnerabilities, giving that mathematical expression of risk a very material and tangible form for which responders will have to provide solutions. Sometimes the most obvious of realities is best expressed in commonplace sentences, and such is the harsh reality that ill-intended actors and adversaries do not wait for alignment of definitions and terms to act, and in fact, thrive on inaction or inconsistency caused by the lack of policies and definition alignments. It is a level of asymmetry that both policymakers and practitioners need to be very well aware of so that even in the absence of agreed policy frameworks, solutions can still be found and applied to counter hybrid threats and protect critical infrastructure across the board. This is especially valid for the Allies, who share a joint commitment and responsibility to defend one another against threats, including hybrid threats.

This enshrined asymmetry is further exacerbated by how the "prevention-response to action" binomial functions. On one side you may have governments and sophisticated protective and response systems in place involving costly resources, on the other the threat vectors can be as simple as one lone-wolf actor managing to negatively impact or access of control systems of a critical entity/infrastructure (e.g. port management systems, government websites and databases, hospital management systems, water management systems, etc.). While virtually all the costs and risks are supported by the prevention and response side, seldom do the ill-intended actors or adversaries support anything in comparison. The latter needs time and a factor of magnitude less resources to eventually materialize on one of the vulnerabilities of a critical infrastructure. It is a game of dice with cheap buy-ins and high payoffs in which the house may eventually lose.

Despite the doom-and-gloom sentencing above, real life demonstrated that societies and systems prosper and that even following very destructive or debilitating events, systems and infrastructures recover, undergo betterment, and further their resistance to other similar threats. Attempts to attack critical infrastructures are taking place daily in such numbers that they might seem fictional, especially true for the cyber domain. For an illustrative example, the National Directorate of Cyber Security of Romania stated that Romania is confronted daily with over 25,000 cyber-attacks across multiple services [20]. Regardless, the news of successful attacks is few and far between, and this is in good measure the result of existing protective and resilient solutions in place.

In the real world, vulnerabilities and threats are constantly found, assessed, tested, and patched across sectors. This signifies that besides the troves of literature on categorizing critical infrastructure threats, vulnerabilities, and risks, there are also troves of applied sectoral solutions that succeed in protecting and keeping critical infrastructures functioning daily. These are essentially functioning sectoral toolkits. But in selected cases, sectoral solutions also hold cross-sectoral applicability, opening the discussion of protecting critical infrastructure and countering hybrid threats through recycling and expanding the applicability of sectoral good practices, and solutions. This inter-disciplinary approach holds the distinctive advantage of potentially reducing costs and time to develop and apply solutions by matching and mirroring existing options and delivering a universal toolkit, that although not finely tuned to the specific needs of a sector, represents a protection-reaction multi-tool in the pockets of all policymakers, decision-makers, practitioners across sectors.

2.5 Countermeasures to Hybrid Threats Across Sectors

The working group identified several areas, based on experience and expertise, which are provided below. The list is not exhaustive, as there are always nuances, but is representative, as several other hybrid threats may exist.

2.5.1 Cyber and Digital Infrastructure Security

Cyberspace encompasses the entire interconnected digital domain, including the tools and technologies that enable communication, information exchange, and digital transactions, while digital infrastructure refers to the specific physical components and software systems that support the delivery of digital services and products. Together, cyberspace and digital infrastructure form the backbone of the modern digital economy and society. The goal of merging communication and information technology is to interconnect and control numerous vital infrastructures. This relatively recent convergence between communications and information technology, with cyber connectivity to critical infrastructure systems has created opportunities for adversaries to exploit vulnerabilities in the critical infrastructure. The threat paradigm has shifted from kinetic to hybrid. Cyber and digital infrastructure adaptation is a proverbial dual-edged sword, since societies must keep up also with digital transformation, remain connected to innovation, and be competitive and relevant among competitors, yet Cyber and Digital Infrastructure security against hybrid threats poses a significant challenge. The countermeasures include a thorough analysis of several hybrid threat vectors that have the potential to attack, undermine, or compromise such infrastructure, as well as the associated risks and vulnerabilities that must be addressed throughout the transformation process. Furthermore, given the global shortage of cyber security professionals and the growing sophistication of cyber threats, Generative AI (GenAI) provides a strategic advantage to strengthen your cyber defenses. GenAI technology enhances the capacity to manage security, prevent threats, and update policies. It can also be used to deploy an AI-powered security assistant—improving the effectiveness of administrators and SecOps analysts by accelerating security administration, reducing the time needed to respond to live threats, and preventing data loss through GenAI security apps.

2.5.2 Water Security

Access to clean water is the most fundamental need of human society, hence adversaries have shown great interest in exploiting this resource to promote their ideology [9, 10, 21]. Many national and international agencies, viz. National Institute of Health (NIH) and NATO have created baseline requirements of resilience for its allies, with water being one of them. Most drinking water resources are openly accessible and are prone to intentional introduction of contaminants. In addition, implementing Supervisory Control and Data Acquisition (SCADA) systems in water purification enhances operational efficiency, ensuring a continuous supply of clean and safe drinking water, yet such systems face several security challenges due to the increasing sophistication of cyber threats. Addressing hybrid threats in the water sector requires a comprehensive approach that integrates cybersecurity, physical security, resilience to natural disasters, and strategic economic and geopolitical planning [9, 22]. By adopting these best practices, it is possible to safeguard water resources against a wide range of threats, ensuring a stable and secure supply of water. From the traditional point of view, developing and regularly updating comprehensive emergency response plans in coordination with local, state, and federal authorities to ensure quick and effective responses to physical threats should be a preferable option. In the modern sense, investing in cybersecurity for water infrastructure can prevent costly disruptions and protect public health. A successful cyber-attack on a water treatment facility can lead to damages costing millions, but proactive cybersecurity measures can reduce these risks by up to 90%. Furthermore, water and wastewater infrastructure are interdependent on other CIS, causing a complex ripple effect throughout society.

2.5.2.1 Emerging Contaminants in Water and Their Impact on Security

Emerging contaminants in water, often referred to as "contaminants of emerging concern" (CECs), include a diverse range of chemicals and biological agents that are increasingly detected in water bodies but are not yet regulated. These contaminants include pharmaceuticals, personal care products, endocrine-disrupting compounds, pesticides, industrial chemicals, and microplastics. Their presence in water poses significant environmental and public health concerns, which in turn impacts various aspects of security. The security implications include public health, the environment, water resources, and national security. Addressing these threats requires comprehensive policy responses, advanced technological solutions for detection and treatment [10, 15, 23, 24], and increased public awareness to mitigate the risks associated with these contaminants.

2.5.3 Energy Infrastructure Security

Energy critical infrastructure security refers to the protection and resilience of systems and assets that are essential to the production, transmission, and distribution of energy. These infrastructures include power plants, electricity grids, oil and gas pipelines, refineries, and other facilities crucial for the continuous supply of energy. Ensuring their security is vital for national security, economic stability, and the well-being of societies. The threat vectors to energy security include physical threats, cyber threats including hacking, ransomware and state-sponsored attacks, supply-chain vulnerabilities, and technological challenges. Similar to water, energy is the basic lifeline for society and any security breach will result in economic disruption, public safety, national security, and even environmental impact. Environmental security in the energy sector focuses on reducing environmental impact while ensuring a stable and sustainable energy supply.

The strategies for enhancing energy infrastructure security include building resilience, such as redundancy and implementation of localized microgrids, cybersecurity to include advanced monitoring and infrastructure protection plans, policy and regulatory frameworks to include regulations compliance and public-private partnerships, and emergency preparedness, response, and consequence management plans. Implementing best practices in the energy sector involves quantitatively assessing and enhancing renewable energy deployment, improving energy efficiency, and developing carbon capture and storage technologies. Environmental security in the energy sector also involves addressing hybrid threats, and multi-sided risks that combine conventional, irregular, and cyber threats. These threats can disrupt energy supplies, damage infrastructure, and impact the environment. To ensure resilience and sustainability, it is essential to adopt best practices that mitigate these hybrid threats. Due to its dependence on supervisory control and data acquisition (SCADA) systems, conducting frequent vulnerability assessments and penetration testing to identify and mitigate potential weaknesses in energy systems or adopting a defense-in-depth approach that includes firewalls, intrusion detection systems, encryption, and secure communication protocols to protect critical infrastructure.

2.5.4 Ground, Aviation, and Marine Transportation Security

Transportation critical infrastructure security focuses on protecting the systems, assets, and networks essential for the movement of people and goods. This includes roads, bridges, tunnels, railways, airports, seaports, and public transportation systems. The security of transportation infrastructure is crucial for economic stability, public safety, and national security. Threats to transportation cis include physical threats, such as terrorism, sabotage, and natural disasters; cyber threats, such as hacking and malware, ransomware, and data breaches; supply chain disruptions, and technological challenges, viz. aging infrastructure and complexity of integration of modern systems. Any security breach, may it be data, or ransomware, will have cascading effects on the economy, and public trust. The security of transportation critical infrastructure is essential for the functioning of economies, the safety of populations, and the stability of nations. As threats continue to evolve, it is crucial to adopt a comprehensive approach that includes physical and cybersecurity measures, policy frameworks, and resilience strategies. By proactively addressing these challenges, societies can ensure the continued safe and efficient movement of people and goods.

2.5.5 Environmental Security

Nowadays, environmental security refers to a multidimensional concept that varies based on differing interpretations of both terms "environment" and "security." This concept is likely to emerge from the environmental movement and the strategic shifts following the second half of the twentieth century, in parallel with numerous "environment-oriented" movements, either driven by formal international institutions or individuals. Environmental security should be considered as a dynamic process that proactively addresses evolving environmental conditions that could potentially undermine global, regional, and national peace and stability. Achieving environmental security objectives often requires the strategic planning and implementation of preventive programs aimed at preventing and mitigating human-induced environmental changes that adversely affect sustainable living conditions within a given region. Key interpretations of environmental security focus on how environmental changes pose risks to national security, contribute to violent conflicts, endanger critical infrastructure elements, and threaten human security as a whole. Quantitatively, environmental change can be linked to national security through various indicators, such as the increased frequency and intensity of natural disasters, which strain a nation's resources and infrastructure. For instance, between 2000 and 2020, the number of weather-related disasters increased by 83%, from 3656 to 6681, according to the United Nations Office for Disaster Risk Reduction (UNDRR, 2020) [25]. These disasters can destabilize regions, leading to significant economic costs, displacement of populations, and potential political instability. Environmental change also plays a role in violent conflicts, albeit to a lesser extent compared to traditional factors such as political, ethnic, or economic tensions. However, it exacerbates existing vulnerabilities. Research from the Institute for Environmental Security indicates that approximately 40% of intrastate conflicts over the past 60 years have had a significant environmental component, often related to resource scarcity, such as water or arable land [19]. Human security is profoundly affected by environmental changes, especially for the world's poor, who rely heavily on natural resources for their livelihoods. According to the World Bank, environmental degradation and climate change could push an additional 100 million people into poverty by 2030 [26]. This is due to declining agricultural productivity, increasing health risks, and heightened exposure to natural disasters. Environmental security is predominantly perceived as addressing threats to nation-states rather than solely focusing on environmental protection, international relations, or individual security. This national-centric perspective promotes the need for comprehensive policies that should integrate environmental considerations into national security strategies, addressing the many-sided challenges posed by environmental changes.

Environmental security threats represent growing concerns that encompass issues like climate change, human pollution, resource depletion, loss of biodiversity, etc. Countermeasures to address these threats involve a combination of policy measures, technological responses, conservation efforts, and community engagement. In the sense of policy measures, governments play a pivotal role in safeguarding the environment through stringent policies and regulations. International agreements like the Paris Agreement aim to mitigate climate change by limiting global warming. Also, international organizations, such as the United Nations and the World Bank, facilitate collaboration and provide funding for environmental projects. Joint research initiatives and knowledge-sharing platforms enable countries to learn from each other's successes and challenges. National policies often focus on reducing emissions, promoting renewable energy, and enforcing environmental standards. From the technology standpoint, advancements in technology offer promising solutions to environmental challenges. Renewable energy technologies, such as wind, solar, and hydroelectric power, significantly reduce dependence on fossil fuels. Innovations in energy storage, like advanced battery systems, help manage the intermittent of renewable energy sources. Additionally, technology is essential in monitoring and managing natural resources. Satellite imagery and data analytics enable precise tracking of deforestation, pollution, and wildlife populations, facilitating informed decision-making. Conservation is equally important and is vital to maintaining biodiversity and protecting ecosystems. Protected areas, such as national parks and wildlife reserves, safeguard critical habitats from adverse human impact. Restoration projects aim to rehabilitate degraded landscapes, such as reforestation initiatives that not only restore habitats but also sequester carbon dioxide from the atmosphere. Marine conservation efforts focus on protecting coral reefs, reducing overfishing, and combating marine pollution, thus preserving the rich biodiversity of our oceans. In the end, effective environmental security actions rely heavily on community involvement. Local communities are often the first to experience the impacts of environmental hazards and thus play a vital role in environmental security efforts. Community-led initiatives foster a sense of stewardship and collective responsibility. Hence, sustainable practices include sustainable transport, eco-friendly urban planning, and green building designs to minimize environmental footprints while promoting economic and social well-being.

2.5.6 Chem-Bio Radiological Nuclear Security Challenges to CIS

Given remote and easy accessibility to SCADA systems using the internet, the possibility of a CBRN attack on critical infrastructure is significantly reduced, nonetheless, due to the ease of assembly of at least the chemical and biological crude devices, it poses a significant challenge to critical infrastructure due to the potential for large-scale harm, disruption, and long-term consequences to critical infrastructure. Some of the key challenges are due to the complexity of detection and false-positive identification that could lead to unnecessary panic or, conversely, a failure to respond in time to a real threat. Other challenges are effective response to CBRN incidents through interagency coordination, resource allocation due to specialized equipment, mitigation and containment, and its environmental impact. High-value targets are always vulnerable to such attacks by state and non-state actors due to the potential for widespread disruption. The interconnectedness of these systems can exacerbate the effects of an attack. Aging infrastructure: many critical infrastructure systems are aging, which can make them more vulnerable to CBRN attacks, either through direct targeting or as a result of accidental release.

In the event of a successful attack, a CBRN incident can result in mass casualties, overwhelming healthcare systems, and leading to long-term public health crises, not to mention the fear and panic associated with CBRN incidents can lead to significant psychological stress on the population. One of the major concerns is that CBRN preparedness requires continuous training and exercises among first responders, healthcare providers, and infrastructure operators. However, maintaining a high level of readiness can be resource intensive.

Chemical and Biological agents have a long history, dating back to Alchemy [27–29]. With advancements in synthesis and processing methodologies, CBRN agents can be engineered or modified to evade detection or increase lethality, making it difficult to stay ahead of emerging threats due to technological challenges. This requires implementing and maintaining cutting-edge detection and response technologies, which can be costly and require constant updates to address evolving threats. The risk of radiological incidents is low, but possible due to the easy availability of radiological sources at low to medium-security facilities, such as university labs and medical facilities. Addressing the challenges of CBRN security in critical infrastructure requires a multi-faceted approach that includes robust detection and identification systems, coordinated response efforts, effective mitigation and containment strategies, and a strong focus on cybersecurity, public health, and international cooperation. Preparedness through regular training, technological innovation, and public awareness is key to minimizing the risks associated with CBRN threats.

2.5.6.1 Risk Relating to Chem-Bio Threats

The first responders, medical professionals, service personnel, and civilians must be adequately protected from chemical and biological warfare agents. Furthermore, it is essential to develop personal protective systems that can protect their user not only passively, but actively, being able to detect, adsorb, degrade, and decontaminate chemical/biological warfare agents. Biological or chemical warfare agents can be microorganisms, toxins, and chemicals, respectively, which are intended for use in military operations to kill, seriously injure, or incapacitate exposed individuals by exerting their physiological effects. Chemical and biological weapons are the respective agents with their delivery systems. Protective measures against chemical and biological agents are respirators, skin protection, detection, decontamination, and medical countermeasures [14, 30, 31]. Many chemical and biological warfare agents are respiratory and dermally active. Protective fabrics provide barrier protection by absorbing/adsorbing or blocking toxic agents. The development of protective materials includes catalytically active nanofibers, sensors, and selective gas permeating properties. Electrospun nanofibrous materials enable the development of effective systems that offer chemical and biological agents' detection, decontamination, and protection [14, 31].

2.5.7 *Maritime Security*

2.5.7.1 Maritime Coordination and Information Fusion Centers

Maritime security regularly requires the coordinated efforts of a multitude of actors operating in the maritime environment. Thus, the maritime environment comprises also vast and diverse interests and efforts, directly representing individuals, coastal communities, industry, and non-governmental, governmental, regional, and international entities. These rarely align and the stakeholders seldom "speak" the same language in the absence of facilitating mechanisms or communication channels. It is for this reason that establishing Maritime coordination and/or Information Fusion Centers became one of the "gold standards" in good practices for effective and efficient action in maritime security. Maritime coordination centers enable stakeholders to complement each other while addressing maritime security challenges or during routine activities. This leads to increased effectiveness and efficiency for action at sea, fosters mutual understanding and cooperation, and helps mitigate misunderstanding and overlapping efforts. Routinely, maritime coordination centers, to include also maritime operations centers, would comprise representatives from governmental agencies and law enforcement (customs, transportation, coast guard, navy, etc.). Ideally, these centers would also include representation from industry and coastal communities, but even liaison and points of contact mechanisms to engage and account from private industry and coastal communities generally suffice for the purpose.

Information Fusion Centers, while restricted in their ability to act, perform the crucial activity of fusion of various sources of information to offer an overview of the maritime situation, all the while accounting for the different perspectives that require these overall situational products. Offering a fused and clear maritime situational picture to stakeholders helps align their views and understanding, fosters collaboration, enables security forces to be more effective in their actions at sea and allows seafarers a better understanding of the security situation.

Establishing coordination and/or fusion centers, while a fairly common practice in the maritime world, is rooted in the general need for coordination of action and sharing understanding across systems and sectors, while not necessarily sacrificing entity autonomy. These mechanisms facilitate the exchange of information, help align views and understanding, allow better mapping of the various data sets and information existing across systems, permit faster communication, and allow for more effective and efficient use of assets at sea. These are desirable qualities, especially in dealing with hybrid threats, where the apparent reason, effects, and even targets are not easily identifiable, reacquiring increased communication, information exchange raised awareness, and effective action to counter them. There is significant value in setting up cross-sectoral information fusion and/or coordination centers encompassing the critical stakeholders, similar to the ones in the maritime world, promoting a more comprehensive approach to systems' resilience and protection against hybrid threats.

2.5.8 Naval Coordination and Guidance for Shipping and Navigational Information Systems

The maritime spaces are equally utilized for many purposes, including commercial and military ones. And it makes sense at times these two mentioned factors would impact one another, or even depend one on the other. The most recent and visible example would be the situation continuing in the Bab-el-Mandeb Strait and the Red Sea region. It is here where the military users of the sea (navies in particular) advise and protect crucial commercial traffic in and out of the Red Sea. And when it comes to these different users, the interaction does not take place as often as one might think, nor it is regularly exercised. For this purpose, and other situations where military usage of the seas (operations, exercises, etc.) might interfere with the commercial users and maritime trade overall, systems of coordination and advice between military and merchant shipping have been put in place by various stakeholders, some operating ad-hoc in a given area, some permanent and staffed or under the responsibility of different information fusion or operations centers. "This provides the interface between military operations and merchant shipping [32] and it is employed to enhance the safety of participating merchant ships in the operations area while supporting military objectives." [33], (NATO, NCAGS & AWNIS).

In essence, these systems offer (most of them free) maritime situational awareness (MSA) and information services to enable the more strategic beneficiary—the global maritime shipping system—to benefit from information and expertise relevant to and curated usually by military and law enforcement beneficiaries. These systems operate under a need-to-share logic, rather than waiting for a demand, to offer readily available information to all needing it. This allows both the ones aware of needing such information, but also the ones unaware of it to benefit from crucial information.

This type of approach holds value also for identifying and/or making others aware of situations of threat otherwise undetectable by the ones targeted. Setting up threat awareness and information systems that readily offer their services to the broader community, and not only to specific recipients, would increase a system's preparedness against all sorts of threats, and particularly would most likely foster the ability of stakeholders to map hybrid threats and campaigns.

2.5.8.1 Space-Based Remote Sensing/Earth Observation (EO) Technologies

Maritime Situational Awareness (MSA) is a central pillar in enabling maritime security efforts and uninterrupted, actionable, and effective MSA is a desire of many a policy across the globe. Simply put, an ideal MSA requires wide coverage and continuous maritime surveillance as its most powerful enabler. Currently, the systems in place involve a great deal of rules compliance, with vessels self-reporting their position or observed situations and abiding by international regulations, for instance, the IMO established a Long-Range Identification and Tracking (LRIT) system (IMO, LRIT). It is fairly easy to imagine that there are any number of vessels out there at any given moment that would not abide by these regulations once away from any coastal station or platform that might detect them by radar or visually, especially convenient if the respective vessel would want to engage in more unconventional or straight-out illegal practices. Due to the sheer size of the water spaces and the complicated jurisdictions of the seas, especially the high-seas, the problem of persistent surveillance, independent of compliance has led to many developments, not the least important in the area of space-based remote sensing/ earth observation (EO). The technical advances in surveillance sensors' sensitivity, accuracy, and reliance coupled with the privatization of space generating more satellites and cheaper prices allow security practitioners to access maritime surveillance tools reaching far broader than air or sea assets, with a persistence that is increasing constantly. Tools like synthetic aperture radar, optical, infrared, and thermal, or electronic interceptions complement the suite of tools and assets used to collect data and information and increase the overall MSA.

In the maritime world, the level of involvement of space-based EO tools and information is increasing in an accelerated manner, and it has to do with the potential to address a persistent gap in surveillance. This integration of newer technology in the maritime domain is an approach that many other systems could nowadays afford to do, cost and performance-wise, given the adaptability of satellite EO tools to perform various jobs. From environmental data gathering and analysis to critical infrastructure persistent monitoring, the list is long and only increasing. One item of notice would be the upskilling of staff in the utilization of space-based EO products and information, which might pose some challenges, although the current workforce has already several digital skills that can be used for this. For instance, most people would correctly identify details in an optical image caught by a satellite, even to a limited extent, just by experiencing this from digital and navigational applications in everyday life.

2.5.9 Emerging Hybrid Threats

The overall hybrid threat landscape is quite complex if we observe it beyond the conventional threat vectors. An exhaustive discussion about such threat vectors is beyond the scope of this discussion, yet it is necessary to recap at least a few of the emerging hybrid threats that pose serious challenges to critical infrastructures. As an example, an electromagnetic pulse (EMP) [6] event is a short burst of high energy that can render electronic and electrical equipment like computers, cell phones, transformers, transmission lines, and the broader critical communications infrastructure inoperative. It could cause failures throughout several digital infrastructure systems—including banking, energy, transportation, food production and delivery, water, emergency services, and cyberspace—across an entire country or region. Disruption to underwater cables, gas pipelines, and other undersea connectivity means pose hidden vulnerability, in addition to intelligence gathering by state or non-state actors. Drones or unmanned ariel vehicles (UAVs) are now increasingly used in battle theatre against critical infrastructure and are a new frontier for policy considerations. Additionally, trade and investment policies by some nations or consortiums can acquire technologies through espionage, cyber operations, evasion of export control restrictions, and coercive intellectual property sharing requirements to destabilize other economies.

2.6 Policy Briefs, Conclusions and Path Forward

Disruptions to critical infrastructure present a complex and severe challenge with significant consequences for society. Addressing these challenges requires a comprehensive approach that includes modernization, enhanced cybersecurity, cross-sector collaboration, effective public communication, and robust contingency planning. By investing in these areas, societies can improve the resilience of their critical infrastructure and better withstand the threats posed by both natural and human-made disruptions. As per the policy briefs discussed above, the following mitigation strategies would enhance CIS security and bridge the security gaps. These include, (a): Infrastructure Modernization, viz. Investing in Upgrades: Modernizing outdated infrastructure with more resilient, secure, and efficient technologies can reduce vulnerability to disruptions. Smart Grids and Decentralization: Implementing smart grid technologies and decentralized systems can enhance the resilience of energy networks and other critical infrastructures. (b): Cybersecurity Enhancement: viz. Advanced Cyber Defenses: Strengthening cybersecurity measures to protect critical infrastructure from cyber-attacks, including intrusion detection systems, encryption, and regular security audits. Incident Response Planning: Develop robust incident response plans to quickly identify, contain, and recover from cyber-attacks on critical infrastructure. (c): Cross-Sector Collaboration: viz. Public-Private Partnerships: Encouraging collaboration between government and private sector entities to share information, resources, and best practices for infrastructure protection. International Cooperation: Engaging in international cooperation to address global threats to critical infrastructure, such as cybercrime and terrorism. (d): Public Communication and Trust-Building: viz. Transparent Communication: Ensuring transparent and timely communication with the public during infrastructure disruptions to prevent panic and misinformation. Community Engagement: Building public trust through community engagement initiatives, educating the public on emergency preparedness, and involving them in resilience-building efforts. (e): Redundancy and Contingency Planning: Backup Systems: Implementing redundant systems and backup plans to ensure the continuity of essential services during disruptions. Resilience Drills: Conduct regular resilience drills and simulations to test and improve the response to infrastructure disruptions. (f): Developing strategic pathways for safety against chemical and biological attacks: viz.: contamination protection gear for first responders and emergency personnel. Developing advanced sensors with bioinformatics capability to sense, detect, and identify contaminants in real time. In conclusion, by implementing advanced cybersecurity measures, enhancing physical security, building resilience against natural disasters, and adopting strategic economic and geopolitical practices, we can significantly improve the security and sustainability of critical infrastructure systems.

Declarations **Conflict of Interest Statement:** The authors declare no competing interest, financial or otherwise.

Authors' Contributions: All authors have contributed to the preparation of this article.

Funding Sources: The authors acknowledge funding from NATO for the Advanced Research Workshop on "Countering Hybrid Threats Against Critical Infrastructures" Brasov, Romania, May 21–23, 2024 "under the NATO Science for Peace and Security Programme.

Research Data Statement: Most of the data reported is either contained in the document and/or easily available from the authors upon request.

Use of Human Subjects or Animals in Research: Research Involving Animals: N.A. Research Involving Human Subjects, N.A.

References

1. UN (1982) United Nations Convention on the Law of the Sea of 10 December 1982
2. UN (1982) United Nations convention on the law of the sea III. Montego Bay, Jamaica
3. Vaseashta A (2022) Applying resilience to hybrid threats in infrastructure, digital, and social domains using multisectoral, multidisciplinary, and whole-of-government approach. In: Bogdanoski M (ed) NATO science for peace and security series—D: information and communication security. IOS Press
4. Thorisson H, Baiardi F, Angeler D, et al Resilience of critical infrastructure systems to hybrid threats with information disruption. NATO Science for Peace and Security Series—D: Information and Communication Security 55:13–26. https://doi.org/10.3233/NICSP190017
5. US WH presidential policy directive—critical infrastructure security and resilience
6. Vaseashta A (2023) Existential risks associated with dual-use technologies. https://doi.org/10.25740/ZY474YF0050
7. Lohela T (2023) Handbook on maritime hybrid threats. Helsinki
8. AGPRESS Bogdan Ivan: Site-ul Directoratului Național de Securitate Cibernetică, ținta unui atac cibernetic
9. Vaseashta A (2022) Future of water: challenges and potential solution pathways using a nexus of exponential technologies and transdisciplinarity. In: Vaseashta A, Duca G, Travin S (eds) Advances in environmental engineering and green technologies. IGI Global, pp. 37–63
10. Vaseashta A (2021) Introduction to water safety, security and sustainability. In: Vaseashta A, Maftei C (eds) Water safety, security and sustainability. Springer International Publishing, Cham, pp 3–22
11. Vaseashta A, Vaseashta S (2008) A survey of sensor network security. Sensors Transducers 94:91–102
12. Vaseashta A (2014) Advanced sciences convergence-based methods for surveillance of emerging trends in science, technology, and intelligence. Foresight 16:17–36. https://doi.org/10.1108/FS-10-2012-0074
13. Van Der Linden S (2022) Misinformation: susceptibility, spread, and interventions to immunize the public. Nat Med 28:460–467. https://doi.org/10.1038/s41591-022-01713-6
14. Vaseashta A, Bölgen N (2022) Electrospun nanofibers: principles, technology and novel applications. Springer International Publishing, Cham
15. Vaseashta A, Khudaverdyan S, Vaseashta A (2013) Advanced sensors for safety and security. Springer Netherlands Springer e-books Imprint. Springer, Dordrecht
16. Vaseashta A, Gevorgyan G, Kavaz D et al (2021) Exposome, biomonitoring, assessment, and data analytics to quantify universal water quality. In: Vaseashta A, Maftei C (eds) Water safety, security and sustainability. Springer International Publishing, Cham, pp 67–114
17. Ayllón D, Grimm V, Attinger S et al (2018) Cross-disciplinary links in environmental systems science: current state and claimed needs identified in a meta-review of process models. Sci Total Environ 622–623:954–973. https://doi.org/10.1016/j.scitotenv.2017.12.007

18. Steiner N, Andriciuc R (2023) Managementul riscului de inundatii si alte dezastre si cooperarea internaţională pentru reducerea efectelor acestora Bucureşti 2009. Bucuresti
19. Kingham R (2006) Inventory of environment and security policies and practices. Hague, Netherland
20. Necsutu M (2022) Cyber attacks hit Romanian government websites. Reporting Democracy, Chisinau
21. Vaseashta A, Duca G, Travin SO (2022) Handbook of research on water sciences and society. Engineering Science Reference, Hershey
22. Brčeski I, Vaseashta A (2021) Environmental forensic tools for water resources. In: Vaseashta A, Maftei C (eds) Water safety, security and sustainability. Springer International Publishing, Cham, pp 333–370
23. McBride J, Berman N (2024) How U.S. Water Infrastructure Works. US
24. Vaseashta A (2006) Carbon nanotubes based devices and sensors. Mater Manuf Process 21:710–716. https://doi.org/10.1080/10426910600613595
25. UNDRR (2024) The human cost of disasters: an overview of the last 20 years (2000–2019)
26. Townsend R Future of Food
27. Frischknecht F (2003) The history of biological warfare: human experimentation, modern nightmares and lone madmen in the twentieth century. EMBO Rep 4. https://doi.org/10.1038/sj.embor.embor849
28. Clancy J, McVicar A (2007) Handbook of chemical and biological warfare agents, 0th edn. CRC Press
29. Szinicz L (2005) History of chemical and biological warfare agents. Toxicology 214:167–181. https://doi.org/10.1016/j.tox.2005.06.011
30. Demir D, Bolgen N, Vaseashta A (2023) Electrospun nanofibers for biomedical, sensing, and energy harvesting functions. Polymers 15:4253. https://doi.org/10.3390/polym15214253
31. Vaseashta A, Karagülle-Bölgen N (2018) Loaded nanofibers: force protection, filtration, decontamination. In: Petkov P, Tsiulyanu D, Popov C, Kulisch W (eds) Advanced nanotechnologies for detection and defence against CBRN agents. Springer, Dordrecht, pp 241–252
32. USN (2023) U.S. Merchant Marine and World Maritime in Review. US
33. NCAGS & AWNIS Naval Cooperation and Guidance for Shipping

Chapter 3
Technology Innovations to Counter Threat Vectors

Arezki Benfdila, Yuri Dekhtyar, Gor Gevorgyan, Radu Muntean, and Emmanuel Karlo Nyarko

Abstract The rapid development of technology has introduced a wide range of new threat vectors, which present significant challenges across multiple sectors. These threats stem from various sources, including human activities, technological advancements, and natural processes, all of which require innovative approaches to mitigate their impacts. To counter these multifaceted threats, the paper reviews and discusses technological innovations designed to enhance system capacities, such as advanced sensors, Building Information Modelling (BIM), and real-time detection systems. A review of monitoring and data processing systems, including drones, photogrammetry, artificial intelligence, and big data, is provided. The integration of these innovations into counter-threat strategies is essential for addressing the complexity of modern challenges, particularly in areas such as water quality monitoring, anomaly detection, and cybersecurity.

Keywords Threat vectors · Technologies · Innovation · Sensors · BIM · Cyber security

A. Benfdila
Mouloud Mammeri University, Tizi-Ouzou, Algeria
e-mail: benfdila@ummto.dz

Y. Dekhtyar (✉)
Riga Technical University, Riga, Latvia
e-mail: jurijs.dehtjars@rtu.lv

G. Gevorgyan
Scientific Center of Zoology and Hydroecology, National Academy of Sciences of RA, Yerevan, Armenia
e-mail: gor.gevorgyan@sczhe.sci.am

R. Muntean
Transilvania University of Brasov, Brașov, Romania
e-mail: radu.m@unitbv.ro

E. K. Nyarko
Josip Juraj Strossmayer University of Osijek, Osijek, Croatia
e-mail: karlo.nyarko@ferit.hr

D. Radu et al. (eds.), *Countering Hybrid Threats Against Critical Infrastructures*, NATO Science for Peace and Security Series B: Physics and Biophysics, https://doi.org/10.1007/978-94-024-2304-4_3

3.1 Introduction to Threat Vectors

Threat vectors represent the ways or methods that malicious actors use to gain unauthorized access to systems, networks, or data. These vectors serve as entry points for various types of cyberattacks, physical intrusions, or hybrid threats that can jeopardize the security and integrity of critical infrastructure. Understanding threat vectors is critical to developing comprehensive security strategies that enable organizations to identify vulnerabilities, prioritize defenses, and allocate resources effectively. The vectors could be issued both because of human activity and sources.

The threats may be initiated from [1, 2]:

- nature, like earthquakes, floods, and storms can disrupt infrastructure and systems, leading to data loss or downtime;
- human activities provoking incidents include intentional acts like cyberattacks, terrorism, and sabotage, as well as unintentional actions such as human error or negligence;
- technologies, like malware, ransomware, software vulnerabilities, and hardware failures, delivering dangerous fields, materials/matters.

Threat vectors include a wide range of potential attack vectors, including but not limited to [1–3]:

- physical access: unauthorized intrusion into restricted areas or hardware;
- social engineering: manipulation of individuals to disclose sensitive information or perform actions that jeopardize security;
- supply chain attacks: compromising the integrity of equipment, hardware, or software during the manufacturing or distribution process;
- insider threats: malicious acts by individuals with legitimate access to systems or information;
- network-based vectors: exploitation of vulnerabilities in network protocols or configurations;
- software vulnerabilities: exploiting bugs or design flaws in applications or operating systems.

As technology continues to evolve, and systems become more interconnected, new and more sophisticated threat vectors are emerging. As such, maintaining an up-to-date understanding of potential threat vectors is essential for effective security management. Threat vectors can emerge from various sources that expose infrastructure to risks. The sources could be external or internal, technological or human, and involve direct or indirect pathways to compromise security. Threat vectors from natural sources, such as biological hazards, radiation, pollution, or natural disasters, are less commonly associated with traditional cybersecurity risks but can have significant impacts on information systems, infrastructure, and operational continuity. These threats can result from both man-made and natural activities and, often, affect physical infrastructure, environmental conditions, and human health, leading to indirect risks and operational challenges.

3.2 Threat Vectors Related to Human Activities

Human activities are one of the primary sources of cybersecurity threats [4]. Threats from human activities can be issued intentionally or unintentionally. Intentional threats come from malicious actors who want to gain unauthorized access to systems, networks, or data, while unintentional threats are expressed by accidents that can lead to security breaches [5].

Some of the primary threat vectors stemming from human activities are characterized in Table 3.1.

Table 3.1 Characterization of main threat vectors related to human activities

Threat vectors	Threat description	Reference
Social engineering	Tricking people into clicking on malicious links opening attachments via SMS messages and voice calls or using attractive offers and gifts.	[6, 7]
Insider threats	Employees, contractors, or outsourced service providers intentionally/unintentionally cause harm to an organization, e.g. by clicking on phishing links or mishandling sensitive data.	[8]
Physical security vulnerabilities	Unauthorized persons obtaining physical access to equipment/ facilities or gathering information about an organization's activities; loss of devices or physical media containing sensitive data; intentional damage to infrastructure/equipment.	[9]
Supply chain attacks	Supply chains are hacked by software vendors, hardware manufacturers, or service providers.	[10]
Negligence and lack of awareness	Misconfigurations, weak passwords, and lack of awareness cause vulnerabilities, the use of personal devices to access corporate systems without proper safeguards	[11]
Lack of training	Employees are not being adequately trained on security best practices causing them to not recognize phishing attempts, not know how to handle sensitive data securely, or how to respond effectively to incidents.	[12]
Social media threats	Gathering information and data about individuals and organizations through social media platforms.	[13]
Remote work threats	Inadequate security measures in home environments and difficulty in managing and securing devices and data for remote work.	[14]
Third-party vendor risks	Organizations relying on external vendors or partners for services, human errors or security lapses from these third parties can introduce threats	[15]
Open-source software	Can expose organizations to risks due to lack of oversight, outdated code, or vulnerabilities in publicly available software packages	[16]
Legacy systems and obsolete technologies	Legacy software that no longer receives security updates can have known vulnerabilities that attackers can exploit. Older hardware may lack the ability to support modern security features, making it easier for attackers to bypass defenses.	[17]
IoT (internet of things) devices	Poorly secured IoT devices, such as smart cameras, thermostats, or medical devices, can be hijacked and used as entry points into networks.	[18]

3.3 Threat Vectors Related to Technologies

3.3.1 Electrical Power Supply Systems

Electrical power infrastructure, which includes power generation facilities and transmission lines, are crucial components of national infrastructure. The complexity and interconnectedness of modern power systems introduce multiple vulnerabilities that malicious actors may exploit. These vulnerabilities pose risks not only to the stability and security of the electrical infrastructure but also to national security and economic stability since compromising these systems can lead to widespread blackouts, affecting communities and critical services. Key vulnerabilities in electrical power infrastructure include:

- Physical threats: Sabotage or destruction of power plants, substations, or transmission lines [19, 20]. The extensive and often remote nature of electric power infrastructure makes it difficult to protect comprehensively.
- Cyber-attacks: Targeting control systems, Supervisory Control and Data Acquisition (SCADA) networks, or smart grid components. The systems that manage the electric grid are prime targets for cyber-attacks [21–24].
- Insider Threats: Employees with access to critical infrastructure can introduce significant vulnerabilities either through malicious intent or negligence to the electric grid [25, 26].
- Electromagnetic pulse (EMP) attacks: Such attacks can potentially disable large portions of the power grid.
- Natural disasters: Extreme weather events or geomagnetic disturbances are known to affect power distribution [27].

Power generation equipment manufacturers play a critical role in the energy sector since they are responsible for the production and distribution of critical equipment, i.e., machinery and technology necessary for generating electricity and for the power grid. However, these manufacturers can also pose significant threat vectors due to various vulnerabilities that can be exploited by malicious actors:

- Supply chain vulnerabilities: The complicated supply chains of manufacturers can be compromised by attacks on vendors or suppliers. This can lead to vulnerabilities that impact the entire manufacturing process and potentially lead to the production of compromised devices.
- Industrial Control Systems (ICS): ICS, such as Supervisory Control and Data Acquisition (SCADA) systems, are essential to the operation of electrical manufacturing equipment and facilities. By installing backdoors or worms, cyber-attacks on these systems can be initiated after being deployed causing significant disruption, as demonstrated by incidents such as the Stuxnet worm [28], which attacked Iran's nuclear facilities by exploiting ICS vulnerabilities.

3.3.2 Security and Communication Systems

Security and communication systems are fundamental components of critical infrastructures that ensure the protection and smooth operation of various sectors such as finance, healthcare, and transport. It is also important to note that these systems not only facilitate the exchange of information critical to operations but also serve as the first line of defense against various threats. Key components of security and communication systems include:

- physical security systems: access control, video surveillance, and intrusion detection systems.
- communication networks: secure voice and data transmission systems, including encrypted channels for sensitive information [29].
- cybersecurity infrastructure: firewalls, intrusion prevention systems, and SIEM (Security Information and Event Management) solutions [30].
- emergency response systems: integrated platforms to coordinate responses to security incidents or natural disasters.

However, these systems are vulnerable to a number of threat vectors that can compromise their functionality and security. Cybersecurity threats are a major concern for security and communication systems. These threats include hacker attacks, malware, and denial of service (DoS) attacks that can disrupt communication channels and compromise sensitive data. The WannaCry ransomware attack of 2017, which affected numerous organizations worldwide, including the UK's National Health Service, is a prominent example of the potential impact of cyber threats on communication systems [31].

In addition to cyber threats, physical attacks on communications infrastructure, such as mobile phones or cell towers and data centers, can also disrupt services and hinder communications. For example, vandalism to fiber optic cables can lead to significant outages affecting internet and telephone services across large regions [32].

3.3.3 Radiation Technologies

Radiation used for different technologies is classified into two groups: (1); ionizing, and (2) and non-ionizing radiation. The first one is when a matter ionizes its atoms/molecules. The non-ionizing radiation does not produce ions, however, it can excite/oscillate molecules. Radiation of both groups can induce radiobiological effects in a human organism [33].

Ionizing radiation that may be absorbed by a human arrives from space, the environment (including water and products), and human-made sources. Space and environmental radiation depend on geography the human lives. Therefore, the typical doses absorbed by inhabitants vary for different territories and are referred to in national legislation.

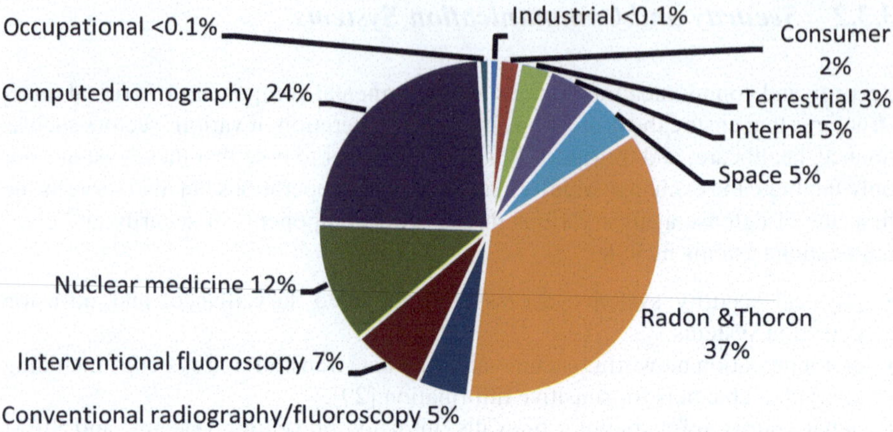

Fig. 3.1 Doses from different sources (the USA) [34]. Medical technologies and backgrounds deliver around 48% and 50% of the dose correspondingly. Doses of medicine do not include radiation therapy

Figure 3.1 gives a sense of the typical ionizing radiation doses from different sources (the example from the USA) [34].

The human-made sources (HMS) deliver radiation to the human organism mainly (see Fig. 3.1) in accordance with medical prescriptions (X-ray diagnostics, positron emission tomography, computed tomography, scintigraphy, radiation treatment, etc). In this case, the medicine follows the ALARA (As Low As Reasonable Achievable) principle.

Alongside HMS may provide humans with radiation occasionally because of the case incidents. Following Fig. 3.1, the highest background dose humans are getting from Radon and its isotope Thoron. Radon is a gas coming to dwellings from the ground the houses are situated on and building construction materials [35].

The non-ionizing radiation (NIR) that affects humans is usually delivered as electromagnetic radiation from the Sun, lighting fixtures, microwave ovens, cell phones, WiFi and Bluetooth transducers, etc. The skin color change is the typical radiobiological effect provoked by the Sunlight at a beach. However, to provide a significant influence of NIR on the human organism the delivered doses are assumed to be high enough that increase the probability of heat and other effects [36].

3.3.4 Pollutions and Contaminations

3.3.4.1 Pollution of Air, Water or Soil

Pollution or contamination of air, water, or soil can cause disruptions to physical infrastructure, personnel safety, and operational integrity. These threats can indirectly affect infrastructure and systems by reducing access to facilities or resources.

Air pollution results from the release of harmful substances into the atmosphere, including gases, particulate matter, and chemicals. These pollutants can originate from industrial processes, vehicle emissions, wildfires, and other human or natural activities [37].

Factories and power plants release pollutants such as sulfur dioxide (SO_2), nitrogen oxides (NOx), carbon monoxide (CO), and volatile organic compounds (VOCs). All these can damage sensitive electronic equipment, reduce air quality, affect human health, and impose the need to implement costly air filtration systems in sensitive facilities [37].

Wildfires release large amounts of particulate matter (PM) and toxins into the air, severely degrading air quality and potentially interfering with infrastructure, causing overheating due to blocked cooling systems and equipment failures.

Ozone pollution, particularly at ground level, contributes to respiratory issues that can affect human operational efficiency and compromise outdoor installations such as telecommunications equipment.

Water pollution involves the contamination of water bodies such as rivers, lakes, and oceans with harmful substances, including chemicals, heavy metals, and biological waste. This pollution can arise from industrial discharges, agricultural runoff, and untreated sewage [38]

Discharge of toxic chemicals, heavy metals (e.g., mercury, lead), and solvents into water bodies by factories and manufacturing plants can affect industries that rely on water-based cooling systems in data centers or power plants. Oil spills, plastic waste, and chemical contaminants from maritime activities can severely impact coastal infrastructure affecting the underwater communication cables or adding delays in maritime operations. Pesticides and fertilizers used in agriculture may contaminate water supplies, leading to toxicity in local ecosystems and affecting water-dependent systems.

Soil contamination occurs when harmful substances such as heavy metals, chemicals, or hazardous waste seep into the ground, affecting land quality and usage. Contaminated soil can result from industrial accidents, improper waste disposal, mining activities, and agricultural practices [39].

Improperly managed industrial waste and landfill sites can lead to soil contamination, affecting nearby infrastructure, compromising building foundations, and health risks to on-site personnel. Long-term environmental damage can make industrial areas unsuitable for habitation or operations, forcing relocations or shutdowns.

Mining activities often lead to soil contamination due to the release of heavy metals, acids, and other harmful by-products causing corrosion of underground cables, affecting network reliability, and increasing the likelihood of cyber incidents.

Noise pollution refers to excessive, unwanted sounds that can interfere with human activities, equipment functionality, and infrastructure stability. This noise originates, mainly, from industrial machinery, transportation systems, and construction activities and can affect worker concentration and efficiency, increasing the likelihood of human error in maintaining secure systems, and delayed response to incidents. A particular case is the airports, railways, and highways which often produce significant noise pollution, that can interfere with sensitive equipment in nearby data centers or control rooms.

Light pollution is excessive or *misdirected* artificial light in the environment, often affecting nighttime visibility and human health. This is particularly common in urban areas with heavy industrial or commercial activity. Excessive lighting can interfere with sensitive instruments and sensors, potentially affecting research facilities, satellite communication systems, and ground-based sensor data, leading to potential data integrity issues.

3.3.4.2 Radioactive Contamination

Radioactive contamination occurs when radioactive materials are released into the environment, often because of nuclear accidents, waste disposal, or medical and industrial use of radioactive materials, leading to long-term evacuation of affected areas, unmonitored facilities, and damage to infrastructure that may be difficult to assess or repair. Also, improper disposal of radioactive waste can lead to contamination of land, water, and air with all the above-mentioned issues.

3.3.4.3 Heat Pollution

Urbanization, despite its numerous advantages, is a major contributor to environmental issues. The microclimate is impacted by urbanization, creating distinct urban climatic environments like the Urban Heat Island, also known as heat pollution, which is a phenomenon where urban areas are noticeably warmer than surrounding rural regions. This phenomenon is primarily due to anthropogenic alteration of the natural landscape, such as an increase of impervious surfaces, reduction of vegetation, industrialization, and poor urban design. It may have detrimental repercussions on the environment, including higher temperatures, worse air and water quality, higher energy use, and infrastructure damage [40]. Heat pollution also poses a serious threat to critical infrastructures, including:

- energy sector: higher cooling demand, straining power networks; thermal expansion causing damage to power lines and transformers; decreased power plant efficiency.
- transportation sector: thermal expansion damages railroad lines and bridges; melting asphalt roads create potholes and safety problems; vehicles, especially air-conditioned ones, operate less well.
- water sector: damage to water treatment facilities and distribution systems; increased water demand for irrigation and cooling; decreased water quality as a result of pollution and increased algae growth.
- communication sector: higher chance of communication failures and outages; decreased performance of electronic components owing to heat [41].

- healthcare sector: an increase in heat-related diseases, including heatstroke, dehydration, and even fatalities; a burden on medical resources to handle heat-related crises [42].

3.3.4.4 Pollution Because of Electricity Production and Transmission

The production and transmission of electricity are crucial to modern society but contribute significantly to environmental pollution. Fossil fuel-based power plants, especially coal-fired power plants, are major sources of air pollution. They emit SO_2, nitrogen oxides (NO_x), particulate matter (PM), and mercury (Hg) [43], which can lead to acid rain, smog and health problems. These power plants also emit significant amounts of carbon dioxide (CO_2), a major greenhouse gas. While natural gas power plants are cleaner than coal, they still emit CO_2 and NO_x and can release methane through leaks [44]. Renewable energy sources are cleaner, but not completely free of pollutants. Toxic materials and energy-intensive processes are used in the manufacture of solar panels and wind turbines [45], while hydroelectric dams can disrupt ecosystems and release methane [46]. Nuclear power plants cause only minimal air pollution but generate radioactive waste that must be disposed of in the long term [47]. On the other hand, the accidental release of radioactive material from nuclear power plants, as seen in disasters such as Chornobyl and Fukushima, has catastrophic effects on the environment and health, contaminating air, water, and soil.

Electricity transmission is also a problem for the environment, though less than production. The construction and maintenance of high-voltage power lines can lead to habitat fragmentation and soil erosion [48]. Electromagnetic fields generated by power lines have raised health concerns, although current evidence does not confirm health effects at low levels of exposure [49]. In urban areas, overhead lines can cause visual pollution, which has led to calls for underground cabling.

3.4 Threat Vectors Related to Nature

3.4.1 Biological Sources

Critical infrastructures that rely on biological resources and systems may be seriously threatened by biological sources in terms of cybersecurity. This is due to the fact that biological data can be difficult to safeguard due to its complexity and diversity, and the fact that many biological systems are integrated with other technologies, which raises the possibility of vulnerabilities [50]. Some key threat vectors stemming from biological sources could be the following:

- biometric data breaches: unauthorized access to sensitive biometric data, such as fingerprints, facial recognition, and DNA sequences; manipulation of biometric data to bypass authentication or identification systems; and misuse of biometric data for unauthorized purposes such as identity theft or surveillance [51].

- healthcare data breaches: unauthorized access to sensitive patient information, such as patient health records and medical device data, in order to disclose patient data, take advantage of security flaws in medical devices to obtain unauthorized control, or encrypt medical records and systems for demanding ransom from organizations [52].
- critical infrastructure threats: using biological agents to destroy critical infrastructures, launching harmful elements against providers of biological materials or equipment, and interfering with vital services that depend on biological resources [53].
- emerging threats: unauthorized biological system experiments and the misuse of synthetic biology might result in the production of dangerous biological agents [54].

3.4.2 Natural Disasters with Radiation or Pollution Effects

Natural disasters such as earthquakes, tsunamis, or volcanic eruptions can release radiation or pollutants into the environment, which in turn can disrupt critical infrastructure, especially in industries relying on safe environmental conditions (e.g., nuclear power, chemical plants, transportation systems, etc.). The combination of natural disasters with radiation or pollution effects poses complex challenges for emergency response, infrastructure resilience, and public safety.

Volcanic eruptions can disrupt air travel, satellite communications, and power grids, leading to cascading effects on IT infrastructure, such as communication loss, data center downtime, power grid failures, etc. The release of toxic gases can pose health risks to employees and force the shutdown of physical operations, evacuation of key personnel, and leaving systems unattended and vulnerable to cyber-attacks or physical intrusion.

A real-life example is the eruption of the Eyjafjallajökull volcano in Iceland in 2010 which released large amounts of ash into the atmosphere, disrupting air travel across Europe for several weeks. Over 100,000 flights were canceled, affecting millions of passengers [55]. The ash cloud contained fine particles that posed a threat to aircraft engines and air traffic systems.

Earthquakes, especially those occurring near nuclear facilities, can result in accidents leading to the release of radioactive material. When safety systems fail due to seismic damage, the uncontrolled release of radiation can contaminate the environment, air, and water, posing long-term health and safety risks [56, 57].

An example is the Fukushima Daiichi Nuclear disaster [58]. A 9.0 magnitude earthquake, followed by a massive tsunami, struck Japan in 2011. The natural disaster caused critical failures at the Fukushima Daiichi Nuclear Power Plant. Cooling systems failed, leading to the release of radioactive materials into the environment. Radioactive isotopes, such as iodine-131, cesium-137, and strontium-90 were released into the atmosphere affecting the air, soil, and water, resulting in long-term environmental contamination and displacement of thousands of residents. The effect was a large-scale evacuation, contamination of local food and water supplies, radiation exposure, and significant economic impacts due to decommissioning and clean-up efforts.

Tsunamis are large sea waves caused by seismic activity that can flood coastal areas, damaging infrastructure and causing significant operational disruption, loss of power and communication, physical damage to servers, and data corruption.

The 2004 Indian Ocean tsunami, triggered by a 9.1–9.3 magnitude earthquake, devastated coastal communities across multiple countries. In addition to widespread death and destruction, the tsunami caused severe environmental pollution as it swept through industrial zones, carrying hazardous materials, chemicals, and waste into the ocean and inland areas.

Floodings caused by heavy rains, hurricanes, or the overflow of rivers can cause widespread flooding, affecting buildings, transportation, and power supply.

Opposite to floodings, prolonged periods of *drought* can have a significant impact on industries that rely on water, such as data centers that use water-based cooling systems causing overheating of IT infrastructure and increased risk of system failures or outages.

Landslides and mudslides are often triggered by heavy rains or earthquakes and can damage physical infrastructure, including buildings, roads, and communication lines, blocking access to key facilities, preventing maintenance or repair, etc.

Besides these natural phenomena, *dust storms* and *hurricanes* bring strong winds, storm surges, and flooding, which can damage any kind of infrastructure, buildings, industrial and agricultural facilities, leading to the release of hazardous and toxic substances, sand and dust, and other particulate matter over long distances, contributing to air pollution and respiratory health issues.

3.5 Technological Innovations as the Hybrid Threat Vectors

3.5.1 Radiation

Although the use of radiation is under strong restrictions and controlled at both the international [59, 60], etc. and national [61, 62], etc. levels, the impact on radiation protection and safety policies has some risks.

To avoid/diminish them, the ethical principles are applied when operating radiation [63]:

- justification: maximize benefits to most people.
- optimization: ALARA
- effective dose limits for members of the public.
- protection of environment.
- protection of human health and human rights.

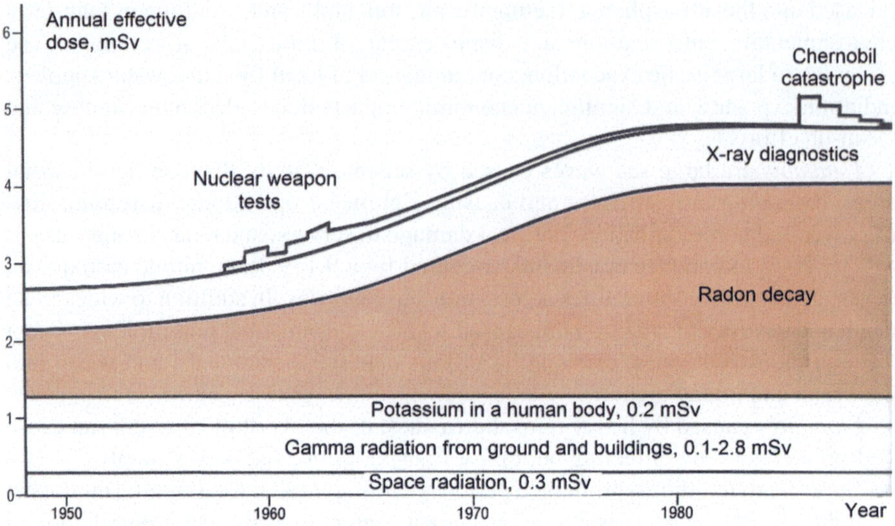

Fig. 3.2 Increment of the background dose [35] (example for inhabitants of Sweden)

The medicine increases contribution to radiation doses from 15% in the early 1980s to 50% today (the USA case) [64]. This relates to the wider application of modern medical technologies using radiation.

The impact of Radon delivering the highest (Fig. 3.1) background dose increases too. In particular, in 1970th alum slinky lightweight concrete that emitted Radon began to be widely used [35]. Because of this, the background dose increased (Fig. 3.2) [35].

3.5.2 Nanotechnologies

When a solid substance scale at least in one dimension is around 1–1000 atoms (nanometric scale) their amount defines the properties of the matter. Modern technologies can fabricate such objects (nanoobjects), assemble/manipulate them, and use them for different applications [65].

The nanoobjects sized to nm in one, two or three dimensions are named as nanofilms, nanowires/nanotubes, and nanoparticles/nanodots, correspondingly. The nanotubes and nanoparticles can travel through a membrane of the cell and impact its survival, proliferation, and differentiation. For instance, DNA and carbon nanotubes are chemically compatible and DNA spontaneously can bind to carbon nanotubes [66]. The biochemical and biological activity of nanoparticles depends on their size [67], etc. and surface electrical charge density [68]. Obviously, the activity of the nanoparticle ensemble is directly proportional to the ratio of the particle surface area to its volume. In other words, smaller particles are characterized by higher

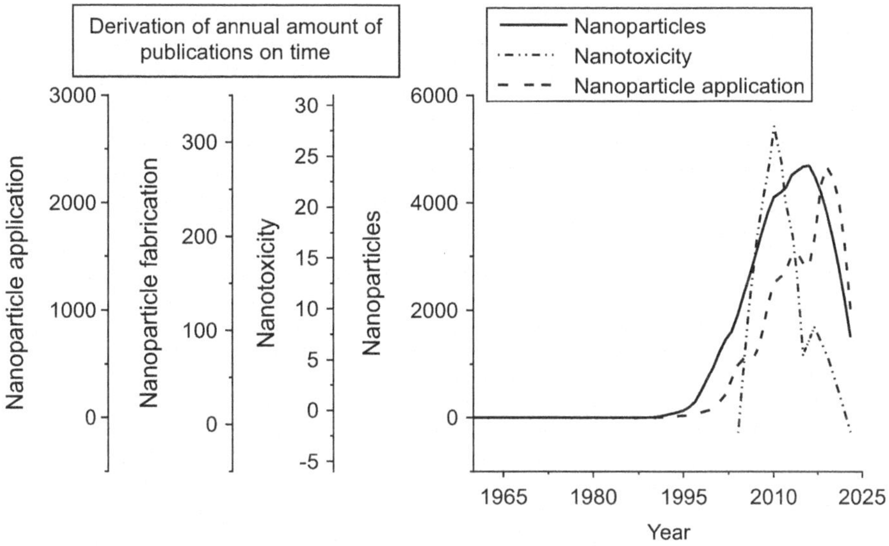

Fig. 3.3 Derivations of annual amount of publications on time (the key words indicated above were used to collect data from SCOPUS)

activity and therefore possible toxicity. The latter paid strong attention after the nanoparticles became widely applied.

The dynamics of the annual amount of publications are evidence in favor of that (Fig. 3.3). In spite of the research in the field of nanoparticles that began in the 1960s, their applications paid a significant interest only after 1995. However, the nanotoxicity importance was indicated with a rapid publication growth a little bit later - after 2000th.

3.6 Technological Innovations to Counter Hybrid Threat Vectors

3.6.1 Capacity of Technologies

Technological innovations play a crucial role in countering different threat vectors, which are increasingly sophisticated and multidimensional nowadays. Many of these advancements are focused on predictive, preemptive, and responsive capabilities, ensuring that organizations and governments can mitigate and neutralize threats across cyber, physical, and informational domains.

Natural threats, such as disasters, environmental pollution, and radiation, require comprehensive strategies that incorporate advanced technologies, forward-thinking policies, and innovative approaches to reduce risk and protect infrastructure, communities, and the environment. As the frequency and intensity of these natural threats increase, governments, organizations, and industries are implementing a variety of measures aimed at preventing, mitigating and responding to these hazards.

One of the first measures to fight against natural threats is building *climate-resilient infrastructure* using advanced sustainable materials, designs and construction methods that can withstand the effects of natural disasters and climate change. Many governments are implementing stricter building codes that require new construction to be resilient to specific natural hazards, such as earthquakes, hurricanes and floods. Designing energy-efficient and disaster-resilient buildings that can maintain functionality during and after disasters, including passive cooling and renewable energy sources like solar panels. Innovations such as self-healing concrete, fiber-reinforced polymers, and earthquake-resistant materials ensure buildings can endure natural threats like earthquakes and floods.

Following the same idea smart cities and IoT sensors to monitor infrastructure conditions in real time and to detect potential issues before they escalate are being developed. For example, sensors can monitor bridge integrity, building stability, and flood levels, as well as air and water pollution, radiation, heat and noise levels, etc.

All these technologies and materials belong to the so-called “Early Warning Systems (EWS) for Natural Disasters”. Early warning systems use sensor networks, satellites and communication technologies to detect and predict natural hazards such as earthquakes, tsunamis, floods, hurricanes, and volcanic eruptions, providing real-time alerts to populations and authorities and enabling timely evacuation and response.

Different sensors are used to monitor and predict threats.

3.6.2 Sensors

To predict threats the corresponding hazards are monitored. The sensors are the fundamental part of observing dangers. To monitor different phenomena a wide range of sensors is employed: electrical, chemical, optical, biological, etc. Because of a huge volume of information should be acquitted and rapidly processed, the sensors are expected to have small sizes, low energy consumption, and fast transition of an informing signal. All of these are achievable when micro and nanotechnologies are applied to fabricate the sensors. Following Moor’s law [69] the elements of the electronic circuits decreased from 20 μm in 1968 to 2–3 nm (corresponds to tens of atoms) in 2024. Miniaturization allows for a decrease in power consumption. For instance, the voltage in mobile computers has been reduced from 5 (1979) to 2.5 V (1997) [70] and the frequency of the processor increased from ~1 MHz (1972) to 5 10 3 MHz (2015) [71].

The sensors have to be protected against external and internal influence. To avoid external impacts the sensors are capsulated, shielded, etc. However, protection against ionizing radiation requires to use of thick and usually heavy containers. Ionizing radiation induces charge carriers in sensor materials and damage/

reconstruct their chemical bonds. These disturb the functionality of sensors and destroy them.

The internal factors, like radiation delivered from radioactive isotopes naturally presented in sensor materials, cannot be restricted. To diminish their influence, the devices are provided with reservation modulus which acts alongside the main one. If the latter fails, the signal is acquired from the reserved unit.

3.6.3 Fundamentals and Design

Sensing devices range from macro to nanoscale dimensions depending on the applications and constraints. Micro-sensors are generally based on the field effect sensing principle that is based on a field effect transistor [72–74] and are called electronic sensors. This type of sensor is embedded in integrated circuits to optimize all performances [75–77].

As field effect devices shrink in size, nanosensors emerged based on nanoelectronics principles [78–80]. There are silicon-based sensors [81], Graphene-based sensors [82] and Carbone Nanotube (CNT) based sensors [83].

Nanosensors are not always related to the nanometric dimensions; they are also related to the base material structure that is a nanomaterial. Nanosensors use nanoparticles, nanomaterials and bulk materials (0D, 1D, 2D, and 3D).

In our work, we are interested in sensors based on the field effect and implemented using nanoelectronics principles. Most of the nanosensors are said to be chemical sensors based on the field effect sensing principle.

Electronic sensors convert physical/chemical phenomena to an electric signal that can be current, voltage, or impedance, which are measurable or extracted through circuit analysis. A device that performs this role is the field effect transistor (FET) shown in Fig. 3.4.

The conducting channel controls the drain-source current (I_{DS}) flow through the applied gate-substrate (V_{GS}) voltage. The current control is subjected to the gate and

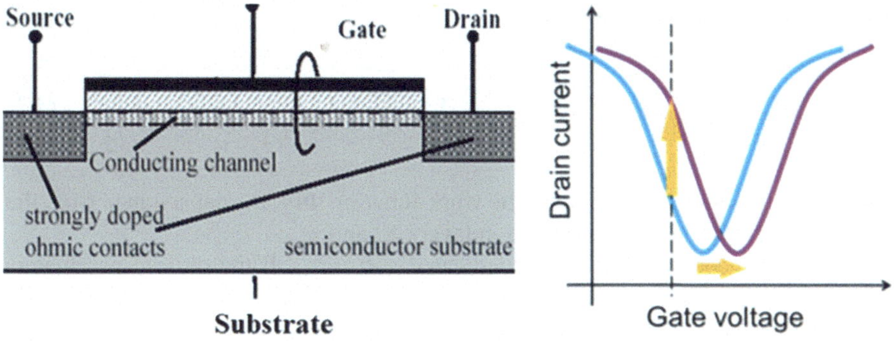

Fig. 3.4 FET transistor Structure (MOSFET) and its typical characteristic current-voltage curves

the oxide structures that act through the field effect (carrier depletion from or attraction to the channel surface) to control the channel resistance by reducing or increasing the carrier density in the conducting channel through the applied gate voltage. The basic current equation is given by: *(the parameters in the equations have their standard meanings. V_T is the threshold voltage).*

$$I_{DS} = \mu_n C_{ox}\frac{W}{L}\left[(V_{GS} - V_{fb} - 2\Phi_{Fi} - \frac{\sqrt{4\varepsilon_{sc} e N_A \Phi_{Fi}}}{C_{ox}}\right]V_{DS}, \text{and}$$

$$V_T = V_{TO} + \gamma\left(\sqrt{2\Phi_{Fi} + V_{SB}}\right) - \sqrt{2\Phi_{Fi}} \tag{3.1}$$

The channel will start conduction when the gate voltage (V_{GS}) is greater than V_T, the threshold voltage. Depending on the quality of the gate, oxide and interface materials, V_T will change accordingly. This variation shown in Fig. 3.1 is used to determine the quality of gate materials.

For the sensing mechanism, SensFETs are used: the gate, the oxide, and the channel all together with gate and drain-source biases are involved. The gate voltage varies with the physical/chemical structure of the complex Metal-Oxide-Semiconductor part of the FET device shown in Fig. 3.1.

For sensing mechanisms, the Metal and oxide are mostly used. The elements to detect are deposited on/in the metal and on/in the oxide. Openings from the metal to the semiconductor or trenches are performed to put the agent to detect.

Gate and oxide-engineered structures allowing the agent introduction leads to a reduction/increase of the threshold voltage V_T this variation is given by ΔV_{Th} and used in conjunction with the graphs in Fig. 3.4 to estimate the detected agent. The gate and the drain voltage are kept constant during the detection operation. Figure 3.5 shows a typical sensor called SensFET and its corresponding MOSFET to illustrate gate engineering.

In Fig. 3.5 configuration, the gate voltage V_{GS} is varied and the reference electrode acts on the solution ions through the field effect. The drain current is plotted with respect to V_{GS} both for presence and absence of agent solution. This case is illustrated by typical measured curves shown in Fig. 3.6. The basic equation giving the curve shift in terms of ΔV_{Th} is:

$$\Delta V_{TO} = \frac{1}{C_{ox}}\sqrt{4e\varepsilon_{Si} N_A \varphi_{Fi}}\,\frac{x_j}{2L}\left[\left(\sqrt{1+\frac{2x_{dS}}{x_j}}-1\right)+\left(\sqrt{1+\frac{2x_{dD}}{x_j}}-1\right)\right] \tag{3.2}$$

The agent parameters are given by the work function ΦFi, the capacitance Cox, the doping concentration NA the depletion layers X, and x.

Using advanced simulators and dedicated software, all the desired parameters can be extracted. The major advantage of such nanosensors is that they can be easily integrated into a chip that handles data processing and computations.

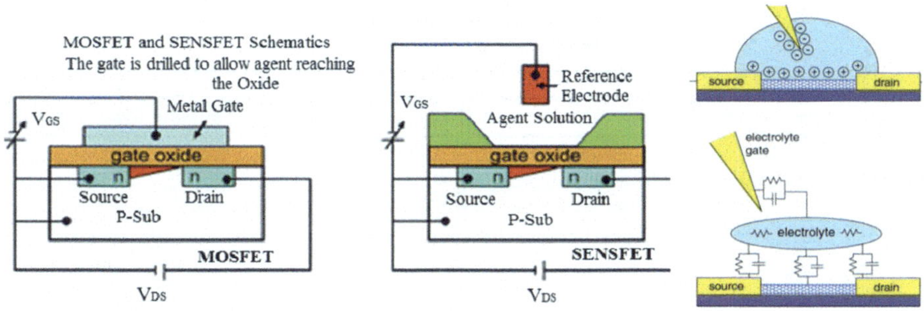

Fig. 3.5 Using the MOSFET as chemical Sensors by engineering its gate

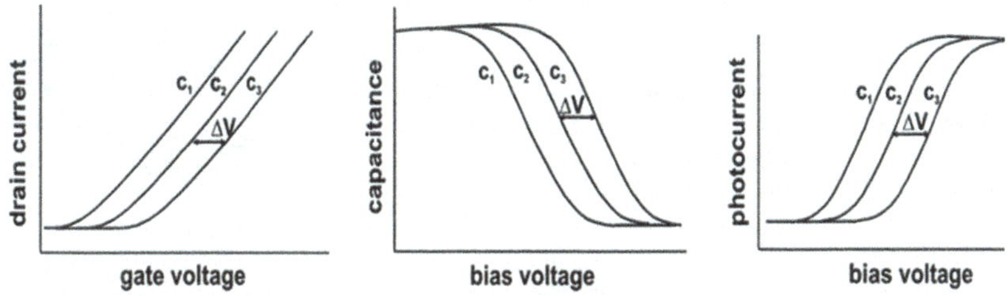

Fig. 3.6 Curve shifts due to threshold voltage for current, capacitance and illumination

The field effect principle applies whether the FET is silicon, graphene and CNT. One must care about the drain current equation parameters which differ from one to the other. There is a variety of configuration, type and structures on nanosensors, some of them are described below.

3.6.4 Fabrication Technologies and Applications

As smart devices are entering all fields of life, nanosensors are the sound of all smart devices and systems. Health care, defense, environment, and industry, are the prominent fields of applications as shown in Fig. 3.7. As nanosensors, they are mostly based on the field effect principle, i.e., the field effect transistor, as shown in Fig. 3.8.

The base transistor material can be silicon (MOSFETs), graphene (GFETs), or CNT (CNFETs). The three devices can be integrated on a chip to build a very efficient functional system.

As we are focusing on nanoelectronics sensors, the most prominent nanosensors are the chemical (ChemFETs) and the Biological (BioFETs) sensors. Figure 3.8 shows an overview of nano-biosensor applications and samples of nanodevices.

Figure 3.8 (right) shows the three major devices used in Biosensing and chemical sensing. The Cell-FET is used in the major biological systems and it is mainly based

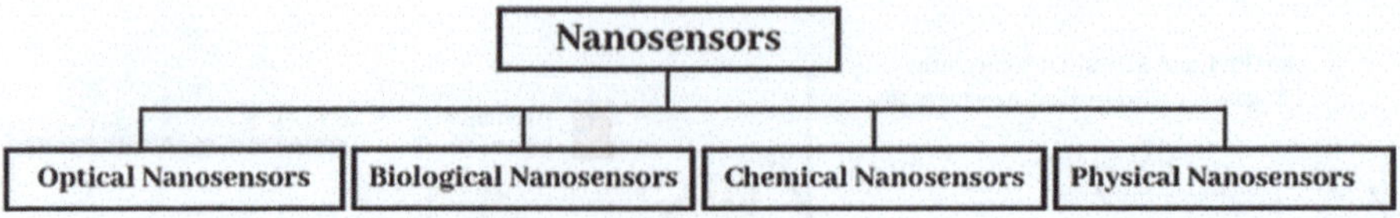

Fig. 3.7 Nanosensors classification and grouping

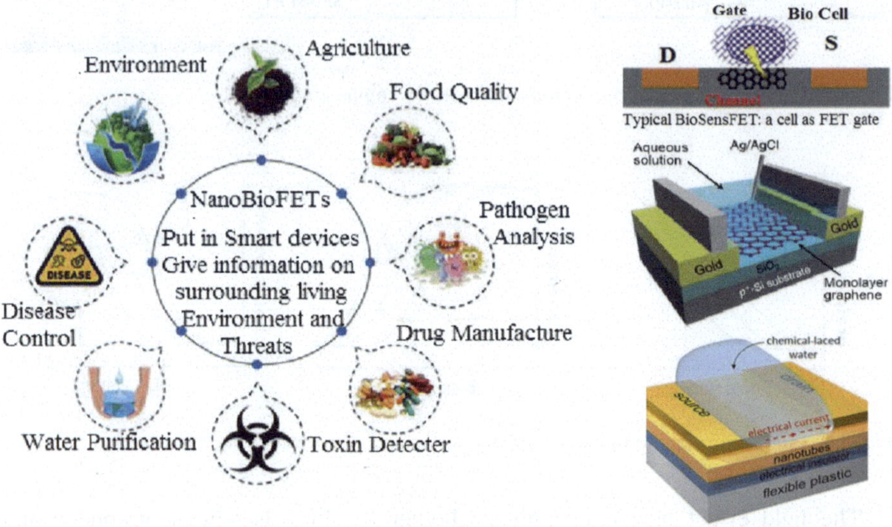

Fig. 3.8 Application and Type of Nanobiosensors (cell Sensors, CNT sensors, and Graphene sensors [76])

on silicon. The Graphene based sensor is used in all aqueous solutions (bio-physiological and Chemical) and the CNT sensor is used for gel solutions.

As it is shown in Fig. 3.8 (right panel), nano-sensor technology depends on the aspect agent to sense. The architecture is always a field effect device; however, the material structure is different. Silicon nanowires (SNW), Graphene nanoribbons (GNR), and carbon nanotubes (CNT) are always used for sensor design and implementation.

Disease control, food analysis, and drug manufacturing generally use CNT and GNR-based sensors. They are integrated into chips of low density. To be efficient, design and manufacturing are done on hybrid integrated circuits. It is known that CNT and Graphene are resistive materials and are not degraded with agents (acid, chemical, or others).

Environment monitoring and water purification use mainly all the technology as in the environment and waters there are many agents (biological, chemical, physical).

To be efficient, nano-sensors are designed as smart hybrid multifunction integrated circuits. Figure 3.9 (left) shows a nanocircuit integrating different devices

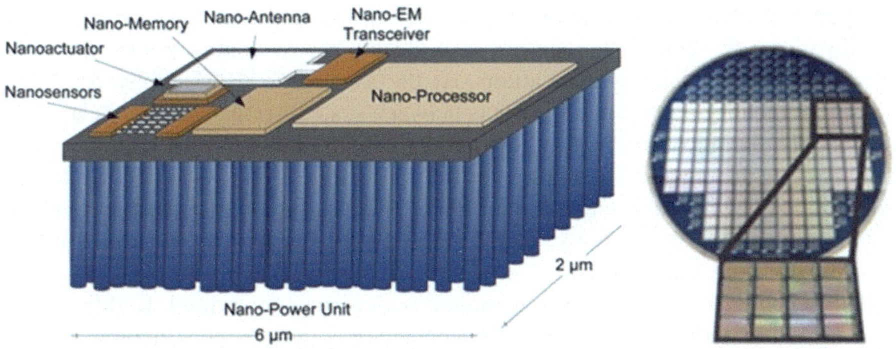

Fig. 3.9 Integrated circuit nanosensors for autonomous and portable uses [84]

ensuring different functions leading to intelligent and smart detection. Data are directly transmitted to a control unit/system.

An array of sensors may be assembled from a specialized unit (sensors). As shown in Fig. 3.9 (right) a chip can hold different sensing units and for the sake of security can use multichannel processing and transmission.

Single-use nano-sensors, especially in biological/analysis, use sensors that are discarded after use. In order to be efficient, they are produced in large numbers as shown on the wafer in Fig. 3.9.

3.6.5 Building Information Modelling

Building Information Modeling (BIM) is a powerful technology used primarily in the architecture, engineering, and construction (AEC) industries. BIM's core strength lies in its ability to integrate data, create digital twins, and facilitate collaboration across all stages of building design, execution, and use. When specifically applied to countering hybrid threats, BIM can enhance both the design and use phases to ensure that buildings are resilient, secure, and equipped to handle both physical and cyber vulnerabilities by providing detailed, real-time information about critical infrastructure and enabling proactive defense, risk assessment, and rapid response.

During the early design stages, BIM facilitates the collaboration of architects, engineers, cybersecurity experts, and security professionals to simulate various emergency scenarios, such as evacuation plans during a natural disaster, a physical attack or system shutdowns during cyberattacks. BIM models allow for the detailed mapping of evacuation routes in complex infrastructures. Real-time data on building occupancy, fire conditions, and other emergency parameters can be integrated into BIM to guide occupants to safe exits during a hybrid attack.

During construction, BIM provides contractors with precise, detailed models that show the exact placement of all structural, electrical, and mechanical components. This precision ensures that security and resilience measures, such as reinforced walls, secure data centers, or protected access control points, are built according to the design specifications. At the same time, BIM allows for clash detection, identifying potential conflicts between systems (e.g., a security camera's line of sight being obstructed by structural elements) before construction begins. This ensures that security measures are fully effective, with no unintended design flaws that could be exploited by attackers. BIM provides an integrated view of all building management systems, including heating, ventilation, air conditioning (HVAC), lighting, and security.

BIM enables the design of infrastructure with resilience in mind. Buildings can be designed to withstand both physical sabotage (e.g., reinforced structures, redundant systems) and cyberattacks (e.g., secured control systems). This level of resilience is critical in mitigating the impacts of hybrid threats. BIM can be used to integrate IoT-enabled sensors into the building design. These sensors can detect potential threats in real time, such as changes in temperature, vibrations (which could signal tampering or physical attacks), and unauthorized access. Real-time monitoring of building systems using BIM provides security personnel with a dynamic view of operational parameters. By integrating AI and machine learning algorithms into BIM, patterns of normal building operation can be analyzed to detect anomalies. For example, if data suggests abnormal network activity (cyber intrusion) or irregular movement in restricted areas (physical threat), BIM systems can provide a real-time alert.

BIM can be integrated with existing defense technologies, such as surveillance systems, access control, and perimeter security. It allows seamless communication and data sharing between these systems, ensuring a coordinated defense against hybrid threats. BIM helps in managing the entire lifecycle of a building, from design to decommissioning, ensuring that all security measures are maintained and updated as new threats evolve.

3.7 Monitoring, Information/Signal Collection and Processing Systems

Effective monitoring and information processing systems are critical to maintaining situational awareness and detecting potential threats to critical infrastructure. These systems enable real-time detection and analysis of potential threats by collecting, analyzing, and interpreting vast amounts of data from multiple sources, providing critical data for timely response and decision-making.

The information systems collecting and processing information about hazards may influence the human being is of great importance. The systems are expected to be precise, deliver online motoring and be available both for professionals and the public. One of the most important innovations in this area is the development of advanced sensor networks. These networks consist of distributed physical and soft sensors that collect data on various environmental and operational parameters. For example, smart sensors in industrial plants can monitor temperature, pressure and other critical variables and detect anomalies that could indicate a security breach or system failure. In addition, the integration of Internet of Things (IoT) devices has revolutionized monitoring and information gathering [85]. IoT devices can continuously collect and transmit data, enabling comprehensive monitoring of critical infrastructure. In the context of energy grids, IoT sensors can provide real-time data on power consumption, device status and potential faults, improving the reliability and safety of power supply systems. In all of these situations, advanced signal processing algorithms must be used to extract meaningful information from complex sensor data.

In addition, the use of unmanned aerial vehicles (UAVs) and drones has expanded the capabilities of surveillance systems. Drones equipped with high-resolution cameras and sensors can inspect infrastructure such as pipelines, power lines, and bridges from the air and detect damage or unauthorized activity. This aerial perspective enables efficient and comprehensive monitoring of large and hard-to-access areas.

Processing the vast amounts of data generated by these monitoring systems requires sophisticated information processing techniques. Big data analyses and machine learning algorithms are indispensable tools for gaining meaningful insights from large data sets. These technologies are used for pattern recognition anomaly detection and predictive analytics that human analysts might miss, providing early warning of potential threats.

Data aggregation platforms and visualization tools are essential for modern monitoring and information processing systems as they enable the efficient collection, integration, and interpretation of large amounts of data from different sources. These tools simplify the complexity of raw data by bringing it together on a central platform, enabling real-time analyses and decisions. Visualization tools convert the collected data into intuitive graphical representations, such as charts, maps, and dashboards, which make it easier for human users to quickly identify trends, patterns, and anomalies. This capability is crucial for improving situational awareness, increasing operational efficiency, and supporting strategic planning.

- emerging trends in this area include:
- blockchain for data integrity: Blockchain technology is being explored to ensure the integrity and security of data captured by surveillance systems. By providing a decentralized and tamper-proof ledger, blockchain can increase the trustworthiness of data, especially in applications that require high security, such as supply chain management, healthcare records, and critical infrastructure monitoring [86].
- edge computing: It is becoming increasingly important to process data closer to the source of collection. By analyzing data at the edge of the network, latency is reduced, and bandwidth usage is minimized. This is particularly important for applications that require real-time responses, such as autonomous vehicles, industrial automation, and remote monitoring in healthcare. Edge computing

also increases security by limiting the amount of sensitive data transmitted over networks [87].

- advanced visualization techniques: Visualization tools are evolving to handle more complex and larger data sets. Techniques such as augmented reality (AR) and virtual reality (VR) are being used to make the visualization of data a real experience. These advanced visualization methods help users to better understand and interact with the data. This makes it easier to identify trends and patterns that may not be apparent using conventional visualization techniques.
- Cognitive Radio Networks (CRNs): These are advanced wireless communication systems designed to optimize the use of the radio frequency spectrum. These networks utilize cognitive radio technology that enables dynamic access to the spectrum, allowing devices to intelligently and autonomously adjust their transmit and receive parameters based on the surrounding radio environment. This adaptive capability significantly improves spectrum efficiency and addresses the problem of spectrum scarcity [88].
- quantum sensing: This is a cutting-edge technology that utilizes the principles of quantum mechanics to measure physical quantities with unprecedented precision and sensitivity. By exploiting phenomena such as superposition, entanglement, and quantum coherence, quantum sensors can outperform classical sensors in various applications, from navigation and imaging to fundamental physics research [89].

Some examples of the publicly available systems informing hazards are listed below:

- radioactivity environmental day-to-day monitoring [90].
- cases of Radon scanning [91].
- geophysical systems focus on seismic activity, magnetic fields etc. [92].

3.7.1 Drones, Photogrammetry

Drones are multipurpose, payload-capable unmanned aerial vehicles (UAVs) that are utilized for data collection. With high-resolution cameras, drones take overlapping pictures of the target area from different perspectives. They may be used in a variety of settings, including both distant and urban ones. Modern drones are capable of carrying a wide range of sensors, such as thermal imaging devices, LiDAR, and cameras. They can investigate dangerous locations without endangering workers and swiftly cover large areas, saving time and resources. Additionally, capable of reaching hard-to-reach places, these instruments provide uncommon viewpoints. In some situations, drone-based solutions might be more economical than conventional surveying procedures [93, 94].

A method called photogrammetry creates a 3D model using the geometric characteristics of the photographs. It analyzes photos and extracts measurements and information from them to produce precise 3D models. These models are commonly shown as point clouds, which are made up of millions of individual points in 3D space. Highly precise point clouds may be produced using photogrammetry; they are frequently on par with conventional surveying techniques [95].

In several sectors, drones and photogrammetry have combined to provide a powerful tool for precise and quick data collection and analysis. Point cloud generation has been revolutionized by drones and photogrammetry. A wide range of sectors may now access and afford point clouds thanks to the combination of photogrammetry and drones. More options for data collection, analysis, and decision-making have been made possible by the capacity to build intricate 3D models from aerial photos. They are essential instruments for tracking, examining, and evaluating critical infrastructures and may be applied successfully in a variety of fields, including construction, land surveying, agriculture, environmental monitoring, and archeology [96].

3.7.2 Big Data, Artificial Intelligence, Machine Learning

Big data, artificial intelligence (AI), and machine learning (ML) have emerged as key technologies in the fight against hybrid threat vectors. These technologies offer advanced capabilities for analyzing large data sets, detecting anomalies, and predicting potential threats with high accuracy. Big data [97] refers to the vast amounts of structured and unstructured data generated by various sources such as sensors, social media, and transaction records. The ability to analyze and interpret this data is critical to identifying patterns and trends in hybrid threats. Big data analysis uses techniques such as data mining, statistical analysis, and predictive modeling to uncover hidden insights and develop security strategies.

Artificial intelligence [98] expands the possibilities of big data analysis by automating complex data analysis processes. AI algorithms can process and analyze data at a speed and scale that exceeds human capabilities, enabling real-time threat detection and response. For example, AI-powered cybersecurity systems can continuously monitor network traffic, detect suspicious activity, and fend off potential cyberattacks before they can cause damage.

In machine learning [99, 100], a sub-area of AI, algorithms are trained on historical data to recognize patterns and make predictions. In the context of hybrid threat vectors, machine learning can be used to develop predictive models that forecast the likelihood of different threats based on historical data. These models can help organizations proactively respond to potential vulnerabilities and improve their preparedness for future threats.

In addition, the combination of big data, AI, and ML enables the development of intelligent systems that are capable of making autonomous decisions. For example, autonomous security systems can recognize and respond to intruders without human intervention, significantly reducing response times and mitigating the impact of threats.

3.7.3 Blocking Systems and Firewalls

Blocking systems and firewalls are fundamental components of cyber security infrastructure designed to protect networks from unauthorized access and malicious activity. These technologies serve as the first line of defense against cyber threats and ensure the integrity and confidentiality of data [101, 102].

Firewalls act as a barrier between trusted internal networks and untrusted external networks, such as the Internet. They monitor and control incoming and outgoing network traffic based on predefined security rules. Modern firewalls use various techniques, including packet filtering, stateful inspection, and deep packet inspection, to recognize and block malicious data traffic.

Intrusion prevention systems (IPS) are a type of blocking system that actively prevents potential threats by detecting and blocking malicious activity. Unlike traditional firewalls, IPSs can analyze the content of network traffic in real time and detect and block threats such as exploits, malware, and denial of service attacks. These systems are often integrated with firewalls to provide comprehensive protection.

Next-generation firewalls (NGFW) combine the capabilities of traditional firewalls with advanced security features such as application awareness, integrated intrusion prevention, and cloud-based threat intelligence. NGFWs can identify and control applications regardless of port, protocol, or evasion tactics, providing enhanced security in complex network environments.

In addition, advances in AI and ML [103, 104] are improving the effectiveness of blocking systems and firewalls. AI-powered firewalls can adapt to the evolving threat landscape by learning from past incidents and predicting future attacks. Machine learning algorithms can analyze patterns of network traffic to detect anomalies and automatically adapt security policies to counter new threats.

3.7.4 Point Clouds

Point clouds are collections of data points defined in three-dimensional space generated by 3D scanning technologies such as LiDAR, photogrammetry, and radar systems. Each point in the cloud represents a specific location in space, usually defined by its X, Y, and Z coordinates. These points can also contain additional information such as color (RGB values), intensity, and timestamps. Point clouds capture the shape, position, and surface properties of objects and environments, providing highly detailed and accurate 3D representations of physical environments.

They are used in various applications, including 3D modeling, geographic information systems (GIS), virtual reality, gaming, and industrial inspection.

Point clouds improve the accuracy and effectiveness of autonomous systems such as mobile robots, unmanned vehicles, and drones by providing detailed spatial information for better navigation and object recognition.

Point clouds can also be used to create realistic environments and, with the aid of virtual or augmented reality platforms, can be used to train first responders or security/military personnel in simulations to improve preparedness by replicating real-world conditions. This immersive training improves response strategies and readiness.

3.7.5 Real-Time Detecting Systems for Anomalies

These systems identify unusual patterns, behaviors, or conditions that deviate from normal operations, allowing for prompt responses to mitigate risks.

Key technologies in real-time anomaly detection systems could be classified as:

1. Statistical techniques, such as autoregressive integrated moving averages (ARIMA), analyze time-series data to detect unusual deviations in metrics like temperature, radiation levels, network traffic, or snow and wind loads.
2. Neural Networks and Deep Learning models like autoencoders, trained to compress and reconstruct data. If the model fails to reconstruct an input with sufficient accuracy, the data point is flagged as an anomaly. This approach is often used in fraud detection and cybersecurity.
3. Edge Computing and IoT (Internet of Things) where edge devices equipped with AI algorithms can detect anomalies locally and in real-time, reducing the need for data transmission to the cloud and enabling faster responses. This is especially useful in industrial systems, smart cities, and environmental monitoring.
4. Cloud Computing and Big Data Analytics provide real-time data streaming and anomaly detection at scale, being particularly effective in detecting anomalies in large-scale networks or distributed systems.

3.7.6 Water Quality Monitoring Systems

Water quality monitoring systems are essential for ensuring the safety of drinking water, environmental protection, and industrial processes. These systems use various technologies to detect, measure, and analyze the physical, chemical, and biological properties of water.

These systems consist of sensors for *real-time monitoring*, measuring the pH, dissolved oxygen (DO), or ion concentrations like nitrates or phosphates, *optical sensors* used to detect turbidity, chlorophyll, and organic compounds and *conductivity sensors* measuring the ability of water to conduct electrical current, which correlates to the concentration of dissolved salts (total dissolved solids or TDS), providing a measure of water salinity.

Remote Sensing and satellite-based monitoring are used to monitor large water bodies by detecting changes in water temperature, turbidity, and pollution levels.

Satellite data is particularly useful for tracking algal blooms and the spread of contaminants over time. Drones equipped with multispectral cameras and sensors can fly over water bodies, collecting data on water quality parameters like surface temperature, turbidity, and chlorophyll levels. They provide a cost-effective way to monitor inaccessible areas.

A new technology in water management uses biosensors for biological contaminants. Biosensors use living organisms or biological molecules to detect specific pollutants, such as bacteria, viruses, and harmful microorganisms like *E. coli*, providing early warnings of microbial contamination in drinking water or natural water bodies. Similarly, *enzyme-based sensors* detect pollutants by measuring the activity of enzymes in the presence of specific chemicals or toxins, particularly pesticides and heavy metals.

AI (Artificial Intelligence) can be used to analyze historical water quality data to predict future contamination events, allowing authorities to take preventive measures. *Machine learning* models can also identify patterns that might be missed by human analysts, providing deeper insights into water quality trends, while *blockchain technology* is being explored for water quality monitoring systems to ensure transparency and traceability of data.

3.8 Conclusion

The introduced threats, their sources, hybrid threat vectors and technologies to protect against them demonstrate a variety of ways and techniques aimed against peaceful human activities. The provision of threats is not limited because novel technologies open new possibilities.

References

1. A Guide to Critical Infrastructure Security and Resilience, Cybersecurity & Infrastructure Security Agency, (2019), https://www.cisa.gov/sites/default/files/publications/Guide-Critical-Infrastructure-Security-Resilience-110819-508v2.pdf, last accessed 2024/07/10
2. Securing Small and Medium-Sized Business (SMB) Supply Chains: A Resource Handbook to Reduce Information and Communication Technology Risks, Cybersecurity & Infrastructure Security Agency, (2019); https://www.cisa.gov/sites/default/files/2023-01/Securing-SMB-Supply-Chains_Resource-Handbook_508.pdf, last accessed 2024/07/10
3. Cybersecurity Supply Chain Risk Management Practices for Systems and Organizations, National Institute of Standards and Technology, (2022); https://nvlpubs.nist.gov/nistpubs/SpecialPublications/NIST.SP.800-161r1.pdf, last accessed 2024/07/10
4. Li Y, Liu Q (2021) A comprehensive review study of cyber-attacks and cyber security; emerging trends and recent developments. Energy Rep 7:8176–8186. https://doi.org/10.1016/j.egyr.2021.08.126
5. Mitnick KD, Simon WL (2011) The art of deception: controlling the human element of security. Wiley

6. Chetioui K, Bah B, Alami AO, Bahnasse A (2022) Overview of social engineering attacks on social networks. Proced Comp Sci 198:656–661. https://doi.org/10.1016/j.procs.2021.12.302
7. Tomsu M, Rosiková V, Vojtek P, Hrabakova M (2023) Spear-phishing and its specifics within social engineering. In: Katalinic B (ed) Proceedings of the 34th DAAAM international symposium. DAAAM International, Viena, pp 0379–0387. https://doi.org/10.2507/34th.daaam.proceedings.052
8. Saxena N, Hayes E, Bertino E, Ojo P, Choo K-KR, Burnap P (2020) Impact and key challenges of insider threats on organizations and critical businesses. Electronics 9:1460. https://doi.org/10.3390/electronics9091460
9. Smith CL, Brooks DJ (2013) Physical security. In: Security science. The theory and practice of security. Elsevier Butterworth-Heinemann, pp 105–127. https://doi.org/10.1016/B978-0-12-394436-8.00005-9
10. Alshamrani A, Myneni S, Chowdhary A, Huang D (2019) A survey on advanced persistent threats: techniques, solutions, challenges, and research opportunities. IEEE Commun Surv Tutor 21:1851–1877. https://doi.org/10.1109/COMST.2019.2891891
11. Talabis M, Martin J (2012) Information security risk assessment: data analysis. In: Information security risk assessment toolkit. Elsevier Syngress, pp 105–146. https://doi.org/10.1016/B978-1-59-749735-0.00004-X
12. Kizza, Joseph Migga, 'Security Assessment, Analysis, and Assurance', 2024, pp. 153–80 https://doi.org/10.1007/978-3-031-47549-8_7.
13. Collier H, Morton C (2024) Teenagers: a social media threat vector. In: du Toit J, van Niekerk B (eds) Proceedings of the 19th international conference on cyber warfare and security, vol 19. Academic Conferences International Limited, Kidmore End, pp 55–61. https://doi.org/10.34190/iccws.19.1.1980
14. Nwankpa JK, Datta PM (2023) Remote vigilance: the roles of cyber awareness and cybersecurity policies among remote workers. Comput Secur 130:103266. https://doi.org/10.1016/j.cose.2023.103266
15. Ilori O (2024) Nelly Tochi Nwosu, and Henry Nwapali Ndidi Naiho, 'third-party vendor risks in IT security: a comprehensive audit review and mitigation strategies'. World J Adv Res Rev 22(3):213–224. https://doi.org/10.30574/wjarr.2024.22.3.1727
16. Hughes C, Robinson N (2024) Effective vulnerability management: managing risk in the vulnerable digital ecosystem. Wiley. https://doi.org/10.1002/9781394277155
17. George AS, Baskar T, Balaji Srikaanth P (2024) Cyber threats to critical infrastructure: assessing vulnerabilities across key sectors. Partner Univ Int Innov J 2(1):51–75. https://doi.org/10.5281/zenodo.10639463
18. Shipena EFN, Gamundani AM (2024) A review of Internet of Things (IoT) smart office data security threats. Stud Manag Finan Account xxxvi:375–381. https://doi.org/10.1108/S1479-351220240000036034
19. Physical Security of the U.S. Power Grid: High-Voltage Transformer Substations, Congressional Research Service, (2015), https://crsreports.congress.gov/product/pdf/R/R43604/6, last accessed 2024/07/10
20. Assault on California Power Station Raises Alarm on Potential for Terrorism - Center for Security Policy, https://centerforsecuritypolicy.org/assault-on-california-power-station-raises-alarm-on-potential-for-terrorism/, last accessed 2024/07/07
21. Cybersecurity Strategy, U.S. Department of Energy, (2024), https://www.energy.gov/cio/articles/doe-cybersecurity-strategy-2024. Accessed 10 Jul 2024
22. Fang X, Misra S, Xue G, Yang D (2012) Smart grid—the new and improved power grid: a survey. IEEE Commun Surv Tutor 14(4):944–980. https://doi.org/10.1109/SURV.2011.101911.00087
23. Dileep G (2020) A survey on smart grid technologies and applications. Renew Energy 146:2589–2625. https://doi.org/10.1016/j.renene.2019.08.092
24. From Innovation to Practice: Re-Designing Energy Delivery Systems to Survive Cyber Attacks, U.S. Department of Energy. (2018). https://www.energy.gov/ceser/articles/

ceds-rd-innovation-practice-redesigning-energy-delivery-systems-survive-cyber, last accessed 2024/07/10
25. Insider Threat Program Maturity Framework, National Insider Threat Task Force, (2018), https://www.dni.gov/files/NCSC/documents/nittf/20181024_NITTF_MaturityFramework_web.pdf, last accessed 2024/07/10.
26. Cappelli DM, Moore AP, Trzeciak RF (2012) The CERT guide to insider threats. Addison-Wesley Professional
27. Campbell RJ, Clark CE, Austin DA, Repair or rebuild: options for electric power in Puerto Rico, https://crsreports.congress.gov/product/pdf/R/R45023, last accessed 2024/07/07.
28. Langner R (2011) Stuxnet: dissecting a cyberwarfare weapon. IEEE Sec Priv 9(3):49–51. https://doi.org/10.1109/MSP.2011.67
29. Katz J, Lindell Y (2020) Introduction to modern cryptography, 3rd edn. CRC Press
30. Edwards J, Weaver G (2024) Artificial intelligence in cybersecurity, in the cybersecurity guide to governance, risk, and compliance. Wiley, pp 497–510. https://doi.org/10.1002/9781394250226.ch28
31. Investigation: WannaCry cyber-attack and the NHS, https://www.nao.org.uk/wp-content/uploads/2017/10/Investigation-WannaCry-cyber-attack-and-the-NHS.pdf, last accessed 2024/07/07.
32. Vandalism of fiber-optic cables in California points to a deeper problem with Internet security, https://www.digitaltrends.com/web/vandalism-of-fiber-optic-internet-cables-raises-security-concerns/, last accessed 2024/07/07
33. John B (2003) Little. Principal cellular and tissue effects of radiation. In: cancer medicine. Ed Kufe DW, Pollock RE, Weichselbaum RR, et al. Hamilton (ON), BC Decker
34. https://www.epa.gov/radiation/radiation-sources-and-doses; 2024.12.08
35. Clavensjö B, Åkerblom G (1994) The radon book. Swedish Council for Building Research, Stockholm, Measures against radon, p 129
36. https://www.fda.gov/radiation-emitting-products/cell-phones/radio-frequency-radiation-and-cell-phones#; https://www.fda.gov/radiation-emitting-products/cell-phones/scientific-evidence-cell-phone-safety; 2024.10.08
37. Akomolafe OO, Olorunsogo T, Anyanwu EC, Osasona F, Ogugua JO, Daraojimba OH (2024) Air quality and public health: a review of urban pollution sources and mitigation measures. Eng Sci Technol J 5(2):259–271. https://doi.org/10.51594/estj.v5i2.751
38. Singh V (2024) 'Water pollution', in textbook of environment and ecology. Springer Nature, Singapore, pp 253–266. https://doi.org/10.1007/978-981-99-8846-4_17
39. Singh V (2024) Soil pollution. In: Textbook of environment and ecology. Springer Nature, Singapore, pp 267–274. https://doi.org/10.1007/978-981-99-8846-4_18
40. Abulibdeh A (2021) Analysis of urban Heat Island characteristics and mitigation strategies for eight arid and semi-arid gulf region cities. Environ Earth Sci 80:259. https://doi.org/10.1007/s12665-021-09540-7
41. Dwivedi A, Soni R (2024) Impacts of urban Heat Island effect on critical urban infrastructure: a review of studies published between 2012 and 2022. Environ Rev. https://doi.org/10.1139/er-2023-0108
42. Artiola JF, Reynolds KA, Brusseau ML (2019) Urban and household pollution. In: Brusseau ML, Pepper IL, Gerba CP (eds) Environmental and pollution science, 3rd edn. Elsevier Academic Press, pp 311–326. https://doi.org/10.1016/B978-0-12-814719-1.00018-5
43. Guttikunda SK, Jawahar P (2014) Atmospheric emissions and pollution from the coal-fired thermal power plants in India. Atmos Environ 92:449–460
44. Alvarez RA et al (2018) Assessment of methane emissions from the U.S. oil and gas supply chain. Science 361(6398):186–188
45. Fthenakis V, Kim HC (2011) Photovoltaics: life-cycle analyses. Sol Energy 85(8):1609–1628
46. Fearnside PM (2015) Emissions from tropical hydropower and the IPCC. Environ Sci Pol 50:225–239
47. Nuclear Energy Agency (NEA) (2022) Management and disposal of high-level radioactive waste: global Progress and solutions. OECD Publishing, Paris

48. Biasotto LD, Kindel A (2018) Power lines and impacts on biodiversity: a systematic review. Environ Impact Assess Rev 71:110–119
49. World Health Organization: Extremely Low-Frequency Fields. Environmental Health Criteria Monograph No. 238. Geneva: WHO Press. (2007)
50. Schabacker DS, Levy L-A, Evans NJ, Fowler JM, Dickey EA (2019) Assessing cybersecurity vulnerabilities and infrastructure resilience. Front Bioeng Biotechnol 7:61. https://doi.org/10.3389/fbioe.2019.00061
51. Abdullahi SM, Sun S, Wang B, Wei N, Wang H (2024) Biometric template attacks and recent protection mechanisms: a survey. Inform Fusion 103:102144. https://doi.org/10.1016/j.inffus.2023.102144
52. Nemec Zlatolas L, Welzer T, Lhotska L (2024) Data breaches in healthcare: security mechanisms for attack mitigation. Clust Comput. https://doi.org/10.1007/s10586-024-04507-2
53. Das S, Kataria VK (2010) Bioterrorism: a public health perspective. Med J Arm Force India 66:255–260. https://doi.org/10.1016/S0377-1237(10)80051-6
54. Bohua L, Yuexin W, Yakun O, Kunlan Z, Huan L, Ruipeng L (2023) Ethical framework on risk governance of synthetic biology. J Biosafe Biosec 5:45–56. https://doi.org/10.1016/j.jobb.2023.03.002
55. Gudmundsson, MT, R Pedersen, K Vogfjörd, S Thorbjarnardóttir, B Jakobsdóttir, and J Matthew, Eruptions of Eyjafjallajökull Volcano, Iceland, Eos, 2011, eruption of the Eyjafjallajökull volcano in Iceland in 2010.
56. Fujinawa Y, Takahashi K (1998) Electromagnetic radiations associated with major earthquakes. Phys Earth Planet Inter 105:249–259
57. Consumer and Clinical Radiation Protection Bureau, Environmental and Radiation Health Sciences Directorate, Healthy Environments and Consumer Safety Branch Health Canada (2015) Limits of human exposure to radiofrequency electromagnetic energy in the frequency range from 3 KHz to 300 GHz. Safe Code 6:24
58. Lochbaum D, Lyman E, Stranahan SQ (2014) Fukushima: the story of a nuclear disaster. Choice Rev Online. https://doi.org/10.5860/choice.186184
59. https://www.iaea.org/topics/health; 2024.12.08
60. https://energy.ec.europa.eu/topics/nuclear-energy/radiological-and-nuclear-technology-health/medical-uses-radiation_en; 2024.12.08
61. https://ncrponline.org/program-areas/pac-4-radiation-protection-in-medicine/; 2024.12.08.
62. https://eurados.sckcen.be/en/important-links/radiation-protection-organisations-and-authorities; 2024.12.08
63. Shrader-Frechette K, Persson L (1997) Ethical issues in radiation protection. Health Phys 73(2):378–382
64. https://www.health.harvard.edu/cancer/radiation-risk-from-medical-imaging#:~:text=The%20proportion%20of%20total%20radiation,on%20Radiation%20Protection%20and%20Measurements; 12.08.2024
65. Bhushan B (ed) (2010) Springer handbook of nanotechnology. Springer
66. Johnson RR, Johnson ATC, Klein ML (2008) Probing the structure of DNA−carbon nanotube hybrids with molecular dynamics. Nano Lett 8(1):69–75. https://www.physics.upenn.edu/~robertjo/gallery/; 2024.12.08
67. Skomorokhova EA, Sankova TP, Orlov IA, Savelev AN, Magazenkova DN, Pliss MG, Skvortsov AN, Sosnin IM, Kirilenko DA, Grishchuk IV, Sakhenberg EI, Polishchuk EV, Brunkov PN, Romanov AE, Puchkova LV, Ilyechova EY (2020) Size-dependent bioactivity of silver nanoparticles: antibacterial properties, influence on copper status in mice, and whole-body turnover. Nanotechnol Sci Appl 13:137–157
68. Dekhtyar Y, Abols D, Avotina L, Stoppel A, Balakin S, Khroustalyova G, Opitz J, Sorokins H, Beshchasna N, Tamane P, Rapoport A (2023) Effects of diamond nanoparticles immobilization on the surface of yeast cells: a phenomenological study. Fermentation 9(2):162
69. https://en.wikipedia.org/wiki/Moore%27s_law; 12.08.2024
70. https://en.wikipedia.org/wiki/Low-power_electronics; 12.08.2024.
71. https://www.karlrupp.net/2015/06/40-years-of-microprocessor-trend-data; 12.08.2024

72. Jafari B, Gholizadeh E, Zhoulideh M, Adibnia E, Ghafarias M, Noori M, Golmohammadi S (2023) Highly sensitive label-free biosensor: graphene/CaF2 multilayer fo gas, cancer, virus, and diabetes detection with enhanced quality factor and figure of merit. Sci Rep 13:16184
73. Krivitsky V, Zverzhinetsky M, Patolsky F (2020) Redox-reactive field-effect transistor nanodevices for the direct monitoring of small metabolites in biofluids toward implantable nanosensors arrays. ACS Nano 14:3587–3594
74. Benfdila A (2014) Sensing using the field effect principle: the MOSFET. NATO ASI, Sozopol
75. Nguyen TT-H, Nguyen CM, Huynh MA, Vu HH, Nguyen T-K, Nguyen N-T (2023) Field effect transistor based wearable biosensors for healthcare monitoring. J Nanobiotechnol 21:411–440
76. Novikov S, Lebedeva N, Satrapinski A, Walden J, Davydov V, Lebedev A (2016) Graphene-based sensor for environmental monitoring of NO_2. Sens Actuat B Chem 236:1054–1060
77. Abdel-Karim R, Reda Y, Abdel-Fattah A (2020) Nanostructured materials-based nanosensors. J Electrochem Soc 167:037554
78. Chen S, Sun Y, Fan X, Xu Y, Chen S, Zhang X, Man B, Yang C, Du J (2023) Review on two-dimensional material-based field-effect transistor biosensors: accomplishments, mechanisms, and perspectives. J Nanobiotechnol 21:144–176
79. Baharfar M, Lin J, Kilani M, Zhao L, Zhang Q, Mao G (2023) Gas nanosensors for health and safety applications in mining. Nanoscale Adv 5:5997–6016
80. Ain ZQ, Fang X, Luo Z, Ullahc S, Fatima S, Batoole S, Qiua B, Shahzadf F (2022) Graphene based nanohybrid aptasensors in environmental monitoring: concepts, design and future outlook. Crit Rev Anal Chem. https://doi.org/10.1080/10408347.2022.2025758
81. Dhahi TS, Yousif Dafhalla AK, Tayfour OE, Mubarakali A, Alqahtani AS, Tayfour Ahmed AE, Elobaid ME, Adam T, Gopinath SCB (2024) Advances in nanosensors for monitoring and optimal performance enhancement in photovoltaic cells. iScience 27(4):109347
82. Manimekala T, Sivasubramanian R, Dharmalingam G (2022) Nanomaterial-based biosensors using field-effect transistors: a review. J Electron Mater 51:1950–1973
83. Tamersit K (2019) An ultra-sensitive gas nanosensor based on asymmetric dual-gate graphene nanoribbon field-effect transistor: proposal and investigation. J Comput Electron 18:846–855
84. Castro KPR, Colombo RNP, Iost RM, da Silva BGR, Crespilho FN (2023) Lowdimensionality carbon-based biosensors: the new era of emerging technologies in bioanalytical chemistry. Anal Bioanal Chem 415:3879–3895
85. Gubbi J et al (2013) Internet of Things (IoT): a vision, architectural elements, and future directions. Futur Gener Comput Syst 29(7):1645–1660. https://doi.org/10.1016/j.future.2013.01.010
86. Roopesh M (2024) Blockchain technology's role in securing data and preventing cyber-attacks: a detailed review. Acad J Sci Technol Eng Math Educ 4(03):16–31. https://doi.org/10.69593/ajsteme.v4i03.86
87. Carvalho G, Cabral B, Pereira V et al (2021) Edge computing: current trends, research challenges and future directions. Computing 103:993–1023. https://doi.org/10.1007/s00607-020-00896-5
88. Muzaffar MU, Sharqi R (2024) A review of spectrum sensing in modern cognitive radio networks. Telecommun Syst 85:347–363. https://doi.org/10.1007/s11235-023-01079-1
89. Bongs K, Bennett S, Lohmann A (2023) Quantum sensors will start a revolution—if we deploy them right. Nature 617(7962):672–675. https://doi.org/10.1038/d41586-023-01663-0
90. https://remap.jrc.ec.europa.eu/Advanced.aspx; 12.08.2024 https://remap.jrc.ec.europa.eu/Atlas.aspx; 12.08.2024. https://map.safecast.org; 12.08.2024. https://www.siliconrepublic.com/climate/radioactive-map-earth; 12.08.2024.
91. https://www.ukradon.org/information/ukmaps; 12.08.2024. https://www.arpansa.gov.au/understanding-radiation/radiation-sources/more-radiation-sources/radon-map; 12.08.2024. https://remon.jrc.ec.europa.eu/About/Atlas-of-Natural-Radiation/Digital-Atlas/Indoor-radon-AM/Indoor-radon-concentration; 12.08.2024. https://www.epa.gov/system/files/documents/2024-05/radon-zones-map_text_link.pdf; 12.08.2024.

92. https://earthquake.usgs.gov/earthquakes/map/?extent=-74.21198,-207.42188&extent=84.33698,278.4375&listOnlyShown=true; 12.08.2024. https://supermag.jhuapl.edu/line/?fidelity=low&start=2001-01-01T00%3A00%3A00.000Z&interval=1%3A00%3A00&stations=BEL,CHD&tab=stations; 125.08.2024.
93. Mohsan SAH, Othman NQH, Li Y, Alsharif MH, Khan MA (2023) Unmanned aerial vehicles (UAVs): practical aspects, applications, open challenges, security issues, and future trends. Intell Serv Robot 16:109–137. https://doi.org/10.1007/s11370-022-00452-4
94. Laghari AA, Jumani AK, Laghari RA, Nawaz H (2023) Unmanned aerial vehicles: a review. Cogn Robot 3:8–22. https://doi.org/10.1016/j.cogr.2022.12.004
95. Aber JS, Marzolff I, Ries JB (2010) Photogrammetry. In: Aber JS, Marzolff I, Ries JB (eds) Small-format aerial photography. Elsevier, pp 23–39. https://doi.org/10.1016/B978-0-444-53260-2.10003-1
96. Colomina I, Molina P (2014) Unmanned aerial systems for photogrammetry and remote sensing: a review. ISPRS J Photogramm Remote Sens 92:79–97. https://doi.org/10.1016/j.isprsjprs.2014.02.013
97. Hashem IAT et al (2015) The rise of "big data" on cloud computing: review and open research issues. Inf Syst 47:98–115. https://doi.org/10.1016/j.is.2014.07.006
98. Russell S, Norvig P (2016) Artificial intelligence: a modern approach, 4th edn. Pearson
99. Halbouni A et al (2022) Machine learning and deep learning approaches for cyberSecurity: a review. IEEE Access 10:19572–19585. https://doi.org/10.1109/ACCESS.2022.3151248
100. Goodfellow I, Bengio Y, Courville A (2016) Deep learning. MIT Press
101. Kaufman C et al (2022) Network security: private communication in a public world, 3rd edn. Addison-Wesley Professional
102. Scarfone, K. and Hoffman, P. (2009), Guidelines on firewalls and firewall policy, special publication (NIST SP), National Institute of Standards and Technology, Gaithersburg, MD., https://tsapps.nist.gov/publication/get_pdf.cfm?pub_id=901083. Accessed 7 Jul 2024
103. Sommer R, Paxson V (2010) Outside the closed world: on using machine learning for network intrusion detection. In: IEEE symposium on security and privacy, pp 305–316. https://doi.org/10.1109/SP.2010.25
104. Mirsadeghi SMH et al (2023) Learning from few cyber-attacks: addressing the class imbalance problem in machine learning-based intrusion detection in software-defined networking. IEEE Access 11:140428–140442. https://doi.org/10.1109/ACCESS.2023.3341755

Chapter 4
From Phantoms to Firewalls: Securing Critical Infrastructures in the Age of Hybrid Threats

Emmanuel Karlo Nyarko

Abstract The rapid development of artificial intelligence (AI) and machine learning (ML) technologies has ushered in a new era of challenges and opportunities for cybersecurity, particularly in the context of hybrid warfare and critical infrastructure protection. This paper explores the dual nature of ML in both offensive and defensive capacities within the evolving landscape of hybrid threats. A brief overview of key ML techniques is provided, including supervised and unsupervised learning, reinforcement learning, deep learning and natural language processing. This article then examines five primary battlefields where AI/ML plays a critical role: Social Media Manipulation, Deep Fakes and Synthetic Media, Targeted Phishing Attacks, Cyber Warfare, and Autonomous Weapons Systems. The background, development, original intentions, and current offensive and defensive applications of the relevant ML technologies are discussed for each area. It also addresses the ethical considerations and challenges of integrating AI into warfare and security, including accountability issues, escalation risks, privacy concerns and the need for human oversight.

Keywords Hybrid threats · Machine learning · Artificial intelligence · Cybersecurity

4.1 Introduction

The twenty-first century is characterised by an unprecedented flow of information and connectivity that drive global economic prosperity and social progress. However, it is precisely this reliance on interconnected critical infrastructure—the lifeblood of modern life—that creates a critical vulnerability, as various hybrid threats can attack

E. K. Nyarko (✉)
Josip Juraj Strossmayer University of Osijek, Faculty of Electrical Engineering, Computer Science and Information Technology Osijek, Osijek, Croatia
e-mail: karlo.nyarko@ferit.hr

D. Radu et al. (eds.), *Countering Hybrid Threats Against Critical Infrastructures*, NATO Science for Peace and Security Series B: Physics and Biophysics, https://doi.org/10.1007/978-94-024-2304-4_4

these systems. Hybrid threats exploit the grey area between peace and war using conventional and unconventional means, including cyberattacks, disinformation and psychological operations [1]. The complexity and scale of these threats require robust defences, including the strategic use of artificial intelligence (AI) and machine learning (ML) in cyber security. The advent of AI, specifically ML, has ushered in a new era of opportunities and challenges in cybersecurity. This paper explores the dual nature of AI/ML technologies in the context of hybrid warfare [2] and their implications for securing vital systems against emerging threats.

As both nations and non-state actors leverage these advanced technologies for various purposes, including potentially malicious activities, it is critical to understand the underlying mechanisms and develop robust defence strategies. This article aims to highlight some of the AI/ML techniques used offensively and defensively, as well as the ethical considerations arising from their use.

4.2 Understanding AI and ML

A basic understanding of AI and ML technologies is essential before looking at the specifics of hybrid threats and their countermeasures. This section provides an overview of this field's most important concepts and methods.

AI refers to the development of computer systems that can perform tasks that usually require human intelligence. These tasks include perception, computer vision (CV), understanding natural language or natural language processing (NLP), recognising patterns, solving problems and making decisions. AI encompasses a wide range of technologies and approaches, including robotics, expert systems, and neural networks, and can be divided into two main categories: narrow AI (specialised for specific tasks) and general AI (with broader capabilities similar to human intelligence).

ML is a subcategory of AI that uses algorithms and statistical modelling to enable computers to learn from data and make predictions or decisions. Rather than being explicitly programmed for each task, ML systems improve their performance by being exposed to more data over time. ML systems improve their performance through experience, recognising patterns, and adapting their actions. ML includes different types of learning, such as supervised learning (SL), unsupervised learning (USL), reinforcement learning (RL) and deep learning (DL).

SL is a type of ML in which the algorithm is trained on a labelled data set and learns to map input data to known output values or labels [3, 4]. This approach is widely used in classification and regression tasks and forms the basis for many predictive models in cyber security applications such as spam and fraud detection.

In contrast to SL, USL algorithms work with unlabelled data and attempt to find patterns or intrinsic structures within the data set [5]. Initially, these techniques were used in market research for customer segmentation. Today, they are particularly useful for clustering and anomaly detection, critical for identifying unusual behaviour in network traffic or grouping similar cyber threats.

In RL, an agent is trained to make a series of decisions in an environment by interacting with the environment and optimising its actions to achieve the best possible results, i.e. to maximise a reward signal. This approach is crucial for adaptive decision-making systems [6]. RL was developed for robotics and gaming and is now used in autonomous systems and complex decision-making processes. This approach has shown promise in developing adaptive defence strategies and simulating attack scenarios to improve preparedness.

DL, a branch of ML, uses artificial neural networks with many layers (deep networks) to learn hierarchical representations of data and, thus, to model complex patterns in data [4, 7]. Initially focused on image and speech recognition, DL has revolutionised many areas of AI, including computer vision (CV) and natural language processing (NLP), which has significant implications for offensive and defensive capabilities in hybrid warfare.

NLP deals with the interaction between computers and human language and enables the analysis and synthesis of natural language texts. It includes tasks such as language comprehension, generation and translation. Originally intended to enable machine translation and sentiment analysis, NLP techniques are now used in chatbots, content filtering and automated summarisation [8, 9]. In the context of hybrid threats, NLP plays a crucial role in analysing and creating text-based content that can be used for both disinformation campaigns and detection.

AI and ML are often used interchangeably due to the close relationship between the two concepts. ML is an important and practical approach within the broader field of AI. As ML has driven many recent advances in AI, the distinction between the two terms has become blurred. When people talk about AI in modern contexts, they often implicitly refer to the ML techniques and applications currently in widespread use, such as image recognition and natural language processing. This relationship is best illustrated in Fig. 4.1. Note that not all components of AI and ML are shown in Fig. 4.1.

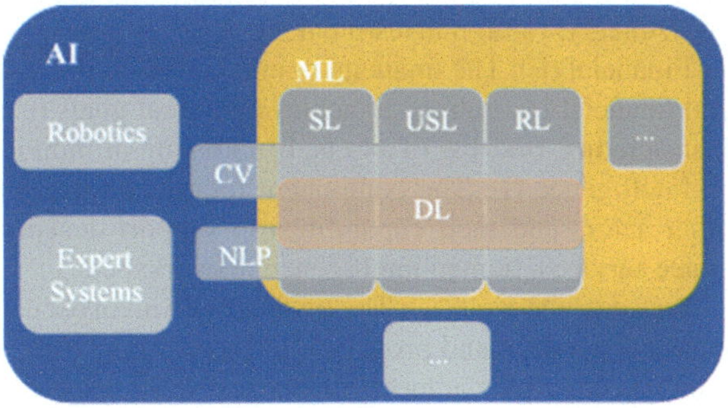

Fig. 4.1 The relationship between AI and ML

4.3 Hybrid Threats and Instruments of Power

In an increasingly complex global security environment, the concepts of hybrid threats and instruments of power (IOP) have gained considerable attention. This section introduces these concepts and examines the relationship between them. In particular, the focus will be on hybrid threats and the DIMEFIL (Diplomacy, Information, Military, Economic, Financial, Intelligence, Law enforcement) framework of national power instruments, providing insights into how they interact and influence modern security challenges.

4.3.1 Instruments of Power (IOP)

Contrary to popular belief, national power is not limited to military power but consists of several other tools or instruments that can be combined depending on the scenario. The effectiveness of these instruments depends on various factors, such as the relative power of the actors involved, the specific situation and international norms. A common framework for understanding these instruments is DIMEFIL [10], an expanded version of the traditional DIME framework (Diplomacy, Information, Military, Economy) [11], which provides a more nuanced understanding of national power in the twenty-first century. This framework comprises seven important instruments:

Diplomatic (D): This domain includes negotiations, alliances, and international agreements to achieve goals (e.g. summits, trade agreements, sanctions, etc.).

Information (I): This aspect focuses on shaping the narrative through media, public relations, and even cyber tools to influence how others perceive a situation (e.g. media, internet, social networks, etc.).

Military (M): This encompasses the use and threat of violence to achieve goals. This can range from peacekeeping missions to full-blown war.

Economic (E): This element involves the strategic application of economic instruments such as trade, aid, and investment to exert influence (e.g. tariffs, embargoes, energy, foreign investment, etc.)

Financial (F): The financial domain refers to the manipulation of currency, credit and financial institutions to achieve political goals (e.g. SWIFT, denial of currency loans/credits, freezing of assets, blacklisting of individuals, etc.)

Intelligence (I): The component focuses on the gathering of information about other actors and their capabilities to support decision-making (e.g. spies, intelligence services, human intelligence networks, etc.)

Law enforcement (L): The final domain involves using international law enforcement cooperation and prosecution threats to pressure other actors (e.g. extradition, joint investigations, international arrest warrants, etc.). It should be noted that non-state actors can also use some of these tools to exert influence.

4.3.2 Hybrid Threats

Hybrid threats represent a complex and multi-layered approach to warfare and conflict that combines conventional and unconventional methods. According to [12], hybrid threats are "a full range of different modes of warfare including conventional capabilities, irregular tactics, and formations, terrorist acts including indiscriminate violence and coercion, and criminal disorder". These threats are characterized by their adaptability, ambiguity, and ability to exploit vulnerabilities in different domains:

1. Synchronisation of multiple instruments of power
2. Exploitation of ambiguity and deniability
3. Use of kinetic and non-kinetic means
4. Exploitation of vulnerabilities in society
5. Blurring the boundaries between war and peace

Hybrid threats pose significant challenges to traditional security paradigms as they often operate below the threshold of conventional warfare and exploit gaps in existing defence strategies [13].

On the other hand, a hybrid conflict refers to a situation where hybrid threats have escalated into a state of conflict that combines conventional military operations with the continued use of irregular tactics. The hybrid conflict lies between "classic" war and peace, as it is deliberately designed to remain below the threshold of conventional military conflict, established red lines and open interstate war [14]. These tactics often operate in the shadows, making it difficult to pinpoint the true source of the attacks. It blends various instruments of power, often operating below the threshold of open armed conflict, to exploit vulnerabilities and create ambiguity [1]. The goal is to create a sense of instability and chaos within the target nation, weakening their government and hindering their ability to function effectively.

Hybrid warfare blends conventional warfare with irregular tactics and cyber operations. It often involves a combination of military actions, cyberattacks, and disinformation campaigns. A typical example is the Russian annexation of Crimea in 2014, which utilized a mix of military force, cyber operations, and media propaganda to create a pretext for annexation and destabilize Ukraine [15].

4.3.3 Relationship Between Hybrid Threats and DIMEFIL

The relationship between hybrid threats and the DIMEFIL framework is complex and multi-layered. By their nature, hybrid threats seek to utilize and combine different instruments of power to achieve strategic objectives. This relationship can be understood through several key aspects:

1. Synergistic exploitation: Hybrid threats often deploy multiple DIMEFIL elements simultaneously, creating synergistic effects that amplify their impact. For

example, a hybrid campaign may combine economic pressure, information warfare, and covert military action to destabilize a target country.
2. Asymmetric application: Hybrid actors usually use DIMEFIL instruments asymmetrically, focusing on areas where they have a comparative advantage or where the target is particularly vulnerable. This approach allows them to maximize impact while minimizing risk and resource expenditure.
3. Threshold manipulation: By carefully calibrating different DIMEFIL elements, hybrid threats can operate in the "grey zone" between peace and war, making it difficult for targets to respond effectively within the existing legal and operational framework.
4. Adaptable strategies: Hybrid threats exhibit high adaptability, switching between different DIMEFIL tools as circumstances change or as countermeasures are taken. This flexibility poses a challenge to traditional, relatively rigid defence structures.
5. Impact on society: The comprehensive nature of hybrid threats and the DIMEFIL framework means that the impact of hybrid operations can be felt across all sectors of society, requiring a whole-of-government and whole-of-society response.

The following section presents an example scenario of a hybrid conflict to show the possible offensive and defensive ML techniques that can be used in such a scenario.

4.4 Possible AI/ML Techniques in Hybrid Warfare

This section looks at some AI/ML techniques commonly used in hybrid warfare scenarios and discusses their development, original purpose and current applications in offensive and defensive contexts.

Imagine a scenario where country A tries to weaken country B without resorting to open military conflict. The following IOPs could be used:

Diplomatic: Country A maintains relations with separatist groups in Country B and supports them politically and financially. They *launch a smear campaign* against the target country's leadership on the international stage, portraying them as corrupt or oppressive.

Information: Country A *spreads disinformation and propaganda via social media and online channels* to sow discord and mistrust in country B. They could *hack media* in the target country to manipulate reporting and public opinion.

Military: Country A provides covert military training and support to proxy forces in the target country to enable them to carry out insurgent attacks. They could *use cyberwarfare tactics* to disrupt the target country's critical infrastructure (e.g., power grids, financial networks, communication networks, railways, etc.) and military communications.

Economic: Country A manipulates economic levers to harm the target nation. This could include manipulating exchange rates or imposing trade sanctions. They could encourage companies to leave country B, paralysing its economy.

Financial: Country A uses financial means to undermine the currency and financial stability of the target country. This could include the manipulation of international markets or the financing of black-market activities in country B.

Law enforcement: Country A could also use tactics such as asset freezes to target key individuals in country B's leadership.

Based on the above scenario, some of the possible AI/ML techniques that can be used are:

1. Manipulation of social media to spread divisive content and exacerbate existing social tensions.
2. Deepfake videos of political leaders making inflammatory statements.
3. Targeted phishing attacks on government officials
4. Cyberattacks on critical infrastructure, such as power grids or financial systems.
5. Autonomous drones for covert surveillance and possible limited attacks.

4.4.1 Social Media Manipulation

Social media manipulation techniques developed in parallel with the rapid growth of social networking platforms in the early 2000s. Initially, these methods were developed for targeted advertising and to optimize user activity. Companies like Facebook and Twitter invested heavily in AI/ML algorithms to analyze user behavior and preferences to keep users engaged and offer them relevant content.

Social media bots are automated software programs designed to simulate human activity and run on social media platforms. They can perform tasks such as posting content, liking or sharing posts, following users, and even participating in conversations. The original intention or purpose is to improve the user experience and increase the effectiveness of digital marketing campaigns. By understanding user preferences and behavioral patterns, platforms can tailor content and advertising to individual users, increasing engagement and advertising revenue.

By implementing AI components in social media bots, human-like interaction can be achieved. The AI components currently used in social media bots are NLP and ML.

ML is extensively utilized to create personalized experiences. One prominent application is in content recommendation systems, where user data and preferences are analyzed to suggest relevant content or products. This capability allows social media bots to generate a personalized feed for each user. Additionally, ML is employed in targeted marketing, where personalized marketing messages are crafted based on user demographics, location, and interests.

NLP is crucial in enabling intelligent interactions. Through NLP, bots can "understand" and respond to natural language, facilitating engaging and human-like conversations with users. Sentiment analysis, another application of NLP, allows bots to analyze the sentiment of user comments or messages, enabling them to tailor responses to positive or negative feedback. Furthermore, NLP can be used for content creation and generating creative text formats such as poems, scripts, pieces of music, or social media posts.

Social media bots offer companies and users a wide range of benefits when used responsibly. Many companies deploy chatbots for customer service, addressing frequently asked questions, providing 24/7 support, or automating simple tasks like scheduling appointments. These bots free up human agents for more complex issues and offer customers a convenient and instant way to get help. Bots can also be programmed for content moderation, recognizing and flagging harmful content such as hate speech, bullying, or misinformation. This helps create a safer and more civilized online environment by filtering out inappropriate content before it reaches a broad audience.

Similarly, educational bots are designed to deliver educational content entertainingly and interactively. For instance, a language learning bot can engage users in conversations in their target language. In contrast, a science bot can provide information and quizzes on various scientific topics, making learning more exciting and accessible. Support bots, particularly mental health chatbots, offer anonymous support and resources for individuals struggling with anxiety, depression, or other mental health issues. While these bots cannot replace professional help, they provide a safe space to vent, obtain basic information, and find mental health professionals. Community-building bots are used to create online communities centred around specific interests. For example, a book club bot can facilitate discussions about assigned readings, and a fitness bot can offer exercise challenges and motivation to a group. These bots help people connect with others who share their passions. Additionally, bots can be customised for news aggregation and alerts, following specific topics and providing real-time updates or alerts. This allows users to stay informed on important topics without experiencing information overload.

It is crucial to remember that transparency is critical when interacting with social media bots. Users should be aware that they are not communicating with a real person.

4.4.1.1 Offensive Use of AI/ML

The same techniques have been repurposed in hybrid warfare to spread disinformation and manipulate public opinion. These bots can be programmed to amplify particular messages, create fake trending topics, target specific demographics with customised content, create echo chambers and foment social unrest [16, 17]. Typical examples of social media manipulation are Cambridge Analytica [18–20], the 2016 US presidential election [21–23], the 2016 Brexit referendum campaign [24–27], the military crackdown in Myanmar in 2017 [28–30] and the COVID-19 pandemic [31–33].

4.4.1.2 Defensive Use of AI/ML

As shown in the previous text, social media manipulation has become an important issue, especially in political campaigns, public health, and social stability. The spread of misinformation, fake news, and coordinated, inauthentic behavior on social media platforms threatens democratic processes and public trust. AI and ML techniques are increasingly being used to detect, mitigate, and prevent attempts at manipulation on these platforms [34–38]. These techniques analyze activity patterns, content, and network structures:

- ML is utilized to analyze activity patterns and enable the identification and flagging of bots or automated accounts. By examining large volumes of user activity data, ML can detect patterns that deviate from typical human behavior, such as unusually high posting frequency, rapid following or unfollowing of users, or activity at atypical times.
- NLP is applied for content analysis to detect coordinated, inauthentic behavior. NLP can scrutinize posted content to identify characteristics like repetitive language, nonsensical wording, or content unrelated to current topics, differentiating bots from genuine users who may be repetitive and authentically engaged.
- Network analysis involves examining the connections between accounts using graph theory. Since bots often display suspicious network patterns, such as following each other in large numbers or having many inactive followers, this analysis can help identify bot rings or coordinated inauthentic behavior.

Platforms and cybersecurity companies are constantly evolving these defences to stay ahead of evolving threats.

4.4.2 Synthetic Media and Deep Fakes

Synthetic data, which includes images, videos and audio data, is generated using advanced AI techniques to overcome the challenges of privacy, efficiency and bias in ML. Historically, the need for synthetic data emerged as a solution to the problems of scarcity, imbalance and privacy associated with real-world data. The generation of synthetic data facilitates the creation of diverse and representative datasets, improving the robustness and fairness of AI models.

Generating synthetic data offers several significant advantages. One of the primary benefits is data augmentation. In scenarios where real-world data is either insufficient or unbalanced, synthetic data can be created to enhance existing datasets, enriching them and improving the performance of ML models. For instance, in the development of self-driving car technologies, synthetic data can simulate rare weather conditions that may not be adequately represented in the collected real-world data. Additionally, synthetic data can be customized to mitigate biases inherent in real-world datasets, providing more balanced and representative data for training AI models.

Another advantage is the efficiency and speed of generating synthetic data. Collecting and processing actual data is often time-consuming and expensive, particularly for large or sensitive datasets such as those in healthcare or finance. In contrast, synthetic data can be generated quickly and cost-effectively, accelerating the development cycles of AI projects.

Synthetic data also plays a crucial role in testing and development. It provides a safe and controlled environment for testing and refining ML models before they are deployed in real-world scenarios. These pre-deployment tests help identify and resolve potential issues without compromising sensitive information.

Furthermore, synthetic data addresses concerns related to data privacy. Real datasets often contain sensitive information that could be misused if improperly handled. Companies can develop and test AI models without risking data confidentiality by creating synthetic data that mimics real data while protecting individual privacy. This is particularly advantageous in fields where sensitive data is processed, such as healthcare and finance.

Synthetic data generation relies heavily on DL techniques, particularly those that deal with complex data structures. Generative models are a central component of this process, as they are trained to learn the underlying probability distribution of the training and use this knowledge to generate new data points very similar to the original data.

Generative Adversarial Networks (GANs) are a prominent example of such models. GANs comprise two neural networks, the generator and the discriminator, which work together. The generator produces synthetic data, while the discriminator evaluates the authenticity of this data and attempts to distinguish between real and synthetic samples. Continuous feedback and refinement make the generator increasingly adept at generating realistic synthetic data [39].

Variational autoencoders (VAEs) are another class of generative models used to create synthetic data. VAEs work by compressing data into a latent space, a low-dimensional representation that captures the essential features of the data. The decoder then reconstructs or generates new data points from this latent space, adhering to the learnt patterns [40].

In audio generation, WaveNet architectures have proven to be particularly effective. WaveNet models predict subsequent audio patterns by analyzing sequences of audio information and generating new patterns that closely mimic the target speaker's voice [41].

In addition, text-to-speech (TTS) systems and speech recognition models play a role in generating and processing synthetic speech. TTS systems based on ML can generate synthetic speech based on text scripts and can be customized to emulate specific voice characteristics [42, 43]. Speech recognition models, especially those using recurrent neural networks (RNNs) with long short-term memory (LSTM) units, are trained on large datasets to recognize and replicate complex speech patterns [44].

Synthetic media is used for numerous positive applications, particularly in the creative and entertainment industries. One notable application is in the field of special effects. In the film and gaming industries, synthetic media has significantly enhanced special effects and provided a more realistic and immersive experience. For example, historical re-enactments can be brought to life with unprecedented realism, and fantastical creatures can be created in previously impossible ways.

Another important application of synthetic media is language accessibility and learning. By automatically dubbing films with actors who speak different languages, synthetic voice generation can cater to different audiences, improving accessibility and facilitating language learning.

In education and training, synthetic media enables personalized educational experiences. Historical figures, for example, can convey content in a way that appeals to younger audiences and makes learning more exciting and effective.

Synthetic media also encourages creative expression. Artists and filmmakers use this technology to explore new storytelling techniques and create unique characters or scenarios that are impossible using traditional methods. This opens up new avenues for creativity and innovation in the arts.

4.4.2.1 Offensive Use of AI/ML

Despite its positive applications, synthetic data, especially in the form of deepfakes, is associated with several negative uses, mainly due to its potential for misuse. Deepfakes use AI to create realistic but fake images, videos or audio recordings, often with malicious intent. One of the most worrying harmful applications is the spread of misinformation and propaganda [45, 46]. Deep fakes can create convincing videos or audio recordings of public figures saying or doing things they have never done. This creates confusion, undermines trust in institutions, and manipulates public opinion, especially during critical events such as elections [47]. Similarly, deepfakes can be used in cyber warfare to disrupt communications, spread misinformation, and sow discord within a population.

Deepfakes also pose a threat to personal and professional reputations. Malicious actors may create fake videos or audio recordings of individuals engaging in inappropriate or illegal activities, seriously damaging their reputation, career or personal life [48]. This form of cyberbullying and harassment is increasingly common, with targets ranging from celebrities to ordinary people. Under the threat of publishing such materials, individuals may be co-opted or coerced to engage in illegal activities.

In the financial sector, deepfakes can be used for fraudulent activities. For example, a deepfake of a CEO could be used to trick employees into authorizing fraudulent transactions [49, 50].

Another serious problem is the creation of revenge porn with deep fakes, where a victim's face is superimposed on explicit content without their consent, causing significant psychological and emotional damage [51].

4.4.2.2 Defensive Use of AI/ML

Given the potential damage that deepfakes can cause, various detection and countermeasures have been developed, many of which are also based on AI [52–59]:

- One of these methods is anomaly detection, in which ML models are trained on large data sets of real videos. These models learn the subtle statistical patterns that occur in real videos, such as natural skin texture, blinking patterns and lighting variations. When these models analyse a video suspected of being a deepfake, they look for anomalies that deviate from these expected patterns, such as inconsistencies in skin tone or unnatural eye movements.
- Another method is facial landmark detection and analysis. AI models can recognise and analyse certain facial features, such as the position of the eyes, nose and mouth, and track their movements in a video. Unnatural movements or inconsistencies in the relative positions of these landmarks can indicate manipulation.
- Analysing temporal consistency is particularly useful for detecting less sophisticated forgeries. This technique analyses how light interacts with objects and facial features across multiple video frames. Inconsistencies in lighting, shadows or objects that appear or disappear abruptly can indicate manipulation.
- AI-supported audio analysis can also recognise deepfakes by identifying the signatures of certain vocoders used to generate synthetic speech. In addition, variations in voice patterns and pitch, as well as inconsistencies in background noise, can help to recognise fakes.
- Interestingly, the same Generative Adversarial Networks (GANs) used to generate deepfakes can also be used to recognise them. By training a GAN with real and fake videos, the discriminative component of the GAN learns to recognise subtle flaws in deepfakes. At the same time, the generator refines its ability to create convincing fakes. This adversarial training results in a powerful detection tool that evolves in parallel with deep fake techniques.
- Finally, manual verification by trained experts is still valuable for recognising forgeries. Experts can recognise inconsistencies in facial features, lighting, skin texture and hair movement that may go unnoticed by automated systems. They can also analyse blink patterns, reflections in the eyes and subtle head and neck movements for signs of tampering.

New technologies such as blockchain and digital watermarking are being explored to authenticate original content as it is created, making it easier to verify media authenticity [60].

To summarise, synthetic data and deep fakes have significant positive potential in various areas, but their misuse poses serious ethical and security challenges. Ongoing research and the development of sophisticated detection techniques are crucial to mitigate the risks associated with these technologies.

4.4.3 *Targeted Phishing Attacks*

Phishing is a fraudulent attempt to trick someone into revealing sensitive information or clicking on malicious links. Phishing scams are typical social engineering attacks in which attackers exploit human psychology to manipulate their victims. A targeted or spear phishing attack is a cyber-attack targeting an individual or organisation. Unlike generic phishing emails that target a wider network, spear phishing attacks are carefully crafted to appear legitimate and trustworthy to the recipient. Whilst spear phishing attacks are highly effective, they also have significant disadvantages. These include the high costs, the time-consuming preparation and the limited scalability due to individualisation. The personalisation required for each recipient makes spear phishing a resource-intensive method, but its precision often results in a higher success rate than traditional phishing techniques.

For targeted phishing to be successful, attackers often invest much effort in researching their target first. They may gather information from social media profiles, company websites, or even data breaches to find details such as the recipient's name, job title, current projects, or even the names of colleagues. Using the information gathered, the attackers personalise the phishing email to make it appear more credible or legitimate. This personalisation can include Addressing the recipient by name and title, referencing specific projects or departments within the target company, or using language or formatting that mimics legitimate emails the recipient might receive (increased credibility). The attackers could impersonate someone the recipient knows or trusts, such as a colleague, supervisor or even a customer (exploitation of trust). The email or message could create a sense of urgency, pressure or emotional triggers to force the recipient to act quickly without thinking critically. This may include fake deadlines, threats of account suspension or promises of financial rewards. The email usually contains a malicious link or attachment. If the recipient clicks on the link, they could be taken to a fake website designed to steal login credentials or download malware onto their device. The attachment could be a document with malware that infects the system when opened.

4.4.3.1 Offensive Use of AI/ML

ML algorithms can be used to analyse vast amounts of personal data (often obtained by mining social media, public data or security breaches) to identify and target individuals [61]. Based on the collected data and using NLP, the human writing style can be mimicked, and the content can be tailored to the target persons. This can increase the success rate of bypassing traditional spam filters and make phishing attempts more convincing. Research has shown that the entire phishing process can be automated using Large Language Models (LLMs), reducing the cost of phishing attacks by more than 95% while achieving the same or higher success rates [62].

4.4.3.2 Defensive Use of AI/ML

To protect against phishing attacks, the first line of defence is user education and awareness. Users need to be regularly trained on recognising phishing attacks, including common tactics attackers use (e.g. fake login pages, urgent requests for personal information). Simulated phishing attacks must be conducted in-house so that users can practise recognising and responding to phishing attempts. As phishing threats constantly evolve, AI-powered methods improve the ability to recognise and prevent such threats:

- ML algorithms are trained on labelled data sets of phishing and legitimate emails. These models can classify new emails based on features such as content, sender information and URL structures, improving the accuracy of phishing detection [63–67].
- DL models [68, 69] and NLP techniques [62, 70] are crucial for analysing email content, detecting suspicious language patterns and identifying common phishing phrases. Sentiment analysis can also assess the tone of emails and flag those that contain fear-inducing or urgent phrases typical of phishing attempts.
- RL algorithms can continuously adapt to new phishing strategies by receiving feedback from the environment (e.g. when a phishing email is successfully blocked or overlooked). This approach helps to develop defences in response to new attack vectors [71].

4.4.4 *Cyberwarfare*

Cyber warfare has roots in early computer network operations but has been significantly enhanced by integrating AI/ML technologies. The development of AI-driven cyberwarfare capabilities began in the late 2000s as nation-states recognised the strategic importance of cyber domains.

Cyber warfare is the strategic use of cyber-attacks by a nation-state or international organisation to disrupt, damage or destroy another nation's critical infrastructure [72] to achieve geopolitical objectives without traditional kinetic warfare. It is a war waged in the digital realm, targeting computer systems and networks. These attacks typically target critical infrastructure such as power grids, financial systems and military networks to cause significant disruption, economic damage, espionage or strategic disadvantage. The increasing reliance on digital infrastructure worldwide has increased the potential impact of cyber warfare, making it a significant national security concern.

Cyberwarfare can be characterised by several key features:

- First, it is usually carried out by state actors such as nation-states or international organisations, although non-state actors with state support may also be involved.

- The primary targets of cyber warfare are often critical infrastructures, thereby disrupting and paralysing a target nation's essential services, such as power grids, financial systems, communication networks or even military operations.
- Cyber warfare is considered a significant threat to national security as it has the potential to paralyse a country's infrastructure and cause widespread economic damage.
- The lines between cyber warfare, cyber espionage (theft of information) and cybercrime (motivated by financial gain) are sometimes blurred, making it challenging to categorise and respond [73, 74]. However, cyber warfare usually has a broader strategic goal of damaging a nation's critical systems. Cyberwarfare is a relatively new concept, and there is no universally accepted definition of what constitutes a cyberwar action [75]. The lack of a clear definition makes dealing with the threats posed by cyber warfare even more complex.

The evolving nature of cyber warfare poses challenges to international law and attribution. There are no specific international laws governing cyber warfare, although some experts argue that existing laws of armed conflict could apply [76]. Attribution of cyberattacks remains a significant challenge due to the sophisticated techniques used to disguise identities.

4.4.4.1 Offensive Use of AI/ML

AI has significantly improved offensive capabilities in cyber warfare [2, 77–80]. One important improvement lies in automated attack campaigns. AI facilitates the automation of various attack processes, including reconnaissance, vulnerability scanning and exploit selection. By automating these tasks, attackers can conduct large-scale cyber campaigns more efficiently, overwhelming traditional defences and increasing the likelihood of a successful intrusion.

Another improvement is advanced phishing and social engineering. AI can create highly personalized phishing emails on any scale. Thanks to Natural Language Generation (NLG), these emails can mimic human writing styles, tailor content to specific targets, and bypass traditional spam filters. This makes phishing attempts more convincing and more challenging to detect.

ML algorithms also improve vulnerability detection by identifying zero-day vulnerabilities in software and systems. This capability allows attackers to develop zero-day exploits that target previously unknown vulnerabilities for which there is no immediate protection, increasing the effectiveness of the attack.

In addition, AI-driven bots are becoming increasingly sophisticated. These self-learning bots can infiltrate networks, steal data, and disrupt operations. Their ability to learn and adapt makes detecting and neutralizing them challenging.

Some high-profile incidents that illustrate the real-world impact of cyber warfare include the Stuxnet worm discovered in 2010 [81, 82], the attacks on the Ukrainian power grid in 2015/2016 [83, 84], and the attack on the Colonial Pipeline in 2021 [85, 86].

4.4.4.2 Defensive Use of AI/ML

Conversely, AI is crucial to develop robust cyber security defences.

- One important application is anomaly detection, where USL and SL algorithms identify anomalies in network traffic, deviations from normal user behavior and system logs that could indicate attacks [87, 88].
- Intrusion Detection and Prevention Systems (IDS/IPS) benefit from DL techniques that analyze patterns in network traffic to detect and prevent malicious activity. RL further enhances these systems by enabling AI agents to learn optimal defense strategies through simulated cyberattacks [89–91].
- User and Entity Behaviour Analytics (UEBA) use SL models to monitor user behavior patterns and help detect insider threats, compromised accounts, and unauthorised access attempts. Graph analysis techniques like community detection and connection prediction are also used to detect suspicious connections and potential insider threats [92–94].
- NLP enhances threat intelligence and predictions, which analyses threat data, security reports, and social media discussions to identify new threats and attacker tactics. Time series prediction models such as ARIMA and LSTM predict future cyberattacks by analyzing historical attack trends and vulnerabilities [95, 96].
- AI-powered systems are increasingly being used for early threat detection. These systems can identify early indicators of cyberattacks, such as unauthorized access attempts and data encryption patterns, which enables rapid mitigation and reduces the overall impact of the attack [80, 97].
- Another important application is automated patch management, where ML algorithms prioritize and automate the patching of software vulnerabilities. This proactive approach minimizes the window of opportunity for attackers to exploit known vulnerabilities, ensuring a robust cyber security defence [98].
- Finally, AI can create decoy systems and networks known as honeypots that lure attackers away from critical infrastructure. These decoys waste attackers' time and resources and provide valuable information about their tactics [99, 100].

For detailed information on various tactics and techniques used by attackers in the different phases of a cyberattack, such as reconnaissance, initial access, execution, and more, please refer to the online resource [101]. It provides a knowledge base of attackers' tactics and techniques based on real-world observations. It also serves as a valuable resource for improving cybersecurity practices by providing a structured way to analyze and mitigate threats. Another online site [102] provides insight and analysis on state-sponsored cyber operations worldwide. It tracks and reports various cyberattacks attributed to nation-states and provides a comprehensive overview of these cyber operations' tactics, objectives, and impact.

4.4.5 Autonomous Weapon Systems

Autonomous systems (AS) have quickly become integral to various industries, from transport to healthcare. These systems, characterized by their ability to perceive, decide, and act autonomously, are now also used for military applications, giving rise to autonomous weapon systems (AWS). An autonomous system can perform tasks and make decisions independently, without human intervention, by perceiving its environment, processing information, and performing actions to achieve specific goals [103, 104]. Autonomous systems are designed to operate in complex and dynamic environments and to adapt to changes and uncertainties.

Autonomous systems have several essential properties, one of which is perception. These systems rely on sensors such as cameras, LiDAR, radar, and ultrasonic sensors to collect data about their surroundings. This data is crucial for sensing and interpreting the environment and enables the systems to make informed decisions. Decision-making is another essential feature. Autonomous systems can analyze the collected sensor data in real-time and make informed decisions based on learned behaviors through AI algorithms or predefined rules. Adaptability is also an important characteristic. These systems can learn from experience and adapt to new or changing environments and conditions, which increases their functionality and efficiency. Fundamental to the operation of autonomous systems are control and actuation. These systems can execute decisions through actuators, such as motors or robotic limbs, to perform physical actions. Finally, communication is also an important feature. Autonomous systems can communicate with other systems or human operators, often in real-time, to coordinate actions or report status. This capability ensures seamless integration and interaction within their operational environment.

Autonomous systems are used in various sectors, e.g. in transport (self-driving cars, unmanned aerial vehicles (UAVs or drones) [105], autonomous underwater vehicles (AUV), ...), healthcare (robotic surgical assistants, rehabilitation robots,...) [106, 107], agriculture (automated planting, harvesting, crop monitoring systems, ...) [108, 109], industry (industrial robots, autonomous mobile warehouse robots, ...) [103], household (robotic vacuum cleaners, robotic lawn mowers, robotic pool cleaners, ...), space exploration (autonomous rovers and spacecraft, ...) [110–112], to name but a few.

Such systems are used in various sectors due to their many advantages and benefits, such as increased efficiency and productivity, greater safety through the reduction of human error, the ability to work in hazardous or inaccessible environments, and the ability to work continuously without fatigue.

There are still many challenges for autonomous systems, especially ensuring reliability and robustness in diverse and unpredictable conditions, resolving ethical and legal issues related to decision-making and accountability [113, 114], addressing cybersecurity risks and protecting against malicious attacks, and developing advanced algorithms for real-time processing and decision-making.

Autonomous Weapon Systems (AWS) have emerged from advances in robotics, computer vision, and decision algorithms. Research into AWS began in the early 2000s, with significant advances made in the 2010s as AI capabilities improved [115–117]. The primary motivation for the development of AWS was to reduce human casualties in military operations and enable more precise and efficient warfare. AWS was seen as a way to remove human limitations such as fatigue, emotions and reaction time from combat situations. AWS are characterized by several key features:

- One of the main features is the autonomous selection and engagement of targets. AWS can autonomously identify, select, and engage targets based on pre-programmed criteria and real-time data analysis.
- Advanced sensor integration is another vital feature. AWS are often equipped with multiple sensors such as radar, infrared, cameras, and sonar to detect the environment and identify potential targets. Sensor fusion techniques combine data from multiple sensors to improve the accuracy and reliability of target identification.
- Decision-making algorithms also play an essential role in AWS. These systems use complex algorithms, including ML and AI, to analyze sensor data, classify objects and make deployment decisions. These algorithms are designed to mimic human decision-making processes but with greater speed and potentially greater accuracy.
- Adaptability and the ability to learn are also key features. AWS can adapt to changing environments and unforeseen scenarios. Some systems may incorporate machine learning components that allow them to improve their performance over time-based on previous experience.
- Precision and accuracy are critical to minimize collateral damage and ensure mission effectiveness. Therefore, AWS are often equipped with guidance and control systems to aim the munition precisely at the target.
- Communication and networking capabilities are essential for coordinated operations. AWS can communicate with other systems and platforms and is integrated with command and control systems to receive updates, new mission parameters, and situational awareness data.
- Speed and efficiency are key advantages of AWS. These systems operate at high speed, make quick decisions, and act faster than human capabilities. This improves operational efficiency as tasks are executed without the delays associated with human decision-making processes.
- The human-machine interface is another important feature. Although AWS are autonomous, they often have interfaces for human monitoring and intervention when required. Operators can monitor system status, receive alerts, and override decisions when necessary.
- Resilience and robustness are critical to AWS and make them more robust than other autonomous systems. They are designed to operate reliably in diverse and demanding environments, including extreme weather conditions, electronic war-

fare and physical attacks. They have redundancy and fail-safe mechanisms to cope with malfunctions or unexpected conditions.
- Finally, ethical and legal compliance is a fundamental feature. AWS must be designed to comply with international laws of war and rules of engagement [118]. They have mechanisms to ensure compliance with ethical standards, such as the distinction between combatants and non-combatants.

AWS are being used more and more frequently due to its numerous advantages. The most important of these are reduced human risk, as the exposure of human soldiers to dangerous situations is minimized; force multiplier, as military capabilities can be expanded without increasing personnel; and increased operational efficiency, as decision-making and the execution of military operations can be faster.

However, the main challenges are [119–121]: ethical concerns, particularly around the question of whether it is acceptable to delegate life and death decisions to machines; accountability and control due to difficulties that may arise when responsibilities for AWS actions need to be assigned [122]; and reliability and safety due to concerns about system malfunction and potential vulnerability to cyber-attack [123].

Current AWS exhibit varying levels of autonomy. For example, loitering munitions loiter around a target area and can autonomously select and engage targets [124]. Autonomous drones are UAVs capable of highly autonomous surveillance and attack missions [125]. Robotic ground vehicles are platforms for various activities, including transport platforms and remote weapon stations [126]. Autonomous naval vessels, on the other hand, are unmanned surface or underwater vehicles for various maritime operations [127]. A significant milestone in military aviation is the historic flight of an experimental AI-controlled F-16 fighter jet in May 2024 that took Air Force Secretary Frank Kendall in an hour-long flight involving lightning-fast maneuvers at more than 550 miles an hour [128].

The development of autonomous systems, particularly AWS, represents a significant technological advance with far-reaching implications for various sectors, including the military. While these systems offer potential benefits regarding efficiency and reduced human risk, they pose significant ethical and technical challenges. As research and development in this area continues, addressing these challenges through robust international dialogue, ethical frameworks, and technological safeguards is critical.

4.5 Challenges and Ethical Considerations

Rapid advancements and deployment of AI/ML technologies in hybrid warfare present numerous challenges and ethical dilemmas.

4.5.1 Accountability and Attribution

One of the primary challenges posed by AI-driven warfare is the need for accountability and attribution [122]. Using autonomous systems in military operations complicates tracing actions back to specific actors, exacerbating tensions and hindering conflict resolution. The problem is compounded by the anonymity that AI systems can provide, making it challenging to hold perpetrators accountable. This lack of clear attribution could destabilize international relations and complicate diplomatic de-escalation efforts.

4.5.2 Escalation Risks

AI-enabled operations can be conducted at unprecedented speed and scale, which raises the risk of rapid escalation in conflict situations. The ability of AI systems to make decisions and execute actions in milliseconds may outpace human decision-making processes, leading to unintended consequences. These dynamics could transform localized conflicts into full-scale conventional warfare, escalating situations before diplomatic interventions occur.

4.5.3 Privacy and Civil Liberties

The deployment of AI for surveillance and data analysis in offensive and defensive operations presents significant privacy concerns. These technologies' pervasive nature allows for extensive population monitoring, potentially infringing on civil liberties. The balance between national security and individual rights becomes increasingly tenuous as AI surveillance capabilities expand, raising ethical questions about the limits of such practices [129].

4.5.4 Bias and Fairness

AI systems are susceptible to biases inherent in the training data, which can result in unfair targeting or discriminatory decision-making in conflict scenarios. These biases may perpetuate existing inequalities or introduce new forms of injustice, particularly in environments involving life-and-death decisions. Ensuring fairness and mitigating bias in AI systems used in warfare is a critical ethical challenge [130].

4.5.5 Human Control and Judgment

The increasing autonomy granted to AI systems in military contexts raises profound ethical questions about human control. As machines are increasingly tasked with making life-or-death decisions, the appropriate level of human oversight becomes a pressing concern. Delegating such decisions to AI systems challenges traditional notions of accountability and moral responsibility in warfare, necessitating re-examining existing ethical frameworks [131].

4.5.6 Arms Race Dynamics

The rapid advancement of AI in military applications has the potential to ignite an arms race, with nations competitively developing increasingly sophisticated AI weapons systems. This arms race could destabilize global security, as the desire for technological superiority may override long-term peace and stability considerations. The competitive nature of AI development in warfare contexts necessitates international dialogue and cooperative governance to prevent escalatory dynamics [132].

4.5.7 Dual-Use Concerns

AI and ML technologies often have dual-use potential, which can be applied in civilian and military contexts. This dual-use nature complicates regulatory efforts, as technologies developed for civilian purposes can be repurposed for military applications. Managing the dual-use nature of AI requires careful consideration of both ethical implications and regulatory frameworks to ensure that such technologies are not misused [2].

4.5.8 Long-Term Consequences

The long-term implications of AI deployment in warfare are primarily uncertain but could be profound. The continuous evolution of AI systems may lead to unforeseen consequences for global security and human society. Ethical considerations must, therefore, extend beyond immediate applications to consider the broader, long-term impacts of integrating AI into military operations [133].

Addressing these challenges requires a multidisciplinary approach involving policymakers, ethicists, technologists, and military strategists. International cooperation and the development of robust governance frameworks will be crucial in managing the risks and harnessing the potential benefits of AI/ML technologies in the context of hybrid warfare and cybersecurity.

4.6 Conclusion

The global security landscape has fundamentally changed with the integration of AI/ML technologies into hybrid warfare and cyber security. As demonstrated in this paper, these advanced technologies present unprecedented opportunities and complex challenges in securing critical infrastructure against evolving threats.

The delicate balance between innovation and security is emphasized by the dual use of AI/ML techniques—from social media manipulation to autonomous weapon systems. While these technologies provide powerful tools to defend against cyber threats and enhance military capabilities, they also enable sophisticated attacks that can undermine societal stability and national security.

A proactive and adaptive approach to cyber security is required due to rapid technological advances in this area. As offensive capabilities are developed, defensive measures must also be enhanced. This arms race between offense and defence underlines the critical importance of ongoing research and development of AI/ML-driven security solutions.

Furthermore, the ethical considerations around the use of AI in warfare and security cannot be overstated. Accountability, privacy, and human control must be carefully considered to ensure these technologies align with societal values and international norms.

Looking ahead, it is clear that AI will play an increasingly central role in shaping the future of global security. A multi-stakeholder approach involving governments, industry, academia, and civil society is essential to navigate this complex landscape. The risks must be managed, and the potential benefits of these transformative technologies are realized through international cooperation and the development of robust governance frameworks.

As further advances are made in this new era of hybrid threats, our ability to secure critical infrastructure will depend on technological innovation and our ability to address the ethical, legal, and societal implications of AI-driven warfare and cybersecurity. A future where AI serves as a bulwark for security and stability rather than a tool for disruption and conflict can be pursued by fostering a comprehensive and nuanced understanding of these challenges.

The journey from phantoms to firewalls in securing critical infrastructure still needs to be completed. Vigilance, adaptability, and a commitment to responsible innovation are required. In navigating this complex terrain, our success will be measured not only by our technological achievements but also by our ability to utilize these powerful tools in the service of a safer and more secure world.

References

1. Hoffman FG (2007) Conflict in the 21st century: the rise of hybrid wars. Potomac Institute for Policy Studies, Arlington
2. Brundage M et al (2018) The malicious use of artificial intelligence: forecasting, prevention, and mitigation. arXiv. preprint arXiv:1802.07228
3. Mitchell TM (1997) Machine learning. McGraw Hill
4. Goodfellow I, Bengio Y, Courville A (2016) Deep learning. MIT Press
5. Hastie T, Tibshirani R, Friedman J (2009) The elements of statistical learning: data mining, inference, and prediction. Springer, New York
6. Sutton RS, Barto AG (2018) Reinforcement learning: an introduction. MIT Press
7. LeCun Y, Bengio Y, Hinton G (2015) Deep learning. Nature 521:436–444. https://doi.org/10.1038/nature14539
8. Manning CD, Schütze H (1999) Foundations of statistical natural language processing. MIT Press, Cambridge
9. Jurafsky D, Martin JH (2022) Speech and language processing, 2nd edn. Pearson Education
10. Rodriguez CA, Walton TC, Chu H (2020) Putting the "FIL" into "DIME": growing joint understanding of the instruments of power. Joint Force Quart 97
11. Nye JS Jr (2004) Soft power: the means to success in world politics. Public Affairs, New York
12. Hoffman FG (2009) Hybrid warfare and challenges. Joint Force Quart 52:34–39
13. Cullen PJ, Reichborn-Kjennerud E (2017) MCDC countering hybrid warfare project: understanding hybrid warfare. A Multinational Capability Development Campaign project
14. Oskarrson K, Barnett R (2017) The effectiveness of DIMEFIL instruments of power in the gray zone. Open Publications (2016-01) 1(2)
15. Renz B, Smith H (2016) Russia and hybrid warfare—going beyond the label. Aleksanteri Papers
16. Lazer DM et al (2018) The science of fake news. Science 359(6380):1094–1096. https://doi.org/10.1126/science.aao2998
17. Bradshaw S, Howard PN (2018) The global organisation of social media disinformation campaigns. J Int Aff 71(1.5):23–32
18. Cadwalladr C, Graham-Harrison E (2018) The Cambridge Analytica files. https://www.theguardian.com/news/series/cambridge-analytica-files. Last accessed 10 July 2024
19. Isaak J, Hanna MJ (2018) User data privacy: Facebook, Cambridge Analytica, and privacy protection. Computer 51(8):56–59. https://doi.org/10.1109/MC.2018.3191268
20. Bongiovi JR (2019) The age of surveillance capitalism: the fight for a human future at the new frontier of power. Soc Forces 98(2):1–4. https://doi.org/10.1093/sf/soz037
21. Hunt A, Gentzkow M (2017) Social media and fake news in the 2016 election. J Econ Perspect 31(2):211–236. https://doi.org/10.1257/jep.31.2.211
22. Howard PN, Ganesh B, Liotsiou D, Kelly J, Francois C (2019) The IRA, social media and political polarization in the United States, 2012–2018. Computational Propaganda Research Project. University of Oxford
23. Traugott M (2019) Cyberwar: how Russian hackers and trolls helped elect a president—what we don't, can't, and do know. Public Opin Q 83(1):166–168. https://doi.org/10.1093/poq/nfy051
24. Howard PN, Kollanyi B (2016) Bots, #Strongerin, and #Brexit: computational propaganda during the UK-EU referendum. Political Bots, Oxford University Research Archive
25. Cadwalladr C (2017) The great British Brexit robbery: how our democracy was hijacked. https://www.theguardian.com/technology/2017/may/07/the-great-british-brexit-robbery-hijacked-democracy. Last accessed 10 July 2024
26. Bradshaw S, Howard PN (2018) Challenging truth and trust: a global inventory of organized social media manipulation. Working Paper 2018.1. Project on Computational Propaganda, Oxford. demtech.oii.ox.ac.uk. 26 pp
27. Bastos MT, Mercea D (2019) The Brexit botnet and user-generated hyperpartisan news. Soc Sci Comput Rev 37(1):38–54. https://doi.org/10.1177/0894439317734157

28. Fink C (2018) Dangerous speech, anti-Muslim violence, and Facebook in Myanmar. J Int Aff. https://jia.sipa.columbia.edu/news/dangerous-speech-anti-muslim-violence-and-facebook-myanmar. Last accessed 12 Aug 2024
29. Kyaw NN (2020) Social media, hate speech and fake news during Myanmar's political transition. In: Sinpeng A, Tapsell R (eds) From grassroots activism to disinformation: a selection. ISEAS–Yusof Ishak Institute, pp 86–104
30. Schissler M (2024) Beyond hate speech and misinformation: Facebook and the Rohingya genocide in Myanmar. J Genocide Res:1–26. https://doi.org/10.1080/14623528.2024.2375122
31. Cinelli M et al (2020) The COVID-19 social media infodemic. Sci Rep 10:16598. https://doi.org/10.1038/s41598-020-73510-5
32. Roozenbeek J et al (2020) Susceptibility to misinformation about COVID-19 around the world. R Soc Open Sci 7:201199
33. Lee EWJ, Bao H, Wang T, Lim YT (2023) From pandemic to Plandemic: examining the amplification and attenuation of COVID-19 misinformation on social media. Soc Sci Med 328:115979. https://doi.org/10.1016/j.socscimed.2023.115979
34. Kaliyar RK, Goswami A, Narang P (2021) FakeBERT: fake news detection in social media with a BERT-based deep learning approach. Multimed Tools Appl 80:11765–11788. https://doi.org/10.1007/s11042-020-10183-2
35. Zhou X, Zafarani R, Shu K, Liu H (2019) Fake news: fundamental theories, detection strategies and challenges. In: WSDM 19: Proceedings of the 12th ACM International Conference on Web Search and Data Mining, pp 836–837
36. Kudugunta S, Ferrara E (2018) Deep neural networks for bot detection. Inf Sci 467:312–322
37. Jiang J et al (2019) Anomaly detection with graph convolutional networks for insider threat and fraud detection. In: MILCOM 2019–2019 IEEE Military Communications Conference (MILCOM), Norfolk, VA, USA, pp 109–114. https://doi.org/10.1109/MILCOM47813.2019.9020760
38. Xu H, Ma Y, Liu HC et al (2020) Adversarial attacks and defenses in images, graphs and text: a review. Int J Autom Comput 17:151–178. https://doi.org/10.1007/s11633-019-1211-x
39. Goodfellow I et al (2014) Generative adversarial nets. In: Proceedings of the International Conference on Neural Information Processing Systems (NIPS 2014), pp 2672–2680
40. Pinheiro Cinelli L et al (2021) Variational methods for machine learning with applications to deep networks. Springer, pp 111–149. https://doi.org/10.1007/978-3-030-70679-1_5
41. Van den Oord A et al (2016) WaveNet: a generative model for raw audio. arXiv:1609.03499
42. Ren Y, Tan X, Qin T, Zhao S, Zhao Z, Liu T (2019) Almost unsupervised text to speech and automatic speech recognition. In: Proceedings of the 36th International Conference on Machine Learning, Proceedings of Machine Learning Research, vol 97, pp 5410–5419
43. Reddy VM, Vaishnavi T, Kumar KP (2023) Speech-to-text and text-to-speech recognition using deep learning. In: 2nd International Conference on Edge Computing and Applications (ICECAA), pp 657–666. https://doi.org/10.1109/ICECAA58104.2023.10212222
44. Barakat H, Turk O, Demiroglu C (2024) Deep learning-based expressive speech synthesis: a systematic review of approaches, challenges, and resources. Eurasip J Audio Speech Music Process 11. https://doi.org/10.1186/s13636-024-00329-7
45. Westerlund M (2019) The emergence of deepfake technology: a review. Technol Innov Manag Rev 9(11):40–53
46. Chesney R, Citron DK (2019) Deep fakes: a looming challenge for privacy, democracy, and national security. Calif Law Rev 107. https://doi.org/10.2139/ssrn.3213954
47. How a magician who has never voted found himself at the center of an AI political scandal. https://edition.cnn.com/2024/02/23/politics/deepfake-robocall-magician-invs/index.html. Last accessed 20 Aug 2024
48. A school principal faced threats after being accused of offensive language on a recording. Now police say it was a deepfake, https://edition.cnn.com/2024/04/26/us/pikesville-principal-maryland-deepfake-cec/index.html. Last accessed 20 Aug 2024

49. A voice Deepfake was used to scam a CEO out Of $243,000. https://www.forbes.com/sites/jessedamiani/2019/09/03/a-voice-deepfake-was-used-to-scam-a-ceo-out-of-243000/. Last accessed 20 Aug 2024
50. A company lost $25 million after an employee was tricked by deepfakes of his coworkers on a video call. https://www.businessinsider.com/deepfake-coworkers-video-call-company-loses-millions-employee-ai-2024-2. Last accessed 20 Aug 2024
51. Teen girls confront an epidemic of Deepfake nudes in schools. https://www.nytimes.com/2024/04/08/technology/deepfake-ai-nudes-westfield-high-school.html. Last accessed 20 Aug 2024
52. Verdoliva L (2020) Media forensics and DeepFakes: an overview. IEEE J Sel Top Signal Process 14(5):910–932. https://doi.org/10.1109/JSTSP.2020.3002101
53. Tolosana R et al (2020) Deepfakes and beyond: a survey of face manipulation and fake detection. Inf Fusion 64:131–148. https://doi.org/10.1016/j.inffus.2020.06.014
54. Agarwal S, Farid H (2021) Detecting Deep-Fake videos from aural and oral dynamics. In: 2021 IEEE/CVF Conference on Computer Vision and Pattern Recognition Workshops (CVPRW), Nashville, TN, USA, pp 981–989. https://doi.org/10.1109/CVPRW53098.2021.00109
55. Amerini I, Anagnostopoulos A, Maiano L, Ricciardi Celsi L (2021) Deep learning for multimedia forensics. Found Trends Comput Graph Vis 12(4):309–457. https://doi.org/10.1561/0600000096
56. Tolosana R et al (2021) DeepFakes evolution: analysis of facial regions and fake detection performance. In: Pattern recognition. ICPR international workshops and challenges. https://doi.org/10.1007/978-3-030-68821-9_38
57. Zobaed S et al (2021) DeepFakes: detecting forged and synthetic media content using machine learning. In: Artificial intelligence in cyber security: impact and implications, Advanced sciences and technologies for security applications. Springer. https://doi.org/10.1007/978-3-030-88040-8_7
58. Qureshi A, Megías D, Kuribayashi M (2021) Detecting deepfake videos using digital watermarking. In: 2021 Asia-Pacific Signal and Information Processing Association Annual Summit and Conference (APSIPA ASC), Tokyo, Japan, pp 1786–1793
59. Agarwal S, Hu L, Ng E, Darrell T, Li H, Rohrbach A (2023) Watch those words: video falsification detection using word-conditioned facial motion. In: 2023 IEEE/CVF Winter Conference on Applications of Computer Vision (WACV), Waikoloa, HI, USA, pp 4699–4708. https://doi.org/10.1109/WACV56688.2023.00469
60. Deepfakes and Blockchain. https://struckcapital.com/deepfakes-and-blockchain/. Last accessed 25 Aug 2024
61. Hegde AM, Kumar SB, Bhuvantej R, Vyshak R, Sarasvathi V (2023) Spear phishing using machine learning. In: International Conference on Advances in Computing and Data Sciences (ICACDS). https://doi.org/10.1007/978-3-031-37940-6_43
62. Heiding F, Schneier B, Vishwanath A, Bernstein J, Park PS (2024) Devising and detecting phishing emails using large language models. IEEE Access 12:42131–42146. https://doi.org/10.1109/ACCESS.2024.3375882
63. Wu CY, Kuo CC, Yang CS(2019) A phishing detection system based on machine learning. In: 2019 International Conference on Intelligent Computing and its Emerging Applications (ICEA), Tainan, Taiwan, pp 28–32. https://doi.org/10.1109/ICEA.2019.8858325
64. Almejrab RM, Sallabi OM, Bushaala FF, Altajori ABS, Mohamed AA (2023) A classification model for phishing detection system based on machine learning Algorithms. In: 2023 IEEE 11th International Conference on Systems and Control (ICSC), Sousse, Tunisia, pp 131–135. https://doi.org/10.1109/ICSC58660.2023.10449834
65. Jain S, Gupta C (2023) A support vector machine learning technique for detection of phishing websites. In: 2023 6th International Conference on Information Systems and Computer Networks (ISCON), Mathura, India, pp 1–6. https://doi.org/10.1109/ISCON57294.2023.10111968

66. Jayaraj R, Pushpalatha A, Sangeetha K, Kamaleshwar T, Udhaya Shree S, Damodaran D (2024) Intrusion detection based on phishing detection with machine learning. Measurement: Sensors 31. https://doi.org/10.1016/j.measen.2023.101003
67. Anirudh S, Radha Nishant P, Baitha S, Dinesh Kumar K (2024) An ensemble classification model for phishing mail detection. Proc Comput Sci 233:970–978. https://doi.org/10.1016/j.procs.2024.03.286
68. Alsubaei FS, Almazroi AA, Ayub N (2024) Enhancing phishing detection: a novel hybrid deep learning framework for cybercrime forensics. IEEE Access 12:8373–8389. https://doi.org/10.1109/ACCESS.2024.3351946
69. Asiri S, Xiao Y, Alzahrani S, Li T (2024) PhishingRTDS: a real-time detection system for phishing attacks using a Deep Learning model. Comput Secur 141. https://doi.org/10.1016/j.cose.2024.103843
70. Alhogail A, Alsabih A (2021) Applying machine learning and natural language processing to detect phishing email. Comput Secur 110. https://doi.org/10.1016/j.cose.2021.102414
71. Smadi S, Aslam N, Zhang L (2018) Detection of online phishing email using dynamic evolving neural network based on reinforcement learning. Decis Support Syst 107. https://doi.org/10.1016/j.dss.2018.01.001
72. Singer PW, Friedman A (2014) Cybersecurity and cyberwar: what everyone needs to know. Oxford University Press. https://doi.org/10.1093/wentk/9780199918096.001.0001
73. Rid T (2011) Cyber war will not take place. J Strateg Stud 35(1):5–32. https://doi.org/10.1080/01402390.2011.608939
74. Nye JS (2017) Deterrence and dissuasion in cyberspace. Int Secur 41(3):44–71. https://doi.org/10.1162/ISEC_a_00266
75. Clarke RA, Knake R (2010) Cyber war: the next threat to national security and what to do about it. HarperCollins Publishers, New York
76. Schmitt MN (2017) Tallinn manual 2.0 on the international law applicable to cyber operations, 2nd edn. Cambridge University Press
77. Reinhold T, Reuter C (2022) Cyber weapons and artificial intelligence: impact, influence and the challenges for arms control. In: Armament, arms control and artificial intelligence, Studies in peace and security. Springer, pp 145–158. https://doi.org/10.1007/978-3-031-11043-6_11
78. Guembe B, Azeta A, Misra S, Osamor VC, Fernandez-Sanz L, Pospelova V (2022) The emerging threat of Ai-driven cyber attacks: a review. Appl Artif Intell 36(1). https://doi.org/10.1080/08839514.2022.2037254
79. Mohamed N (2023) Current trends in AI and ML for cybersecurity: a state-of-the-art survey. Cogent Eng 10(2). https://doi.org/10.1080/23311916.2023.2272358
80. Kaur R, Gabrijelčič D, Klobučar T (2023) Artificial intelligence for cybersecurity: literature review and future research directions. Inf Fusion 97. https://doi.org/10.1016/j.inffus.2023.101804
81. Kushner D (2013) The real story of Stuxnet. IEEE Spectr 50(3):48–53. https://doi.org/10.1109/MSPEC.2013.6471059
82. Stuxnet explained: the first known cyberweapon (2022) https://www.csoonline.com/article/562691/stuxnet-explained-the-first-known-cyberweapon.html. Last accessed 20 Aug 2024
83. Lessons from Ukraine's energy grid cyber attack (2016) https://www.boozallen.com/s/insight/thought-leadership/lessons-from-ukranians-energy-grid-cyber-attack.html. Last accessed 20 Aug 2024
84. How hackers attacked Ukraine's power grid: implications for industrial IoT security (2016) https://www.zdnet.com/article/how-hackers-attacked-ukraines-power-grid-implications-for-industrial-iot-security/. Last accessed 20 Aug 2024
85. Colonial Pipeline hack explained: everything you need to know (2022) https://www.techtarget.com/whatis/feature/Colonial-Pipeline-hack-explained-Everything-you-need-to-know. Last accessed 19 Aug 2024

86. The attack on Colonial Pipeline: what we've learned & what we've done over the past two years (2023) https://www.cisa.gov/news-events/news/attack-colonial-pipeline-what-weve-learned-what-weve-done-over-past-two-years. Last accessed 19 Aug 2024
87. Chandola V, Banerjee A, Kumar V (2009) Anomaly detection: a survey. ACM Comput Survey (CSUR) 41(3):1–58. https://doi.org/10.1145/1541880.1541882
88. Li Z, Zhu Y, van Leeuwen M (2022) A survey on explainable anomaly detection. arXiv. arXiv:2210.06959
89. Buczak AL, Guven E (2016) A survey of data mining and machine learning methods for cyber security intrusion detection. IEEE Commun Surv Tutor 18(2):1153–1176. https://doi.org/10.1109/COMST.2015.2494502
90. Umer MF, Sher M, Bi Y (2017) Flow-based intrusion detection: techniques and challenges. Comput Secur 70:238–254. https://doi.org/10.1016/j.cose.2017.05.009
91. Apruzzese G, Colajanni M, Ferretti L, Guido A, Marchetti M (2018) On the effectiveness of machine and deep learning for cyber security. In: 2018 10th International Conference on Cyber Conflict (CyCon), Tallinn, Estonia, pp 371–390. https://doi.org/10.23919/CYCON.2018.8405026
92. Shashanka M, Shen MY, Wang J (2016) User and entity behavior analytics for enterprise security. In: IEEE International Conference on Big Data (Big Data), Washington, DC, USA, pp 1867–1874. https://doi.org/10.1109/BigData.2016.7840805
93. What is AI-powered behavioural analysis in cybersecurity. https://www.crowdstrike.com/cybersecurity-101/secops/ai-powered-behavioral-analysis/. Last accessed 20 Aug 2024
94. Advanced threat detection with User and Entity Behaviour Analytics (UEBA) in Microsoft Sentinel. https://learn.microsoft.com/en-us/azure/sentinel/identify-threats-with-entity-behavior-analytics. Last accessed 20 Aug 2024
95. Suh-Lee C, Jo JY, Kim Y (2016) Text mining for security threat detection discovering hidden information in unstructured log messages. In: 2016 IEEE conference on Communications and Network Security (CNS), Philadelphia, PA, USA, pp 252–260. https://doi.org/10.1109/CNS.2016.7860492
96. Mittal S, Das PK, Mulwad V, Joshi A, Finin T (2016) CyberTwitter: using Twitter to generate alerts for cybersecurity threats and vulnerabilities. In: 2016 IEEE/ACM International Conference on Advances in Social Networks Analysis and Mining (ASONAM), San Francisco, CA, USA, pp 860–867. https://doi.org/10.1109/ASONAM.2016.7752338
97. Salem AH, Azzam SM, Emam OE et al (2024) Advancing cybersecurity: a comprehensive review of AI-driven detection techniques. J Big Data 11:105. https://doi.org/10.1186/s40537-024-00957-y
98. Nowakowski J, Keller J (2024) AI-powered patching: the future of automated vulnerability fixes. Google Security Engineering Technical Report. https://research.google/pubs/ai-powered-patching-the-future-of-automated-vulnerability-fixes/. Last accessed 19 Aug 2024
99. Wang C, Lu Z (2018) Cyber deception: overview and the road ahead. IEEE Secur Priv 16(2):80–85. https://doi.org/10.1109/MSP.2018.1870866
100. Bao T, Tambe M, Wang C (eds) (2023) Cyber deception: techniques, strategies, and human aspects. Springer Cham. https://doi.org/10.1007/978-3-031-16613-6
101. MITRE ATT&CK Homepage. https://attack.mitre.org. Last accessed 21 May 2024
102. Cyber operations tracker. https://www.cfr.org/cyber-operations/. Last accessed 21 May 2024
103. Müller M, Müller T, Ashtari Talkhestani B, Marks P, Jazdi N, Weyrich M (2021) Industrial autonomous systems: a survey on definitions, characteristics and abilities. At - Automatisierungstechnik 69(1):3–13. https://doi.org/10.1515/auto-2020-0131
104. Chen J, Sun J, Wang G (2022) From unmanned systems to autonomous intelligent systems. Engineering 12:16–19. https://doi.org/10.1016/j.eng.2021.10.007
105. Olayode IO, Du B, Severino A, Campisi T, Alex FJ (2023) Systematic literature review on the applications, impacts, and public perceptions of autonomous vehicles in road transportation system. J Traffic Transp Eng (English Edition) 10(6):1037–1060. https://doi.org/10.1016/j.jtte.2023.07.006

106. Trusted autonomous systems in healthcare: a policy landscape review (2021) https://www.kcl.ac.uk/policy-institute/assets/trusted-autonomous-systems-in-healthcare.pdf. Last accessed 20 May 2024
107. Morgan AA, Abdi J, Syed MAQ, Kohen GE, Barlow P, Vizcaychipi MP (2022) Robots in healthcare: a scoping review. Curr Robot Rep 3(4):271–280. https://doi.org/10.1007/s43154-022-00095-4
108. Rose DC, Lyon J, de Boon A, Hanheide M, Pearson S (2021) Responsible development of autonomous robotics in agriculture. Nat Food 2(5):306–309. https://doi.org/10.1038/s43016-021-00287-9
109. Emmi L, Fernández R, Gonzalez-de-Santos P, Francia M, Golfarelli M, Vitali G, Sandmann H, Hustedt M, Wollweber M (2023) Exploiting the internet resources for autonomous robots in agriculture. Agriculture 13(5):1005. https://doi.org/10.3390/agriculture13051005
110. Gao Y, Chien S (2017) Review on space robotics: toward top-level science through space exploration. Sci Robot 2(7). https://doi.org/10.1126/scirobotics.aan5074
111. Nesnas IA, Fesq LM, Volpe RA (2021) Autonomy for space robots: past, present, and future. Curr Robot Rep 2:251–263. https://doi.org/10.1007/s43154-021-00057-2
112. Shah V (2024) Next-generation space exploration: AI-enhanced autonomous navigation systems. J Environ Sci Technol 3(1)
113. Yazdanpanah V, Gerding EH, Stein S et al (2023) Reasoning about responsibility in autonomous systems: challenges and opportunities. AI & Soc 38:1453–1464. https://doi.org/10.1007/s00146-022-01607-8
114. Vicente DM, Pereira RS, Leal AA (eds) (2024) Legal aspects of autonomous systems: a comparative approach. Springer Cham. https://doi.org/10.1007/978-3-031-47946-5
115. Scharre P (2018) Army of none: autonomous weapons and the future of war. W.W. Norton & Company
116. Taddeo M, Blanchard A (2022) A comparative analysis of the definitions of autonomous weapons systems. Sci Eng Ethics 28(5):37. https://doi.org/10.1007/s11948-022-00392-3
117. Wood NG (2023) Autonomous weapon systems: a clarification. J Mil Ethics 22(1):18–32. https://doi.org/10.1080/15027570.2023.2214402
118. McFarland T (2020) Autonomous weapon systems and the law of armed conflict: compatibility with international humanitarian law. Cambridge University Press
119. Arkin R (2009) Governing lethal behavior in autonomous robots, 1st edn. Chapman & Hall/CRC. https://doi.org/10.1201/9781420085952
120. Allenby BR (2015) The applied ethics of emerging military and security technologies, 1st edn. Routledge. https://doi.org/10.4324/9781315241364
121. Amoroso D, Tamburrini G (2020) Autonomous weapons systems and meaningful human control: ethical and legal issues. Curr Robot Rep 1:187–194. https://doi.org/10.1007/s43154-020-00024-3
122. Lin P, Bekey G, Abney K (2008) Autonomous military robotics: risk, ethics, and design. California Polytechnic State University, San Luis Obispo
123. The weaponization of increasingly autonomous technologies: autonomous weapon systems and cyber operations (2017) https://unidir.org/files/publication/pdfs/autonomous-weapon-systems-and-cyber-operations-en-690.pdf. Last accessed 21 May 2024
124. Unveiling the power of loitering munition: a game-changer in modern combat (2023) https://defensebridge.com/article/unveiling-the-power-of-loitering-munition-a-game-changer-in-modern-combat.html. Last accessed 22 May 2024
125. Rise of the drones: 11 exciting UAS companies innovating in defense (2023) https://www.karveinternational.com/insights/rise-of-the-drones-11-exciting-uas-companies-innovating-in-defense. Last accessed 22 May 2024
126. Top 10 military robots and unmanned ground vehicles in the world (2020) https://roboticsbiz.com/top-10-military-robots-and-unmanned-ground-vehicles-in-the-world/. Last accessed 22 May 2024

127. US Navy's four unmanned ships return from Pacific deployment (2024) https://www.defensenews.com/naval/2024/01/16/us-navys-four-unmanned-ships-return-from-pacific-deployment/. Last accessed 22 May 2024
128. AI-controlled fighter jet takes Air Force secretary on historic ride (2024) https://www.airforcetimes.com/news/your-air-force/2024/05/03/ai-controlled-fighter-jet-takes-air-force-secretary-on-historic-ride/. Last accessed 23 May 2024
129. Richards NM (2013) The dangers of surveillance. Harv Law Rev 126(7):1934–1965
130. Barocas S, Selbst AD (2016) Big data's disparate impact. Calif Law Rev 104:671
131. Horowitz MC, Scharre P (2015) Meaningful human control in weapon systems: a primer. Center for a New American Security. https://www.cnas.org/publications/reports/meaningful-human-control-in-weapon-systems-a-primer. Last accessed 22 May 2024
132. Geist EM (2016) It's already too late to stop the AI arms race—we must manage it instead. Bull At Sci 72(5):318–321. https://doi.org/10.1080/00963402.2016.1216672
133. Bostrom N (2014) Superintelligence: paths, dangers, strategies. Oxford University Press

Chapter 5
Forecasting and Mitigating Epidemic Spread During Natural Disasters Triggered by Seismic Events

Mirsada Hukić, Almir Karabegović, Mirzć Ponjavic, and Naida Ademović

Abstract Natural disasters such as earthquakes, floods, and tornadoes can trigger infectious disease outbreaks due to factors such as the destruction of communal infrastructure, forced migration, environmental changes, inadequate access to clean food and water, increased exposure to pre-existing pathogens, and the introduction of new pathogens. The real-time tracking and analysis of the spread of infectious diseases and the characteristics of seismological disasters are facilitated by the Epidemiological Locational Intelligence System (ELIS), a tool developed by the Academy of Sciences and Arts of Bosnia and Herzegovina. Initially created to oversee the COVID-19 pandemic, ELIS's effectiveness has been demonstrated to depend on efficient information exchange among local entities responsible for collecting, updating, and managing epidemiological data. The system's ability to promptly recommend preventive actions is enhanced by access to spatial information about population dynamics, environmental conditions such as climate, vegetation, pollution, seismic activity, community infrastructure, and healthcare services. A recent enhancement to ELIS includes a model designed to predict the likelihood and spread of an epidemic in the aftermath of an earthquake. This enhancement involved analyzing statistical data on seismic activity in Bosnia and Herzegovina as well as

M. Hukić
Department of Medical Sciences, Academy of Sciences and Arts of Bosnia and Herzegovina, Sarajevo, Bosnia and Herzegovina

A. Karabegović (✉)
University of Sarajevo-Faculty of Electrical Engineering, Sarajevo, Bosnia and Herzegovina
e-mail: akarabegovic@etf.unsa.ba

M. Ponjavic
Faculty of Engineering and Natural Sciences, International Burch University Sarajevo, Ilidža, Bosnia and Herzegovina

N. Ademović
University of Sarajevo-Faculty of Civil Engineering, Sarajevo, Bosnia and Herzegovina

D. Radu et al. (eds.), *Countering Hybrid Threats Against Critical Infrastructures*, NATO Science for Peace and Security Series B: Physics and Biophysics, https://doi.org/10.1007/978-94-024-2304-4_5

epidemiological data related to infectious diseases, hot spot areas of endemic disease, and immunization rates of vaccinated vaccine-preventable diseases. Based on the model's forecasts, it provides and establishes a set of preventive and control actions.

Keywords Forecasting epidemic · Mitigating epidemic spread · Natural disasters · Earthquake · Landslide areas · Water supply infrastructure

5.1 Introduction

If our goal is effective disease control, then early detection of disease outbreaks is extremely important. An obvious characteristic of every epidemic is a degree of spatial diffusion of cases, but still, little or no spatial data is used in the routine surveillance of infectious diseases [1–3]. Spatio-temporal surveillance algorithms are being developed for this very reason to facilitate early detection of disease outbreaks that show spatial clustering. This especially applies to those associated with person-to-person transmission or a localized source of infection [4, 5].

Geographic Information Systems (GIS) technologies have played an important role in the fight against COVID-19. This could be seen in many aspects, including rapid aggregation of data from multiple sources, spatial tracking of confirmed cases, rapid visualization of epidemic information, transmission prediction, spatial segmentation of epidemic risk and degree of prevention, material resource management, social-emotional management, and elimination of panic [6].

In Bosnia and Herzegovina (BiH), the Epidemic Location Intelligence System (ELIS) was established to monitor the spread of the COVID-19 virus to combat the epidemic of COVID-19. This was the result of a project initiated by the Committee for Microbiology and Related Disciplines of the Department of Medical Sciences of the Academy of Sciences and Arts of Bosnia and Herzegovina (ANU-BiH). At that time, a multidisciplinary team was formed to implement the system.

The system, as shown in Fig. 5.1, aimed to collect, process, and analyze data on the epidemic, conduct research on the epidemic of COVID-19, and facilitate better communication and the exchange of information with public health institutions. An important element of this system was evaluating the state system for monitoring the epidemic and the control measures taken. Finally, a mechanism was established for reporting the epidemic situation to state institutions and the public [7].

The most recent improvement of ELIS is the development of a model for predicting the occurrence of an epidemic under earthquake conditions. For this purpose, an analysis of statistical data on seismic phenomena in Bosnia and Herzegovina was carried out, with the definition of preventive and control measures to suppress the epidemic, based on a model for predicting its occurrence and spread.

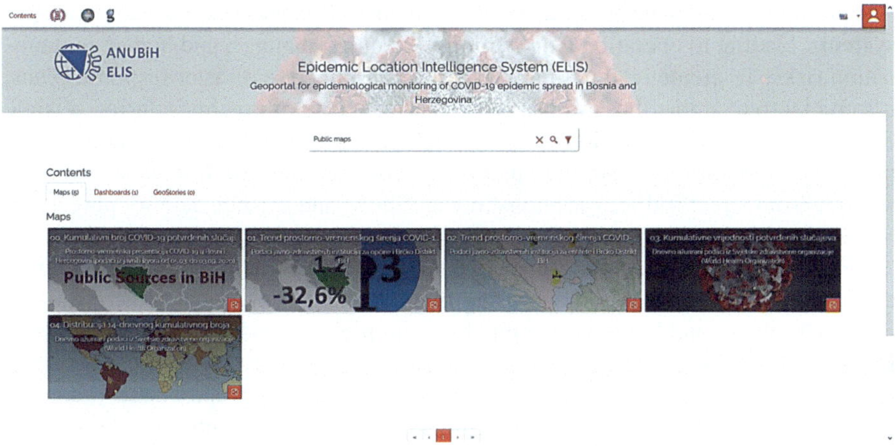

Fig. 5.1 A geoportal for geo-visualization, communication, and data manipulation (http://epidemija.gis.ba/)

5.2 The Impacts of Earthquakes

Earthquakes are catastrophic natural phenomena that cannot be prevented. However, their impacts are profound and multifaceted, affecting both the natural environment and human societies in numerous ways. The consequences of earthquakes can be broadly categorized into infrastructure destruction, environmental changes, human casualties, population displacement, healthcare challenges, and public health crises [8].

Destruction of infrastructure is one of the most immediate and visible effects of an earthquake. The seismic waves generated during an earthquake can cause severe damage to critical infrastructure systems, including power grids, water supply systems, transportation networks, information and communication systems, sanitation systems, healthcare facilities, and financial institutions. Power system disruptions can result in widespread blackouts that can linger for long periods of time, making rescue and relief operations more difficult. Damaged reservoirs and burst pipelines can cause shortages in water supplies as well as possible contamination of the water supply. When sanitation systems fail, there can be serious consequences for public health because improper disposal of trash and uncontaminated water raises the possibility of disease epidemics. Damage to transportation networks, including highways, railroads, airports, and bridges, usually isolates impacted areas and impedes the delivery of emergency services and necessary supplies. Furthermore, the breakdown of communication and information networks might make it more difficult for emergency responders to coordinate, which could postpone vital operations. Furthermore, the collapse of financial institutions can impede economic activity, postponing recovery and making conditions worse.

Earthquakes, as natural disasters, can put immense pressure on **water supply systems**, leading to operational disruptions, structural damage, and higher contamination risks. To create resilient infrastructure and effective disaster response plans, it's crucial to examine these impacts. The performance of pipelines during seismic events is influenced by various factors, including the material and rigidity of the pipes, soil conditions, and the quality of construction. Additionally, dynamic stresses from street traffic, variations in water flow, and seismic activity play a significant role [9].

Earthquakes cause compressive and tensile stresses in pipelines, leading to cracks, fractures, and water loss due to ground displacement and pipeline misalignment [10]. Water and hydrocarbon pipelines are highly vulnerable to seismic activity, even at low intensities, due to their extensive reach and varying soil properties [11]. Protection strategies include expansion joints at rigid pipe connections and designing pipeline bridges to withstand seismic forces. It has introduced a seismic safety system to improve pipeline resilience in water supply networks [12].

Water storage facilities are essential lifelines, but many of them in British Columbia are susceptible to earthquake damage as they were constructed before current seismic codes were established [13]. Harrison [13] recommended measures to improve the seismic resilience of concrete reservoirs. After the 2001 Bhuj earthquake, efforts in India focused on retrofitting vulnerable structures, including an elevated water tank over 40 years old. A case study confirmed that a simple retrofitting method effectively maintained the tank's functionality post-earthquake [14]. Without intervention, such damage can cause significant water loss and contamination.

Eskandari [15] highlights key damages to water pipelines from past earthquakes. In the 1976 Mindanao earthquake, a collapsed bridge burst a major water line. The 1986 El Salvador earthquake damaged 80 km of water and 65 km of sewage pipelines, while the 1997 Venezuelan earthquake caused significant damage to subterranean pipelines and drinking water lines. The 2003 Bam earthquake damaged 11 wells and six key water pipeline locations. The 1994 Northridge earthquake caused widespread fractures and contamination. The 1995 Kobe earthquake led to soil liquefaction, damaged 86 reservoirs, and ruptured major pipes [16]. Turkey's 1999 earthquakes destroyed 45% of the pipeline network [17], and the 2003 Bam earthquake affected groundwater contamination levels [18].

Similarly, the 2011 Tohoku earthquake in Japan severely damaged the country's water infrastructure, resulting in large-scale pipeline breakage and leaks as well as severe debris and sewage contamination. In Japan, the Great East Earthquake of 2011 claimed almost 16,000 lives. For 3 months, it resulted in serious malfunctions in over 120 wastewater treatment plants, liquefaction of soil, and failures in 102 km of pipelines, depriving many families of clean water and sanitation and causing untreated sewage to be discharged into water bodies [19].

These occurrences show how important it is to have water infrastructure systems that are resilient enough to withstand earthquakes and prevent contamination. Effective mitigation measures are necessary to reduce the impact of earthquakes on water supply systems and guarantee the security and dependability of water supplies in the wake of such catastrophes. Seismic-resistant design and routine maintenance are key strategies in addressing these issues.

Earthquakes can lead to long-term **environmental changes** that have significant impacts. Landslides, tsunamis, and soil liquefaction triggered by earthquakes can drastically alter the landscape. Ground failure can manifest in three main ways: faulting, landslides, and liquefaction [20]. Landslides and soil liquefaction, in particular, can severely damage water systems. Soil liquefaction occurs when saturated soil loses its strength and acts like a liquid, potentially causing pipelines and other critical infrastructure to collapse. Landslides, on the other hand, can bury and damage water pipes and treatment facilities. Both of these effects can severely disrupt water infrastructure and complicate recovery efforts after a seismic event.

Serious issues are raised by earthquakes, such **as human fatalities and psychological effects.** These incidents may cause serious mental anguish, injuries, and fatalities. Anxiety, sadness, and post-traumatic stress disorder (PTSD) are among the long-term mental health conditions that might result from the terror and panic felt. Food supply disruptions can cause famine, and shortages of essentials can make survivors' suffering worse. Furthermore, because individuals impacted may find it difficult to cope with the aftermath of the tragedy, the psychological effects of an earthquake might hinder recovery efforts.

Population displacement is a major consequence of earthquakes. People are frequently compelled to relocate to safer areas after houses and infrastructure are damaged. These displaced people typically wind up living in cramped housing units or improvised shelters, which frequently lack adequate hygienic conditions. Disease risk can be increased by unsanitary living conditions in these shelters, particularly in the absence of clean water and proper sanitation. Inadequate nutrition, stress, and overcrowding can all contribute to the spread of infectious diseases, including diarrhea, respiratory infections, and vector-borne illnesses. Large-scale population inflows can also put a burden on host communities' resources, causing social unrest and making recovery efforts more difficult.

Earthquakes have the potential to seriously damage **healthcare systems**, making the problems with public health in the impacted areas worse. The ability of medical facilities to treat the ill and injured may be severely limited by their complete or partial destruction. Even when these institutions are still open, they frequently deal with overcrowding, which makes it more difficult to provide prompt and efficient medical care. Supply chain disruptions can result in shortages of vital drugs, medical supplies, and other items, which makes healthcare solutions even more difficult. In addition, the system may be put under further strain if healthcare personnel leave due to illness, death, or displacement, which would make meeting the rising demand for services even more challenging.

The damage to healthcare systems caused by earthquakes exacerbates the public health challenges faced by affected populations. The ability of healthcare facilities to treat the ill and injured may be severely limited by partial or total destruction. Even when facilities remain operational, they are often overwhelmed by the sudden influx of patients, making it difficult to provide timely and adequate medical care. The disruption of supply chains can lead to shortages of essential medicines, medical equipment, and supplies, further hampering the ability of healthcare providers to respond to the needs of the population [21]. Additionally, the loss of healthcare workers due to injury, death, or displacement can further weaken the healthcare system, making it difficult to manage the increased demand for services.

Public health crises are a common outcome in the aftermath of devastating earthquakes. The combination of disruption in infrastructure, population displacement, and a damaged healthcare system creates the perfect conditions for the spread of infectious diseases and the outbreak of an epidemic [22, 23]. The lack of clean water, and poor food safety and hygiene conditions can lead to outbreaks of waterborne diseases such as hepatitis A and typhoid fever as well as foodborne diseases including salmonellosis and shigellosis. Overcrowded living conditions in shelters facilitate the transmission of respiratory infections like influenza, COVID-19, and tuberculosis. The mixing of populations with varying immunity levels accelerates risks of infectious diseases, particularly vaccine-preventable diseases such as diphtheria, whooping cough, measles, etc. Disruption of vector control programs can result in outbreaks of vector-borne diseases including West Nile virus, Lyme disease, Zika virus, rickettsial diseases, tick-borne encephalitis, and malaria, as well as rodent-borne diseases such as hantaviruses and leptospirosis. Temporary shelters located in endemic hotspots for these diseases pose additional risks. Furthermore, the physiological and psychological stress associated with earthquakes can weaken the immune system, making individuals more susceptible to infections. In the aftermath of large-scale earthquakes, the number of injured people is often overwhelming, with many suffering from wounds that, if not properly treated, can lead to hospital-acquired infections in overcrowded hospitals lacking adequate sanitation. When a healthcare system is overwhelmed, it can have difficulty diagnosing and treating illnesses, which may result in higher rates of morbidity and mortality. Limited resources can hinder efforts to control disease outbreaks and respond quickly to epidemics, leading to widespread suffering and long-term health issues for the affected population [24].

5.3 Expert Epidemic Information System

A special type of information associated with specific locations is called geoinformation. A geoinformation system is a system for collecting, analyzing, storing, and presenting various types of data, including geospatial data. Geoinformation system (GIS) is a technology that integrates various techniques such as databases, statistical analysis, digital cartography, image processing, etc. Because of this, GIS is used in various scientific disciplines and professional areas. Geoinformation management requires the ability to apply GIS tools, as well as understanding and knowledge of concepts and methods based on different fields such as informatics, computing, geography, geodesy, cartography, remote sensing, and other relevant fields [2].

Generally speaking, geoportals provide the presentation of spatial data. They usually do this through an interactive map, so that users can easily select and display different thematic layers or perform a data search [6]. Like standard web portals, the essential features that make it useful are availability regardless of time, location, or platform of use. That is why it is used as a tool to support decision-making and facilitate communication with the public through the transparent presentation of data on the Internet. In this way, we have a simpler monitoring and analysis of data changes, including spatial ones. The system enables the presentation of data on the locations of important business-administrative institutions and organizations, business and public infrastructure, public safety, and civil defense, cadastres (real estate, utilities, green areas), communal infrastructure, spatial planning documentation, and more. In Bosnia and Herzegovina, the exchange of all data, including statistical and epidemiological data, is characterized by poor availability and inconsistency in data collection. A particularly pronounced problem is the weak use of standards, lack of coordination, as well as limitations in data distribution due to multiple levels of administrative jurisdiction [3]. This situation makes it difficult to identify, access, and use all data in the country. Most of the epidemic data are georeferenced (address-based), enabling enhanced analytics with location intelligence. Data flow is illustrated in Fig. 5.2, where input data includes both health and spatial data [7]. The upgraded process diagram presented here includes recognizing and mapping areas of disasters, which then become part of geospatial analysis. In the analysis phase, any data not already geocoded must be geocoded, usually based on address, followed by data classification, and statistical and geospatial analysis. Various tables and associated frequency or location maps are generated as output. The system enables the distribution of scientific data on the risk of epidemics following natural disasters, making this information available to both experts and citizens to raise awareness of these risks. The software supports rapid computational analysis and

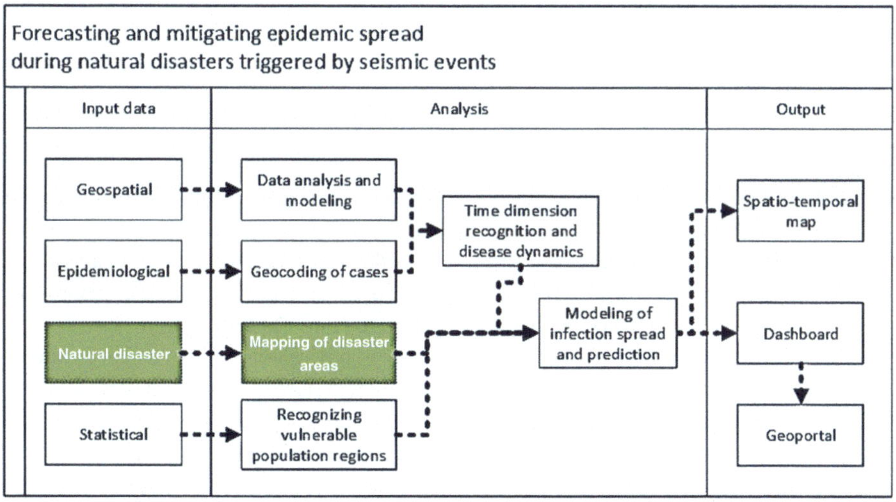

Fig. 5.2 Spatial analysis with epidemic data processing and spatial analysis. The results of this analysis are presented in the form of maps and tables. (Adapted from [7])

the creation of spatial risk assessments by combining scientific hazard data and land use data. The system only requires Internet access and adapts its presentation and tools to the user's needs, whether they are health professionals, decision-makers, or citizens. In this way, ELIS aids in disaster management both before and after the event. Specifically, through planning activities, it supports mitigation efforts aimed at reducing the impact of disasters before they occur. For this, it is necessary to be prepared, which involves developing emergency operations plans. When a disaster strikes, ELIS facilitates the implementation of these plans and monitors the recovery process [4].

The combination of health and geospatial sciences includes various disciplines focused on the study of earth and climate, the risks associated with exposure to their changes, and their impact on health. The field of geohealth studies is precisely related to human health and the environment and is based on the fact that nature, human activities, climate, and public health are all closely related. Also, the influence of all these factors on the spatio-temporal distribution, and especially on the transmission of infectious and other diseases, is evident. The Center for Disease Control and Geo-health Studies has launched research aimed at predicting and mitigating the spread of epidemics due to natural disasters with an emphasis on seismic. Such a task requires interdisciplinary cooperation, where each member from different groups (geohealth studies, clinical methods, epidemiology, laboratory diagnostics, recurrent zoonoses) becomes an active user of ELIS. The research results are presented in the form of interactive maps, together with other related data.

The goal of this kind of cooperation is primarily to combine knowledge and different approaches, but also to apply methods specific to different fields.

5.3.1 Methodology

To adapt the existing expert system to improve it and use it to forecast and mitigate the spread of infectious diseases during natural disasters, particularly in the aftermath of earthquakes, the research, and development included the following aspects:

- Methodology: The methodology applied to develop and introduce the Epidemiological Location System (ELIS) as a modern tool in epidemiology for forecasting and mitigating the spread of infectious diseases during and after earthquakes includes undegraded existing ELIS and collection data related to geography, infrastructure, healthcare capacitates and epidemiology.
- Data Collection: Collecting and analyzing geographical data on earthquake-prone areas, including seismic maps and historical data on previous earthquakes; Including detailed information on key infrastructure, such as water supply networks, communication networks, energy facilities, and industrial sites that produce toxins or hazardous materials; Mapping healthcare facilities, including hospitals, clinics, and emergency services, with data on their capacities, equipment, and current status; Collecting and analyzing historical epidemiological

data, with a focus on areas with high rates of vector-borne and rodent-borne diseases; Collecting data on vaccination and immunization rates among children and adults to assess the risk of disease outbreaks due to the mixing of vaccinated and unvaccinated populations.

- Database Development: Integrating collected data into ELIS through multi-layered databases, enabling spatio-temporal analysis and visualization. Using GIS software for mapping and analyzing data, allows for identification of potentially risky areas, including zones with high rates of infectious diseases and low vaccination coverage.
- ELIS Enhancement: Adaptation of ELIS by adding new data layers, including epidemiological risks and vaccination data, to enable more accurate modeling and prediction of disease spread after earthquakes; Conducting simulations of different earthquake scenarios to assess the system's effectiveness in real-world conditions.
- Data Analysis: Using ELIS for analyzing spatio-temporal patterns of disease spread in relation to the locations of key infrastructure, healthcare facilities, epidemiologically risky areas, and vaccination levels; Assessing the risk of epidemic outbreaks and other public health crises based on collected data and simulations.
- Development of Guidelines: Develop guidelines to assist local authorities, healthcare institutions, and the public in organizing and responding quickly after an earthquake. The guidelines will include recommendations for shelter placement, disease control, and healthcare resource management.
- Validation: Validation methods using historical data on earthquakes, epidemiological risks, and epidemic outbreaks. Engaging local authorities and healthcare institutions in testing and validating the system and guidelines.

5.4 Analyzing Data About Epidemic Spread and Natural Disasters Triggered by Seismic Events

The Epidemic Location Intelligence System (ELIS) is an expert information system designed for the analysis and reporting of epidemic data. It features functionalities such as communication, analysis, and reporting. The ELIS portal, which serves as a web user interface, provides access to various resources for data analysis. Primarily, it offers an environment for users to access data and interact with data. The portal integrates data from external systems.

To perform data analysis and modeling, it is necessary to extract, transform, and load data into a database. Due to the diversity of data sources, its formats, and structures, the system provides appropriate mechanisms for transforming raw data into usable models. The new module of ELIS includes a model for predicting epidemic occurrences in the context of earthquakes. For this purpose, a map has been created that integrates live feed sources for earthquake data. The Euro-Mediterranean

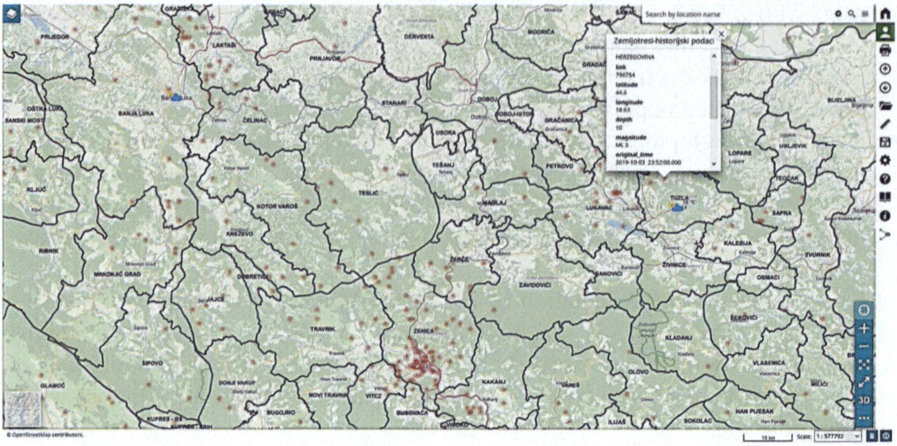

Fig. 5.3 The map presents the integration of the service about recent earthquake information from the Euro-Mediterranean Seismological Centre (EMSC) program with ELIS data

Seismological Centre (EMSC), an international non-profit organization with over 70 member institutes worldwide, was utilized as the primary data source. EMSC manages a system for the rapid collection, determination, and dissemination of earthquake parameters. This system integrates seismic data from various seismological institutes and crowdsourced information from eyewitnesses of earthquakes. This map provides real-time updates on earthquake data. It allows users to zoom in and examine the seismic intensity in the vicinity of significant seismic events, as illustrated in Fig. 5.3.

5.5 Results and Discussion

The outcome of the population analysis is a composite map illustrating both the population density distribution across Bosnia and Herzegovina and the epicenters of significant earthquakes. These earthquakes are highlighted based on their magnitude, as well as the extent of casualties and/or damage they have caused (Fig. 5.4). This analysis underscores the efficacy of the ELIS in providing hazard alerts in regions with limited data availability, aimed at protecting vulnerable populations. Achieving this goal necessitates the development of a robust database designed to enhance disaster prediction capabilities, refine existing predictive models, and extend the lead time for warnings in high-risk areas.

While an earthquake itself is a natural phenomenon, the resulting effects of earthquakes on the surrounding environment, particularly on built structures, can have catastrophic consequences. The primary risks associated with earthquakes include extensive infrastructural damage, which can severely compromise the integrity and functionality of critical infrastructure such as bridges, roads, and utilities.

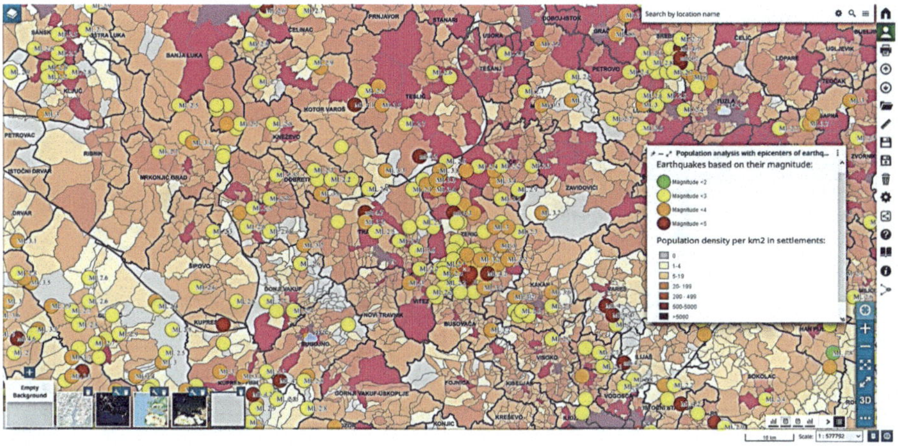

Fig. 5.4 This map illustrates the population density distribution across Bosnia and Herzegovina, along with the locations of significant earthquake epicenters

Additionally, earthquakes can induce mass movements like landslides [15], which exacerbate infrastructural damage and pose further risks to human safety. The seismic activity can also precipitate secondary hazards such as fires, often resulting from ruptured gas lines or electrical malfunctions, and floods, especially if earthquake-induced damage to dams or levees occurs. Moreover, the disruption of services and the creation of temporary, overcrowded shelters can lead to outbreaks of infectious diseases, as compromised sanitation systems and closed living quarters facilitate the spread of pathogens. An illustrative analysis comparing earthquake locations with vulnerable infrastructure is presented in Fig. 5.5.

Each earthquake undergoes an intensity analysis to evaluate the distribution of shaking intensity in the vicinity of the epicenter. This analysis categorizes the intensity into classifications such as severe, strong, moderate, or light, based on the observed ground motion and its impact on structures [25]. The results of this fundamental intensity analysis are depicted in the map presented in Fig. 5.5, which visually represents the spatial distribution of shaking intensity across the affected areas. Such analyses are crucial for assessing the impact of seismic events on specific regions and for guiding the development and implementation of effective mitigation, response, and recovery strategies. A magnitude 3.0 earthquake (Fig. 5.5) is typically felt within a few kilometers and rarely causes significant structural damage. However, its effects can be more severe in areas with pre-existing ground instability. If the region has experienced prior seismic activity, landslides, or heavy rainfall, the already weakened soil or rock layers may further destabilize. In such cases, even a minor tremor could trigger secondary hazards like landslides or localized soil failures. In regions characterized by steep terrain or unconsolidated sediments, these secondary effects can pose substantial risks to nearby infrastructure and communities. Although low-magnitude earthquakes generally do not result in major damage, the geological context in which they occur can significantly amplify their potential impact [26].

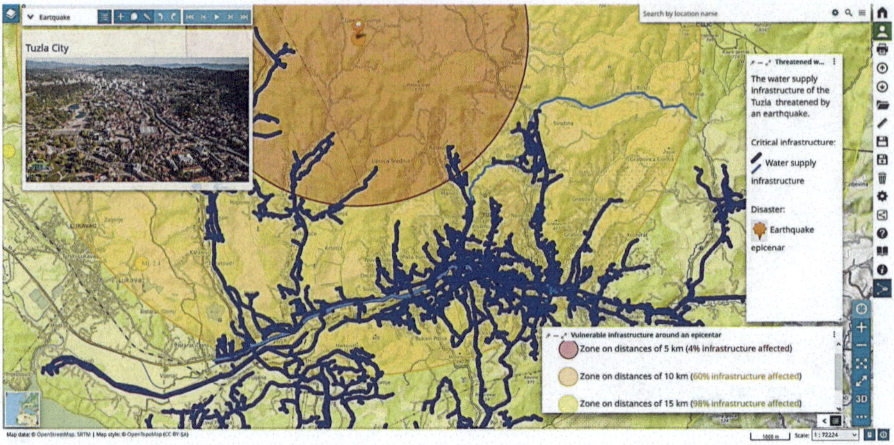

Fig. 5.5 This map shows the percentage of water supply infrastructure within different zones. The epicenter of the closest historical earthquake to the urban area was selected as the reference point, and zones representing distances of 5, 10, and 15 km were created. In the first zone, 4% of the infrastructure is located, in the second zone as much as 60%, while the third zone covers 98% of the observed infrastructure

5.5.1 Application of ELIS for Predicting Epidemic Occurrence in Water Supply Systems as a Consequence of Earthquakes in Bosnia and Herzegovina

5.5.1.1 Case Study of Tuzla in Bosnia and Herzegovina

Temporal variations in seismic activity, derived from historical data and instrumental earthquake measurements, reveal an increased seismic risk in the regions of Banja Luka, Tuzla, Zenica, and Herzegovina [27]. Considering the degree of urbanization in Tuzla Canton, characterized by a high density of multi-story buildings, the presence of hazardous industries such as chemical production and mining, and the known deficiencies in seismic construction practices, it is apparent that Tuzla Canton is a region of significant seismic vulnerability (Fig. 5.5). This risk is exacerbated by the phenomenon of land subsidence in the urban areas of Tuzla, which has resulted from prolonged, uncontrolled extraction of salt deposits. Moreover, the close proximity of several industrial complexes to the main tectonic fault in the municipality of Lukavac increases the seismic risk even further [27]. The study [26] examines how site characteristics influence seismic vulnerability in Tuzla. It highlights that incorporating site-specific effects into seismic calculations results in amplified damage and increased structural vulnerability.

In this specific case, we concentrate on the application of ELIS for Predicting Epidemic Occurrence in Water Supply Systems in the Tuzla region as a Consequence of Earthquakes in Bosnia and Herzegovina. The selection of this city was twofold. First of all this region is prone to earthquake motions [28], one of the most critical

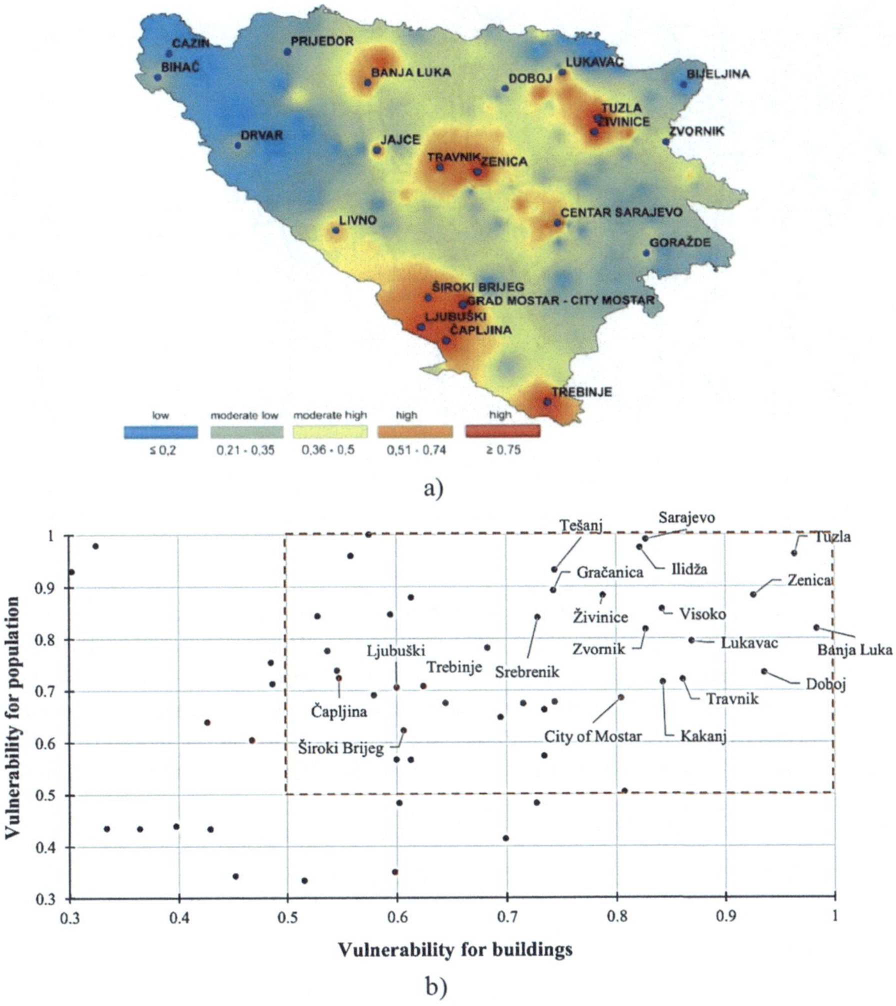

Fig. 5.6 (**a**) Distribution of relative seismic risk in Bosnia and Herzegovina [28]; (**b**) The relationship between building vulnerabilities and population distribution in Bosnia and Herzegovina [28]

regions when referring to seismic risk assessment is located around the cities of Tuzla and Živinice in the northeast of Bosnia and Herzegovina. According to the current hazard map [29], the expected PGA (Peak Ground Acceleration) for Tuzla city is 0.21g. With this PGA value, the estimated relative risk is approximately 75% (Fig. 5.6a), whereas the vulnerability for both population and buildings is above 95% (Fig. 5.6b) [28].

The information obtained from the building damage can be used as a starting point for water supply damage. A study [30] found that there is a strong correlation between the spatial distribution of pipeline damage and building damage. This

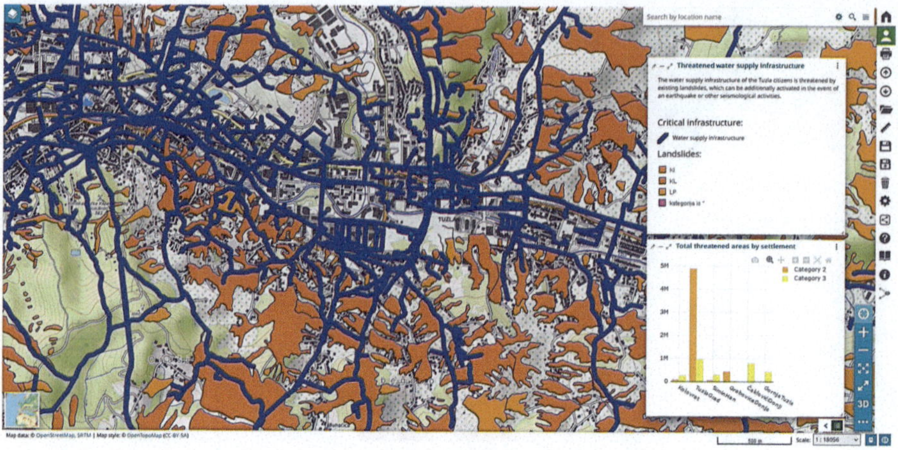

Fig. 5.7 Threatened water supply infrastructure from existing landslide areas potentially triggered by earthquakes, with an emphasis on two categories of landslides: category 3 (LP) High susceptibility to sliding; and Category 2 (KL) Medium susceptibility to sliding

suggests that the patterns of damage were primarily influenced by variations in the intensity of seismic motion across different districts. The comparison of building and pipeline damage rates across Düzce indicates that pipeline damage occurs only when ground motion exceeds the level required to cause building damage [30]. This observation supports findings from earlier studies conducted in Japan and the USA [31] examining several pipeline damages from six different earthquakes (four in Japan, one in Nicaragua, and one in the United States) to estimate the fragility relations for both segmented cast iron and asbestos cement pipelines in terms of PGA. Barebberg [32] used data from three different earthquakes in the US to develop the first fragility relationship for buried cast iron pipelines based on Peak Ground Velocity (PGV) in the late 1980s. The 1991 ASCE study [33] utilized information from pipeline damage in Coalinga in 1983, along with data from a prior study conducted in 1975 [32], to establish fragility relationships based on Peak Ground Acceleration (PGA). Additionally, study conducted by O'Rourke and Ayala [33] used data from other U.S. earthquakes [31] to create a fragility relationship based on PGV. According to reports from the American Lifelines Alliance, PGV and Repair Rate (RR) are commonly used to estimate damage to buried pipes, while PGA is typically used to evaluate damage to pumps and tanks [34, 35].

The minimum Peak Ground Acceleration (PGA) needed to damage water pipelines depends on various factors, such as the pipeline's design, construction, soil conditions, and maintenance levels. Typically, significant damage to water pipelines can occur at PGA values of approximately 0.1g to 0.2g. Following the 1989 Loma Prieta earthquake, and the development of the fragility curves, data from a large number of tanks showed that those exposed to ground motions of 0.10g or less were generally unlikely to suffer damage [34]. In contrast, older or poorly maintained systems might suffer damage at lower PGA values. It's important to consider that even lower levels of ground shaking can cause minor damage or contribute to cumulative effects that result in significant problems over time. In areas where PGV was less than 10 cm/s during the 1994 Northridge earthquake, no pipeline damage was reported [36]. In a different study [37] authors hypothesized that pipeline damage is likely to happen when strong-motion thresholds reach PGV of 15 cm/s and PGA of 100 cm/s^2 when constructing pipeline fragility correlations from the 1995 Hyogoken-nanbu earthquake data.

The threat to water supply infrastructure in the urban areas of Tuzla, especially in the context of a natural disaster, an earthquake, represents a serious challenge to public health and safety. When data on landslides, the location of the water supply network, and other relevant factors are available, it is possible to accurately simulate and predict the potential damage that earthquakes can cause. For example, an earthquake can trigger landslides (Fig. 5.7), which can lead to serious damage to water infrastructure. The position of water pipes and facilities, especially in areas of high risk of landslides, is key information in such scenarios. When damage to these systems is predicted, the impact on populated parts of the City of Tuzla can be additionally simulated. Population density and the availability of alternative water sources become crucial in determining the risk of epidemic outbreaks in areas supplied with water from damaged systems.

An illustration featuring a map of critical landslides in relation to the water supply network can provide clearer insight into which areas are most at risk and facilitate the planning of emergency measures.

5.6 Conclusion

The Epidemiological Intelligence Location System (ELIS) is a valuable tool for the collection and analysis of data from areas of interest, such as geography, basic infrastructure, demographics, population density, health capacity, epidemiological information, soil characteristics, earthquake history, and more.

Based on the magnitude of earthquakes, soil characteristics, landslides, infrastructure maps, water supply networks, and location of urban areas, ELIS can predict the intensity of destruction in affected areas.

By overlaying maps of earthquake impact, the water supply system with the density of population in the areas is possible to assess how many people could be left access to clean water. This population is at increasing risk of infectious diseases, especially waterborne and foodborne infectious.

ELIS plays an important role in predicting risks of infection among the population in affected areas. It communicates with stakeholders and provides solutions for alternative water supply by mapping available water wells and pathways for transportation.

The extensive data on population, epidemiology, infrastructure, communication systems, and industry presented through various maps and data layers in ELIS are invaluable for forecasting and mitigating epidemic spread during natural disasters triggered by seismic events.

References

1. Coetzee S, Ivánová I, Mitasova H, Brovelli M (2020) Open geospatial software and data: a review of the current state and a perspective into the future. ISPRS Int J Geo Inf 9(2):90
2. Helmi AM, Farhan MS, Nasr MM (2018) A framework for integrating geospatial information systems and hybrid cloud computing. Comput Electr Eng 67:145–158
3. Hukic M, Ponjavic M, Karabegovic A (2023) Epidemic Location Intelligence System (ELIS): a multidisciplinary approach for prediction, early detection, tracking and response to disease outbreaks. In: Keković Z, Duev R, Polović J (eds) How to deal with uncertainties in an increasingly complex environment? (The new cartography of risk and crises). Proceedings of the International Scientific Conference, Cavtat-Dubrovnik, pp 27–29
4. Ponjavic M, Karabegovic A, Ferhatbegovic E, Tahirovic E, Uzunovic S, Travar M, Pilav A, Mulic M, Karakas S, Avdic N, Mulabdic Z, Pavic G, Bico M, Vasilj I, Mamic D, Hukic M (2020) Spatio-temporal data visualization for monitoring of control measures in the prevention of the spread of COVID-19 in Bosnia and Herzegovina. Med Glas 17(2). https://doi.org/10.17392/1215-20
5. Hukic M, Ponjavic M, Tahirovic E, Karabegovic A, Ferhatbegovic E, Travar M, Serdarevic F (2021) SARS-CoV-2 virus outbreak and the emergency public health measures in Bosnia and Herzegovina: January–July, 2020. Bosn J Basic Med Sci 21:111–116
6. Karabegovic A, Ponjavic M (2014) Geoportal as interface for data warehouse and business intelligence information system. In: Advances in intelligent systems and computing: advances in business ICT. Springer, pp 27–40
7. Karabegovic A, Ponjavic M, Hukic M (2021) Epidemic location intelligence system as response to the COVID-19 outbreak in Bosnia and Herzegovina. J Appl Syst Innov 4(4):79
8. Ahmed SK, Chandran D, Hussein S, Sv P, Chakraborty S, Islam MR, Dhama K (2023) Environmental health risks after the 2023 Turkey-Syria earthquake and salient mitigating strategies: a critical appraisal. Environ Health Insights 17:11786302231200865. https://doi.org/10.1177/11786302231200865
9. Meniconi S, Brunone B, Frisinghelli M (2018) On the role of minor branches, energy dissipation, and small defects in the transient response of transmission mains. Water 10:187
10. Uckan E, Aksel M, Atas O et al (2024) The performance of transmission pipelines on February 6th, 2023 Kahramanmaras earthquake: a series of case studies. Bull Earthq Eng. https://doi.org/10.1007/s10518-024-01957-2
11. Bross L, Krause S, Wannewitz M, Stock E, Sandholz S, Wienand I (2019) Insecure security: emergency water supply and minimum standards in countries with a high supply reliability. Water 11:732
12. Ancaş AD, Aşchilean I, Profire M, Toma I (2019) System for increasing the seismic safety of pipelines in the water supply and distribution networks. Water 11:1049. https://doi.org/10.3390/w11051049
13. Harrison DB (1999) Seismic upgrading of water storage facilities in British Columbia. In: Proceedings of the 8th Canadian Conference on Earthquake Engineering, Vancouver
14. Nayak CB, Thakare SB (2019) Seismic performance of existing water tank after condition ranking using non-destructive testing. Int J Adv Struct Eng 11(4):395–410. https://doi.org/10.1007/s40091-019-00241-x
15. Eskandari M (2011) Seismic damage estimation of pipelines buried fuel—case study: Kermanshah City. Master's Thesis, Tehran University, Tehran, Iran (In Persian)
16. Schiff AJ (1999) Hyogoken-Nanbu (Kobe) earthquake of 17 January 1995: lifeline performance. ASCE, Reston. ISBN 978-0-7844-0408-9
17. Bragato DDA (2016) Downtime estimation of lifelines after an earthquake. University of California, Berkeley
18. Malakootian M, Nouri J (2010) Chemical variations of ground water affected by the earthquake in bam region Malakootian. Int J Environ Res 4:443–454

19. Qi WK, Sanuba T, Norton M, Li YY (2014) Effect of the great East Japan earthquake and tsunami on sewage facilities and subsequent recovery measures. Water Sustain 4:27–40
20. Eskandari M, Omidvar B, Modiri M, Nekooie MA, Alesheikh AA (2017) Geospatial analysis of earthquake damage probability of water pipelines due to multi-hazard failure. ISPRS Int J Geo Inf 6:169. https://doi.org/10.3390/ijgi6060169
21. Mohammad Y, Bayram H, Kayalar O, Madonna F, Annesi-Maesano I (2023) Earthquake disaster and respiratory health: lessons from Turkey and Syria in 2023. Eur Respir J 62(5):2300534. https://doi.org/10.1183/13993003.00534-2023
22. Okuyama J, Seto S, Fukuda Y, Suzukamo Y, Okazaki T, Furusawa Y, Izumi SI, Ito K, Imamura F (2024) Establishment of a post-disaster healthcare information booklet for the Turkey-Syrian earthquake, based on past disasters. Sci Rep 14(1):1558. https://doi.org/10.1038/s41598-024-52121-4
23. Goyet S, Rayamajhi R, Gyawali BN, Shrestha BR, Lohani GR, Adhikari D, Salvador E, Ofrin R, Vandelaer J, Samuel R (2018) Post-earthquake health-service support, Nepal. Bull World Health Organ 96(4):286–291. https://doi.org/10.2471/BLT.17.205666
24. Ho ML, Lim JZ, Tan MZ, Kok WL, Zhang JR, Tan MY, Tan AC (2016) Humanitarian assistance and disaster relief mission by a Tripartite Medical Team led by the Singapore Armed Forces after the 2015 Nepal earthquake. Singapore Med J 57(8):426–431. https://doi.org/10.11622/smedj.2016132
25. USGS Homepage. https://www.usgs.gov/programs/earthquake-hazards. Last accessed 30 Aug 2024
26. Ademovic N, Hadzima-Nyarko M, Zagora N (2022) Influence of site effects on the seismic vulnerability of masonry and reinforced concrete buildings in Tuzla (Bosnia and Herzegovina). Bull Earthq Eng 20:2643–2681. https://doi.org/10.1007/s10518-022-01321-2
27. Cantonal Administration of Civil Protection Tuzla Canton, Federation of Bosnia and Herzegovina (2016) Threat assessment to natural and other disasters in the Tuzla Canton. (in Bosnian)
28. Ademović N, Kalman Šipoš T, Hadzima-Nyarko M (2020) Rapid assessment of earthquake risk for Bosnia and Herzegovina. Bull Earthq Eng 18:1835–1863. https://doi.org/10.1007/s10518-019-00775-1
29. Institute for Standardization of Bosnia and Herzegovina [BAS] (2018) National Annex BAS EN 1998-1:2018—design of structures for earthquake resistance—Part 1: General rules, seismic actions and rules for buildings
30. Tromans I (2004) Behaviour of buried water supply pipelines in earthquake zones. Ph.D. Thesis, Imperial College of Science, Technology and Medicine, University of London
31. Katayama T, Kubo K, Sato N (1975) Earthquake damage to water and gas distribution systems. In: Proceedings of the National Conference on Earthquake Engineering, Earthquake Engineering Research Institute, Oakland, CA, USA, pp 396–405
32. Barenberg ME (1988) Correlation of pipeline damage with ground motions. J Geotech Eng 114:706–711
33. O'Rourke M, Ayala G (1993) Pipeline damage due to wave propagation. J Geotech Eng 119:1490–1498
34. American Lifelines Alliance (2001) Seismic fragility formulations for water systems, part 2. American Lifelines Alliance
35. Bata MH, Carriveau R, Ting DSK (2022) Urban water supply systems' resilience under earthquake scenario. Sci Rep 12:20555. https://doi.org/10.1038/s41598-022-23126-8
36. O'Rourke TD, Toprak S, Sano Y (1998) Factors affecting water supply damage caused by the Northridge earthquake. In: Proceedings of the Sixth US National Conference on Earthquake Engineering
37. Isoyama R, Ishida E, Yune K, Shirozu T (2000) Seismic damage estimation procedure for water supply pipelines. In: Proceedings of the 12th World Conference on Earthquake Engineering, CD-ROM Paper No. 1762, 8pp

Chapter 6
Empowering Communities: Education for Improving Public Awareness and Response Capacity in Earthquake Emergencies

Konstantinos Papatheodorou, Emmanuil Kirtas, Georgios Panagopoulos, Konstantinos Ntouros, Nikolaos Theodoulidis, Nikolaos Klimis, Can Zulfikar, Dragos Toma-Danilă, Dragos Vintilă, and Vladlen Cardanet

Abstract Earthquakes pose significant challenges to societal structures and critical infrastructure, particularly in seismic-prone regions. They cause widespread damage, population displacement, and infrastructure disruption, conditions that can later be exploited by human actors for adverse purposes. Addressing these challenges

---

K. Papatheodorou (✉)
Surveying & Geoinformatics Engineering Department, International Hellenic University, Serres, Greece
e-mail: conpap@ihu.gr

E. Kirtas · G. Panagopoulos
Civil Engineering Department, International Hellenic University, Serres, Greece

K. Ntouros
International Hellenic University Surveying & Geoinformatics Engineering Department, Serres, Greece

N. Theodoulidis
Research Director, Institute of Engineering Seismology & Earthquake Engineering, Pylaea, Greece

N. Klimis
Civil Engineering Department, Democritus University of Thrace, Xanthi, Greece

C. Zulfikar
Disaster Management Institute, Istanbul Technical University, Istanbul, Türkiye

D. Toma-Danilă
Scientific Researcher, National Institute for Earth Physics, Măgurele, Romania

D. Vintilă
Civil Engineering Department, Ovidius University, Constanța, Romania

V. Cardanet
Scientific Researcher, Institute of Geology & Seismology, Chișinău, Moldova

D. Radu et al. (eds.), *Countering Hybrid Threats Against Critical Infrastructures*, NATO Science for Peace and Security Series B: Physics and Biophysics, https://doi.org/10.1007/978-94-024-2304-4_6

requires a comprehensive approach, including robust public communication, investment in resilience and preparedness measures, and coordination among stakeholders to enhance societal resilience. This paper introduces the development of an Educational Hub (Edu.Hub) as part of the Rapid Earthquake Damage Assessment System (REDAS), which includes three components: the Rapid Earthquake Damage Assessment platform, which provides real-time damage data for planning and emergency response; a smartphone application for public information and communication during emergencies; and the Edu.Hub, which aims to enhance public preparedness and response, influencing the effectiveness of actions taken by authorities. The Edu.Hub (https://www.redact-project.eu/educational-hub/) builds on national guidelines, incorporating innovative ideas, practical tips, and solutions tailored to public needs during emergencies. It addresses public behavior, compliance with emergency protocols, risk comprehension, and preparedness actions. It includes instructions for maintaining communication, self-assessment of preparedness, creating personal and family emergency plans, and navigating to safe areas. Its content is adaptable to various emergency scenarios beyond earthquakes, increasing its utility. The Edu.Hub has been well-received and integrated by regional Civil Protection authorities, demonstrating its positive impact on community resilience against earthquake-related risks.

Keywords Earthquake disaster mitigation · Public response · Earthquake education

6.1 Introduction

Hybrid threats encompass a wide range of tactics, which target facilities and services vital for the essential, necessary operations of Society, to directly or indirectly, exploit vulnerabilities, create serious problems, finally aiming at eroding public trust, and undermining confidence in the State's ability to provide security. The complexity of modern critical infrastructure, which often integrates cyber and physical systems, increases their vulnerability to hybrid threats. Moreover, the interconnection of these systems, indicates that a failure in one component may have negative effects across different components and thus affect the entire infrastructure [1, 2].

Earthquakes are natural disasters that can cause widespread damage to critical infrastructure (CI) including buildings, roads, bridges, utilities, and communication networks. The physical impact of earthquakes can lead to infrastructure failures, including structural collapse, trigger landslides, and liquefactions, and cause secondary effects such as gas leaks, water main breaks, power outages, and fires. These disruptions can both increase vulnerability against hybrid threats impede emergency response efforts and exacerbate the overall impact of the disaster. The relationship between earthquakes and hybrid threats to critical infrastructure is multifaceted, with earthquakes posing a direct physical threat to infrastructure, while hybrid threats, exploit vulnerabilities in both physical and digital domains.

As is therefore evident, a multidisciplinary and multisectoral approach is essential to address these combined threats through enhancing resilience by developing decision-making frameworks, which consider the interdependencies of digital, social, and inftructure domains [3]. Understanding and mitigating these risks requires a comprehensive approach that integrates seismic risk management with strategies to counter hybrid threats. This comprehensive strategy must include risk assessment at an early stage and the integration of safety and security measures across all stages of infrastructure design and operation [4]. Moreover, enhancing situational awareness and preparedness is crucial. This involves understanding the nature of earthquake and hybrid threats and implementing measures to protect vital societal functions from both natural and man-made disruptions [5].

As one of the main actors, the Public plays a crucial role in both earthquake and hybrid-threat Risk mitigation through various mechanisms, including perception, participation, and collaboration with governmental and scientific bodies. Public engagement may include reporting suspicious activities, participating in disaster drills, and adhering to evacuation protocols, which can greatly enhance the collective resilience of communities and support the recovery following a crisis. Public involvement can, therefore, significantly enhance the effectiveness of disaster preparedness and response strategies.

Public perception of earthquake risk is a critical factor in disaster mitigation. Studies show that a lack of knowledge and infrastructure can hinder effective disaster response [6]. Risk perception influences public behavior in disaster scenarios so, enhancing public awareness and self-support capabilities can improve emergency management outcomes [7]. To that end, civic education and communication are essential for fostering a culture of risk awareness [8].

Public participation in disaster mitigation is also vital for sustainable risk reduction. The Theory of Planned Behavior (TPB) framework shows that attitudes, subjective norms, and perceived behavioral control significantly predict public intention to participate in mitigation activities. However, past disaster experiences can negatively impact participation intentions, highlighting the need for effective communication and engagement strategies [9]. Community-based models, such as the socio-seismology demonstrate the potential of public involvement, which can enhance earthquake awareness and encourage protection initiatives [10]. The public can even be engaged in data and information collection (crowdsourcing) through educational initiatives, thus increasing risk awareness and contributing to acquiring important information [11].

While public involvement is essential for effective earthquake risk mitigation, several challenges remain. For instance, increasing public awareness alone is insufficient for genuine participation because some basic skills that must be acquired at the preparedness level are necessary to enhance public engagement [12].

Consequently, the public's response to hybrid threats and earthquakes can significantly influence the safety and resilience of critical infrastructures. Public awareness, preparedness, and effective response can help mitigate the impact of disasters and enhance the overall resilience of society. Conversely, misinformation, panic, and lack of coordination, challenge-response efforts and increase vulnerabilities.

An effort to provide solutions to problems the Public faces during Earthquake emergencies and to enhance its preparedness and response capacity, which can strongly affect the dynamics of any response actions taken by competent authorities, is presented in the following paragraphs. Interestingly, much of its content is adaptable to emergencies stemming from various causes beyond earthquakes, thereby increasing its versatility and utility across a range of emergency scenarios.

6.2 Public Response to Threats and Hazards

Public response to threats and hazards is a multifaceted issue that involves understanding societal behavior, effective communication, and preparedness strategies. The public's response to different types of threats varies, depending on the nature of the threat, whether it is natural, technological, or hostile, and hazards, thus emphasizing the importance of communication, cooperation, and preparedness.

Natural hazards are often perceived by the Public, as unavoidable acts of nature, leading them to a focus on mitigation rather than prevention. In contrast, responses to technological hazards are seen as controllable, prompting efforts to find solutions and reduce risks [13]. As a principle, management of these hazards involves informing the public, deciding on actions, and implementing measures to control or mitigate consequences [13].

Earthquake-imposed crises invoke the entire community with all of its structural components: the operational capacity of services, their response efficiency, and the response of the population, which strongly affects the dynamics and progress of response actions, both during the event and after that. Response of the population is strongly related to the level of communication, of comprehending the situation, and of being trained to respond.

Public Behavior during emergencies is often characterized by cooperation and support rather than panic. People tend to help each other and follow instructions from authorities, although they may question unclear directives [14]. Reasons for that behavior can be found in the lack of adequate education related to disaster mitigation and crowd and personal psychology during crises. While previous research suggested the existence of a crowd "irrationality" during emergencies, which causes "mass panic" [15, 16], more recent research [17] shows that in fact, crowd behavior is usually "normal" and "mass panic" is very rare [18–20]. In some cases the Public, given the critical character of those moments, is not willing to comply and follow the recommendations [21, 22].

Recent findings, in fact, suggest that during emergencies, public behavior can't be pre-determined because it is shaped by the way civil protection authorities (including emergency responders), manage the event [23]. To further improve public response efficiency, the same research suggests that effective communication strategies must be designed, to enhance public compliance and cooperation. Within that scope, public warnings must be authoritative, accurate, and timely, to ensure the population is well-informed about potential risks [24].

Social media can also play an important role in disseminating information but adaptation of communication protocols is required [24].

In any case, trust in communication sources is vital, as people tend to follow reliable information during emergencies. False alarms and perceived threats may lead to emotional behavior, but this is usually structured and strongly influenced by social norms and relationships rather than mass panic.

6.3 Factors Influencing Public Response During Earthquakes and Other Emergencies

Compliance with recommendations and/or instructions is a key factor in preventing the Public from exposing themselves to risks intentionally, and thus affecting response actions taken by competent authorities. To identify motivation and the reasons causing such behavior, an identification of the Public's main concerns during an emergency situation is needed.

Towards that end, two main facts need to be considered: (i) a "critical situation" is strongly related to the viewpoint of the one who is undergoing it [25] and (ii) decisions during an emergency are binary (only one option is always selected and followed) implying that each person may take a different decision based on own perception of risk.

On the other hand, decisions made during emergencies, have to be aligned among all involved parties to minimize the threat. At the same time, any individual faces various concerns, which indicate "social vulnerability" [26]; a set of characteristics that include a person's initial well-being, self-protection, livelihood and resilience, social protection, and the social and political networks and institutions. To be more specific, there are six levels of "social" vulnerability [27] which, based on significance are the: (i) individual; (ii) household; (iii) administrative community; (iv) cultural community; (v) national, and (vi) regional.

As is therefore evident, the Public needs situational awareness at different levels, related to the six social vulnerability levels mainly regarding persons (family members, friends, pets) and assets (house, car) of "interest" to them.

Information regarding the event itself (magnitude, consequences, future trends) can help reduce concerns that cause stress. So, reliable information provided by real-time communication during critical moments can help the public stay calm and make informed decisions, act responsibly, and prevent them from exposing themselves to risks, in a way that leads to improving their capacity to respond efficiently.

On the other hand, information to provide "situational awareness" regarding personal concerns, is a matter of personal responsibility. Each person needs to communicate with others to acquire this information and as communication during crises can't be taken for granted (communication problems often occur due to comms network overloading), provisions for actions that support communication infrastructure viability should also be considered and applied.

Practice from previous events has shown that during emergencies, communication over data may be feasible while voice communication is unavailable. For that reason, the required "situational awareness" should be provided in real-time, by using free services and a set of tools facilitating communication over data, since voice communication is usually unavailable due to the overloading of telecommunication systems [28].

6.4 Problems Identified During Earthquake Emergencies and Proposed Solutions

Problems related to Public response during Earthquake emergencies include [29] panic, traffic congestion (traffic jams), ignorance of the geographical location of safe/refuge areas, unknown duration of being able to stay there, and unknown routes to get there. Especially the latter is greatly enhanced in tourist areas where a large number of visitors may be present.

As already explained in previous paragraphs, "irrational" mass panicking doesn't exist. Panic reported may be attributed to the instinct of self -preservation and can last until people have evacuated the buildings. If such a condition remains, it should be attributed to the main concerns people have during emergencies, which are related to the inability to assess the potential impact of the event and its consequences and are further enhanced by concerns regarding their personal interests.

As a response to those problems, trustworthy information regarding the event, provided by competent authorities is necessary. In addition, motivation to drive the Public to be interested, to devote some time to be better informed and trained, is necessary.

At the same time, individuals need situational awareness regarding the conditions of persons and assets of "interest". Lack of communication at this point, can increase stress and make people expose themselves to the risks, in order to establish communication in person and get the information they need. In the process, they crowd the streets, causing traffic jams especially in case they use their cars for transportation.

Moreover, since the safe area location is unknown, they remain in the same area, occupying the middle of the streets in an attempt to stay as far as possible from building facets. As it therefore seems, that most of the problems reported are interrelated and so are the solutions to those problems.

6.5 Implementation—The REDACt Project Educational Hub

To respond to problems presented in previous paragraphs, proposed solutions have been developed and presented in a sequence in the form of an Educational Hub, hosted on the REDACt project's website (https://www.redact-project.eu/educational-hub/). Having said that, it must be noted that these efforts aim at developing, evaluating, and demonstrating solutions to existing problems and do not, by any means, intend to replace the official websites and material published by competent authorities at National level.

Education to process information and to comprehend the risks is one of the main requirements for improving Public response capacity, so a large part of the educational hub is attributed to providing targeted education, which capitalizes on guidelines issued and/or published by competent National authorities and adds some new ideas, tips, and solutions to potential problems the public faces during emergencies.

The entire content is presented on one page so that users can have easy access by just scrolling through the page. This is especially convenient when using small-screen devices such as smartphones and tablets. To content is presented as a series of sections (Fig. 6.1) leading from the introductory section, to the main part with useful information on what to do and tutorials on how to do it and ends with a map of safe locations which provides navigation capabilities.

Considering that the public needs to comprehend the risks and be motivated and prepared, easy-to-comprehend information justifying the need for an individual to

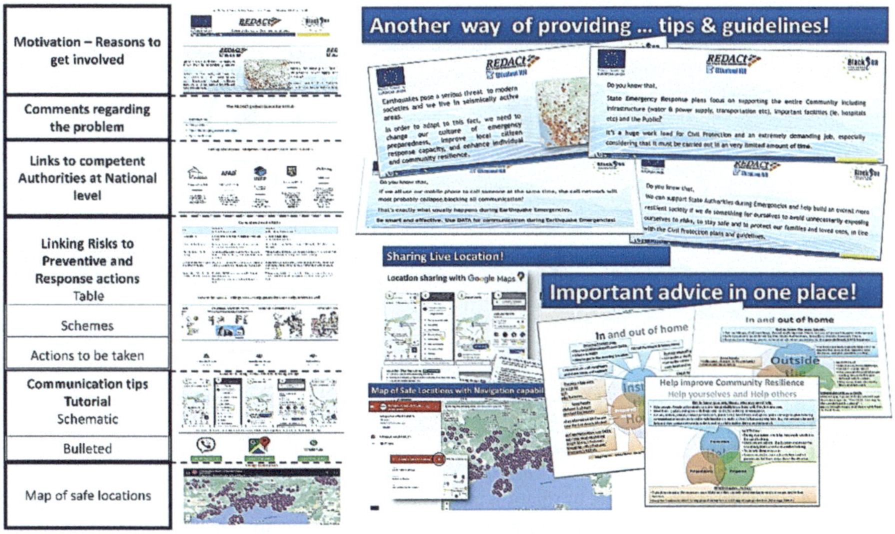

Fig. 6.1 The Educational Hub structure (left section) and indicative content (right section). The description and the respective content can be seen on the left while on the right side, parts of the, embedded in slide shows, messages are displayed

be involved, tips on what to do and what to avoid, and tools to facilitate information sharing and communication are provided (Fig. 6.1).

This information is complemented with a quiz so that any individual, can perform a "self-assessment" regarding their earthquake risk mitigation status (Fig. 6.2). Quiz questions and answers, actually promote the basic principles as set out by competent authorities at the National level, worldwide.

Additionally to those, a series of popularized books aiming at providing a knowledge base that can help the Public better understand the basic principles and the mechanisms from the earthquake to ground motion and to damage to structures, is provided. These include Planning for Emergencies and Safe Citizens, to help the Public develop their emergency plans at various levels (family, individual, work, etc.); From Earthquake Focus to induced Damage, to better understand the mechanisms of this natural phenomenon and its consequences; Earthquake Damage to structures and Infrastructure, which describes causes of damage and can help understand the necessity of Preventive measures; Earthquake-induced Geotechnical failures, which presents the mechanisms of earthquake-induced landslides and liquefactions and their potential consequences.

Links to competent at the National level, State authorities follow, to allow for easy access to already published educational material.

The following section of the EDU-HUB focuses on helping the public to comprehend the risks. Information is provided in the form of a table linking actual risks to the appropriate preventive, preparedness, and response actions. The table is followed by a slideshow, based on already published educational material by the

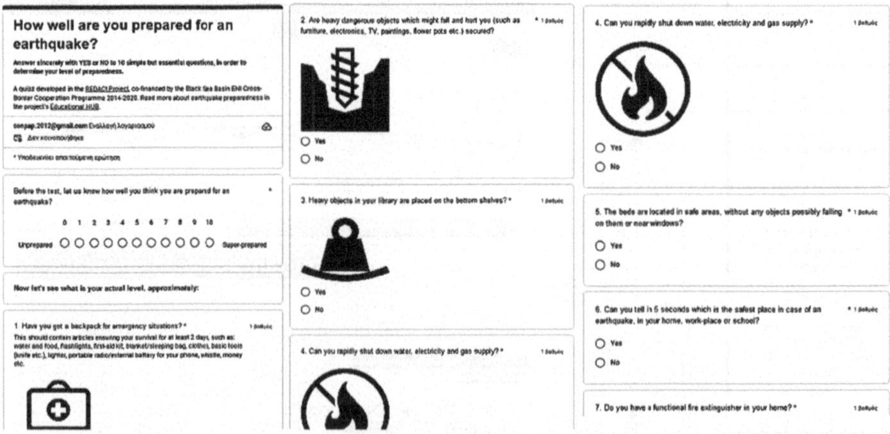

Fig. 6.2 A "self-assessment" risk mitigation test. The participants are first asked about their opinion regarding their status and then, answer a series of questions including basic prevention, preparedness, and response actions that must be taken by citizens. After they answer, they receive their grade regarding their actual risk mitigation status

Hellenic Earthquake Planning and Protection Organization (EPPO), which again attempts to link risks to prevention, preparedness, and response actions. Finally, comprehensive diagrams, grouping actions for Prevention, Preparedness, and Response inside and outside of the house/building are given so that users can find most of the info, gathered in three schemes.

Earthquakes trigger the response of everyone within a very large area. Any attempt to communicate over voice phone calls at that time, may easily overload the telecoms network and render it useless, thus preventing persons in need, of communicating. A key response to this problem is to prefer using data for either text messaging or even VoIP communication. Communication over data is based on the Voice over Internet Protocol (VoIP), which uses a broadband data connection instead of using analog signals to send calls to the Public Switched Telephone Network (PSTN). During VoIP communication, audio is converted into data packets which are then compressed, thus further reducing the data volume transmitted over the IP network, while maintaining voice quality.

Additional benefits stemming from respective capabilities, include Group calls, chat integration, and communication over any kind of electronic device.

Moreover, existing software for VoIP communication can provide additional, very important capabilities for sharing the required, important, personal information as is the "live" geographic location. "Live" location provides the ability not only to locate someone (their exact location at a specific moment) but also to "track" them in real-time as they move and change places. In other words, sharing live location in real-time means that, members of the same group (family, friends) can monitor each other as they move towards safety, while at the same time, they can exchange text messages or even talk over VoIP comms.

This function can reduce or even eliminate anxiety regarding the whereabouts of "persons of interest" and in that way, remove the reasons for trying to reach those persons physically and be exposed to risks, in the process. Interestingly, this capability is offered by comms software most people use for everyday communication, so there's no need for extra training or for installing additional software (Fig. 6.3).

To help the Public locate, choose, and navigate towards safe/refuge areas using any software platform, maps of safe locations with navigation capabilities have been developed (Fig. 6.4). Indicatively they include safe locations of selected areas (a Region in Greece and Constanta city in Romania). Reverse geocoding performed on coordinate data, provided additional information such as the address, the name of a nearby landmark if one exists and more, to help the public identify each safe location. This map can be used at all risk mitigation stages, to plan at the family level, and to use during emergencies for implementing response actions.

It's worth noting that the entire REDACt EDU-HUB material has been produced in English, Greek, Turkish, and Romanian and is accessible over the internet (https://www.redact-project.eu/educational-hub/) by any platform available.

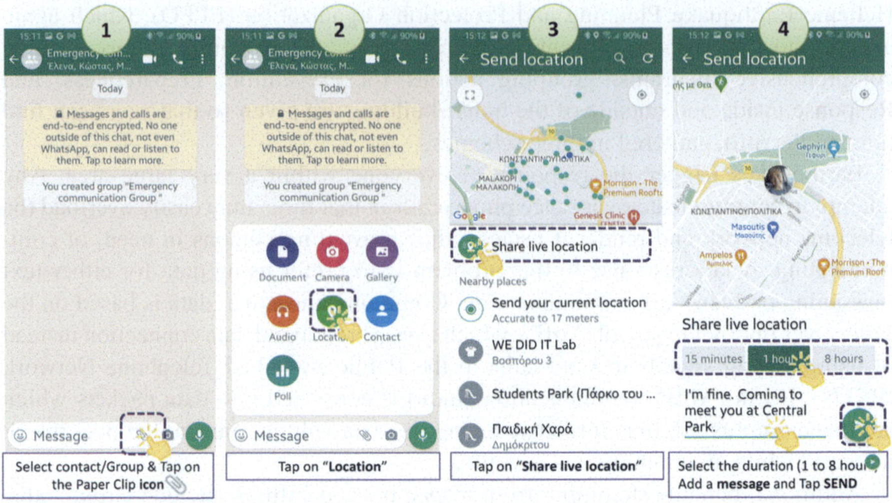

Fig. 6.3 Short tutorials on how to acquire situational awareness by exchanging messages and sharing live geographic locations using Whatsapp

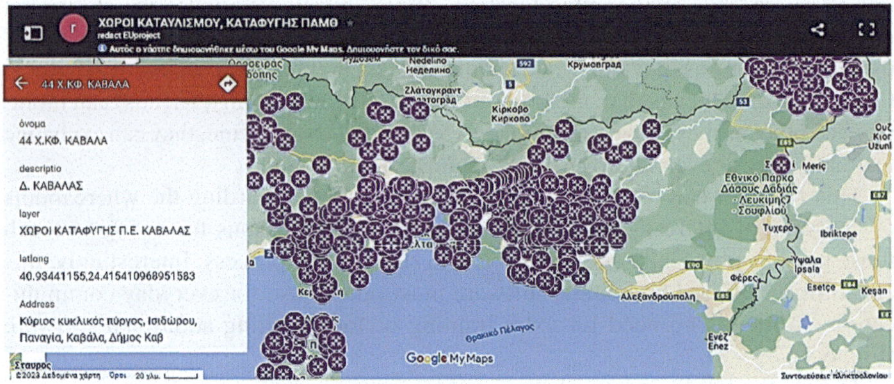

Fig. 6.4 REDACt EDU-HUB map of safe locations for the entire Region of Anatoliki Makedonia and Thraki (Greece). The map has navigation capabilities, as shown in the small inset window

6.6 Conclusions

This study aims to identify the challenges the public faces during earthquake emergencies and proposes, develops, and evaluates potential solutions to address these issues. Existing research highlights several major problems encountered by the public during such events, including panic, traffic congestion, unfamiliarity with safe locations, lack of knowledge about evacuation routes, and uncertainty regarding the duration of time spent at safe locations.

Additional challenges involve reluctance to engage in risk mitigation activities, non-compliance with guidelines, and a lack of education and understanding of the associated risks. Therefore, there is a need to motivate the public to become more interested in learning about and participating in risk mitigation actions, including personal and family-level prevention planning and preparation for emergency response through organization and training.

To address these issues, various methods were developed and presented via an Educational Hub (Edu-Hub), which were subsequently tested and evaluated based on feedback from national and regional authorities. The Edu-Hub extensively uses colorful images and diagrams to minimize reading requirements and to create a more engaging user interface (https://www.redact-project.eu/educational-hub/).

To motivate the public, the introductory section of the Edu-Hub provides realistic descriptions of conditions during emergencies, emphasizes the protective role of Civil Protection authorities, and outlines the complexities of their responsibilities, especially given the limited time available during crises. This section also highlights the importance of individual actions in safeguarding oneself and family members.

To enhance public understanding of the risks, the Hub includes tables and schematic representations linking specific risks to preventive and response actions, both inside and outside buildings. Tips to encourage volunteering at the building or neighborhood level are provided, recognizing the significant role that community engagement plays in enhancing societal resilience, particularly in supporting vulnerable populations such as the elderly and individuals with mobility challenges.

The Edu-Hub addresses the causes of stress that may lead individuals to knowingly expose themselves to risks by offering practical tips, tutorials, and tools. Research indicates that during emergencies, people require "situational awareness" regarding the status of family members, possessions, and other valued elements such as pets. To support this need, the Hub offers advice on maintaining telecommunication capabilities, encouraging the use of data communication instead of voice calls. This recommendation has been adopted and promoted by relevant institutions and integrated into existing software tools.

The Edu-Hub also provides tutorials on using free communication services and on-demand geographic location sharing through popular smartphone apps like WhatsApp, Viber, Google Maps, and Messenger, ensuring easy adaptation for everyday users.

Additionally, a comprehensive map of safe locations, covering entire regions and cities where data were available, was developed. The locations are reversely geocoded to include addresses and descriptions, making them easily identifiable to the public. This map can be utilized at various stages of risk mitigation—both for preventive planning and for navigation during emergencies—and is accessible across platforms via web browsers or specific apps such as Google Maps.

The solutions presented in the REDACt project's Educational Hub have been well-received, with many being integrated by regional Civil Protection authorities, demonstrating the Hub's positive impact on efforts to enhance community resilience against earthquake-related risks. Notably, much of the Edu-Hub's content is adaptable to emergencies arising from diverse causes beyond earthquakes, thus increasing its versatility and utility across a range of emergency scenarios.

References

1. König, S., Schauer S (2019) Cascading threats in critical infrastructures with control systems
2. König S, Gouglidis A, Rass S, Adams N, Smith P, Hutchison D (2020) Analyzing disaster-induced cascading effects in hybrid critical infrastructures: a practical approach. In: Guide to disaster-resilient communication networks. Computer communications and networks. Springer, Cham, pp 769–789. https://doi.org/10.1007/978-3-030-44685-7_31
3. Francesco M (2022) Applying resilience to hybrid threats in infrastructure, digital, and social domains using multisectoral, multidisciplinary, and whole-of-government approach. NATO Sci Peace Sec Ser. https://doi.org/10.3233/nicsp220017
4. Nikolaos P, Joonas L, Jarmo A, Ahmed ZB, Bryan MO, van Bossuyt Douglas L (2019) Early hybrid safety and security risk assessment based on interdisciplinary dependency models. https://doi.org/10.1109/RAMS.2019.8768943
5. Tikanmäki I, Ruoslahti H (2022) How are hybrid terms discussed in the recent scholarly literature? Eur Conf Cyber Warf Sec 21(1):296–304. https://doi.org/10.34190/eccws.21.1.457
6. Yusran A (2022) Analysis of microtremor h/v spectral ratio and public perception for disaster mitigation. Int J Geomate 23(97). https://doi.org/10.21660/2022.97.3311
7. Cai J, Hu S, Sun F, Tang L, Fan G, Hui-ge X (2023) Exploring the relationship between risk perception and public disaster mitigation behavior in geological hazard emergency management: a research study in Wenchuan county. Disas Prevent Resil 2(4):10.20517/dpr.2023.26
8. Maio RA, Ferreira TM, Vicente R (2018) O papel da perceção e comunicação na mitigação do risco de terramoto. Territorium Revista Portuguesa de riscos prevenção e segurança 25(2):69–78. https://doi.org/10.14195/1647-7723_25-2_6
9. Xing H, Que T, Yuxin W, Shiyu H, Li H, Li H, Skitmore M, Talebian N (2023) Public intention to participate in sustainable geohazard mitigation: an empirical study based on an extended theory of planned behavior. Nat Hazards Earth Syst Sci 23:1529–1547. https://doi.org/10.5194/nhess-23-1529-2023
10. Eric C, Dominique B, Steeve S, Claude P, Bétonus P, Sophia U, Laennec H, Alain G, Jean-Marie T, Tony M, Anne D, Françoise C, Jérôme C, Fabrice P, Etienne B, Jean-Paul A, Bernard M, Lépinay D, Julien B, Jean-Luc B, Rémy B, Laure F, Valérie C (2020) A socio-seismology experiment in Haiti. Front Earth Sci 8:542654. https://doi.org/10.3389/FEART.2020.542654
11. Chiara S (2022) Can high-school students contribute to seismic risk mitigation? Lessons learned from the development of a crowd-sourced exposure database. Int J Disast Risk Reduct 69:102755–102755. https://doi.org/10.1016/j.ijdrr.2021.102755
12. Valibeigi M, Feshari M, Zivari F, Motamedi A (2019) How to improve public participation in disaster risk management: a case study of Buein Zahra, a small city in Iran. Jàmbá J Disas Risk Stud 11(1):741–741. https://doi.org/10.4102/JAMBA.V11I1.741
13. Kasperson RE, Kasperson J (2005) Social contours of risk: two volume Set, 1st edn. Routledge. https://doi.org/10.4324/9781849772556
14. Drury J, Arias S, Au-Yeung TC, Barr D, Bell L, Butler T, Carter H, Choudhury S, Eriksson J, Neville F, Philpot R, Radburn M, Reicher S, Ronchi E, Scott C, Telga M, Templeton A (2023) Public behavior in response to perceived hostile threats: an evidence base and guide for practitioners and policymakers. https://doi.org/10.20919/vjvt7448
15. Le Bon G (1895) The crowd: a study of the popular mind. international relations and security network. ETH Zurich. https://www.google.com/url?sa=t&source=web&rct=j&opi=8997844

9&url=https://www.files.ethz.ch/isn/125518/1414_LeBon.pdf&ved=2ahUKEwiMmtWou7qIAxVfBtsEHXIpBHgQFnoECBkQAQ&usg=AOvVaw2WFl0KBlLvJACt_QDBAnim. Accessed 11 Sept 2024
16. Bendersky J (2007) "Panic": the impact of Le Bon's crowd psychology on U.S. military thought. J Hist Behav Sci 43:257–283. https://doi.org/10.1002/jhbs.20239
17. Stott C, Hutchison P, Drury J (2001) "Hooligans" abroad? Inter-group dynamics, social identity and participation in collective "disorder" at the 1998 World Cup Finals. Br J Soc Psychol 40(3):359–384
18. Feinberg WE, Johnson NR (2001) The ties that bind: a macro-level approach to panic. Int J Mass Emerg Disasters 19(3):269–295
19. Drury J, Cocking C, Reicher S (2009a) The nature of collective resilience: survivor reactions to the 2005 London bombings. Int J Mass Emerg Disasters 27(1):66–95
20. Drury J, Cocking C, Reicher S (2009b) Everyone for themselves? A comparative study of crowd solidarity among emergency survivors. Br J Soc Psychol 48(3):487–506
21. Carter H, Drury J, Amlot R, Rubin GJ, Williams R (2013) Perceived responder legitimacy and group identification predict cooperation and compliance in a mass decontamination field exercise. Basic Appl Soc Psychol 35(6):575–585
22. Carter H, Drury J, Rubin GJ, Williams RJ, Amlot R (2014) Effective responder communication improves efficiency and psychological outcomes in a mass decontamination field experiment: implications for public behaviour in the event of a chemical incident. PLoS One 9(3):e89846
23. Holly C, John D, Richard A (2018) Social identity and intergroup relationships in the management of crowds during mass emergencies and disasters: recommendations for emergency planners and responders. Policing, pp 1–14, https://doi.org/10.1093/police/pay013, Crown copyright 2018
24. Rossi C, Frisiello A (2024) Human-centered public warnings, pp 243–274. https://doi.org/10.1002/9781119741374.ch12
25. Habermas J (1975) Legitimation crisis. Beacon Press, Boston, MA
26. Cannon T, Twigg J, Rowell J (2003) Social vulnerability. Sustainable livelihoods and disasters. Report to DFID conflict and humanitarian assistance department (CHAD) and sustainable livelihoods support office. https://www.researchgate.net/publication/254398816_Social_Vulnerability_Sustainable_Livelihoods_and_Disasters. Accessed 6 Jun 2022
27. Schneiderbauer S, Ehrlich D (2006) Social levels and hazard (in)dependence in determining vulnerability. In: Birkmann J (ed) Measuring vulnerability to natural hazards—towards disaster resilient societies. United University Press, Tokyo
28. Papatheodorou K, Emmanuel K, Kazakis P, Paschalis G, Georgios P, Konstantinos N, Nikolaos T, Nikolaos K, Can Z, Toma-Danila D, Vintila D, Cardanet V (2023). Deliverable D.T3.5.1: The REDACt project Educational Hub. Project Rapid Earthquake Damage Assessment Consortium-REDACt [BSB 966]. Contract Nr: MLPDA 88712/26.06.2020. https://www.redact-project.eu/wp-content/uploads/2023/11/D.T3.5.1_The-REDACt-Educational-Hub.pdf
29. Kalliopi S (2021) "Evacuation of Buildings and finding refuge in safe areas for the Population", Manual No.3. Earthquake Planning & Protection Organization of Greece (EPPO), Hellenic Ministry of the Environment, Spatial Planning and Public Works, European Centre for Earthquake Prevention and Forecasting. Athens, 2001

Chapter 7
Assessment of Existing Steel Bridges, a Critical Infrastructure Challenge

Dorin Radu, Radu Băncilă, and Ciprian Bratu

Abstract The rehabilitation of existing steel bridges has become a critical issue as transportation networks face the challenges of increased traffic loads. With many bridges reaching the end of their designed lifespan, bridge owners must now contend with the realities of an aging infrastructure and the need to ensure the continued safety and functionality of these crucial transportation links (Spuler T, Moor G, Meng N (2013, January 1) The role of SHM systems in planning bridge renovation works. https://doi.org/10.2749/222137813808627055).

One key aspect of this challenge is the assessment of bridge condition and load-carrying capacity. Reliable condition assessments are essential to ensure public safety, as well as to prioritize bridges for repair, rehabilitation, and replacement. Researchers have been working to develop methods for improving the accuracy of bridge-capacity evaluation through the use of non-destructive evaluation techniques, which can provide more precise data on the structural integrity of aging bridges. Romania's extensive railway and road infrastructure includes a significant number of historical steel bridges, many of which were built using riveted construction techniques. These bridges not only serve as critical transportation links but also hold immense cultural and historical value. However, as these structures age, they face numerous challenges, including deterioration, increased maintenance requirements, and the need to accommodate modern traffic demands. The paper aims to present assessment methods for investigating the structural integrity of existing steel bridges and to present some study cases for rehabilitation strategies that consider also the impacts of increased traffic loads.

D. Radu (✉)
Transilvania University of Braşov, Braşov, Romania
e-mail: dorin.radu@unitbv.ro

R. Băncilă
Politehnica University of Timişoara, Timişoara, Romania

C. Bratu
Politehnica University of Timişoara, Timişoara, Romania

Doctoral school, Politehnica University of Timişoara, Romania,
Transilvania University of Braşov, Braşov, Romania

D. Radu et al. (eds.), *Countering Hybrid Threats Against Critical Infrastructures*, NATO Science for Peace and Security Series B: Physics and Biophysics, https://doi.org/10.1007/978-94-024-2304-4_7

Keywords Structural integrity · Steel bridges assessment · Critical infrastructure

7.1 Introduction

The transportation network relies heavily on bridges as critical infrastructure, which facilitates the uninterrupted flow of people, goods, and services across the country. However, the aging and deteriorating condition of these bridges poses a significant challenge to maintaining the safety and efficiency of the transportation system [2].

According to the European Commission (2020), approximately 35% of the bridges in Europe are made of steel, making them essential for transportation networks [3]. The evaluation of existing steel bridges is a critical infrastructure issue, as these structures play a vital role in transportation networks and public safety.

The first step in the rehabilitation process is to conduct a thorough assessment of the bridge's current condition. This involves a comprehensive inspection, including visual assessments, non-destructive testing, and structural health monitoring to identify any structural deficiencies, corrosion, or other issues that may compromise the bridge's integrity [1, 2].

The primary objective in assessing the existing steel bridges is to ensure structural safety and integrity. Regular inspections help identify potential issues, such as fatigue cracks, corrosion, and deflection, which can compromise a bridge's performance. Furthermore, assessments are critical for determining whether infrastructure meets current codes and regulations, which often evolve over time. The steel bridges assessment process must take into account several key points: structural integrity, safety concerns, economic considerations and compliance with regulations. Ensuring the structural integrity and longevity of existing steel bridges is a critical concern, particularly when they are exposed to fatigue loading. In addition to condition assessment, strategies for the rehabilitation of existing steel bridges must also consider the impacts of increased traffic loads.

The preservation and maintenance of aging steel bridges in Europe have become increasingly critical as infrastructure faces growing demands from increased traffic loads. Bridges built in the post-war era with a design life of 50–100 years are now subject to significant deterioration, requiring innovative strategies for rehabilitation and structural upgrades. A research study of European railway administrations covering nearly 220,000 bridges across Europe [4] has shown that almost 22% of the bridges are steel structure type. Among them, 28% of the steel bridges are over 100 years old, and 40% are between 50 and 100 years old. Railway bridges are subjected to high stresses caused by extreme vibrations and dynamic deflections, which are increased by the high traffic and the high-speed trains demand.

Evaluation of the existing steel bridges in Europe faces several challenges. Firstly, funding limitations can hinder regular inspections and maintenance activities. Many European countries are experiencing economic constraints, making it difficult to allocate sufficient resources to infrastructure assessment programs [5]. Moreover, variations in regulations and standards across countries may lead to

Table 7.1 Damaged steel structures according with [6]

Structure type	Damaged percentage
Buildings (including industrial buildings and crane supporting structures)	45.10%
Railway bridges	16.10%
Cranes	15%
Road bridges	8.70%
Plant and big machinery used in surface mining	8.00%
Masts and towers	5.80%
Other steel structures	1.30%

inconsistencies in assessment methodologies. The European Union has made strides in harmonizing standards, but regional differences still exist, which can complicate cross-border infrastructure management. Lastly, the dynamic nature of traffic loads, including the increase in heavy vehicle usage, has outpaced original design parameters established decades ago. Engineers must constantly adapt assessment techniques to account for these changes, which may not have been anticipated when the bridges were initially designed.

Considering only steel structures, P. Oehme [6] made a study considering on the cause of damage and type of structures. This study is interesting because steel has been used in a wide variety of different structures. A total of 448 damage cases are reported. Approximately 98% occurred in the period of 1955 to 1984 and 62% in less than 30 years after erection. From this study, we can see (Table 7.1) that bridges (combining railway and roadway) are among the structures most often damaged.

Another way of looking at the study is to separate by the cause, which leads to each of the damages. The study reveals that fatigue is the third in the frequency of cause of all recorded damages; for bridges, fatigue ranks first [6].

7.2 Present Methodologies for Existing Steel Bridges Assessment

7.2.1 Structural Integrity and Reliability Evaluation of Steel Bridges

In 2008 Joint Research Centre (JRC) and the European Convention for Constructional Steelwork (ECCS) published guidelines and methodologies [7] that emphasize systematic evaluation processes for assessing the integrity and reliability of steel bridges. The assessment process is separated into four phases (Fig. 7.1): a *Preliminary Evaluation* phase, in which simple assessment methods, (e.g. visual inspection), are applied in order to gather information about the existing structure; a *Detailed Investigation* phase in which uses quantitative methods (like Non-Destructive Testing) in order to have a more clear and in-depth view of the structure; an *Expert Investigation* phase in which large consequences problems, in terms of

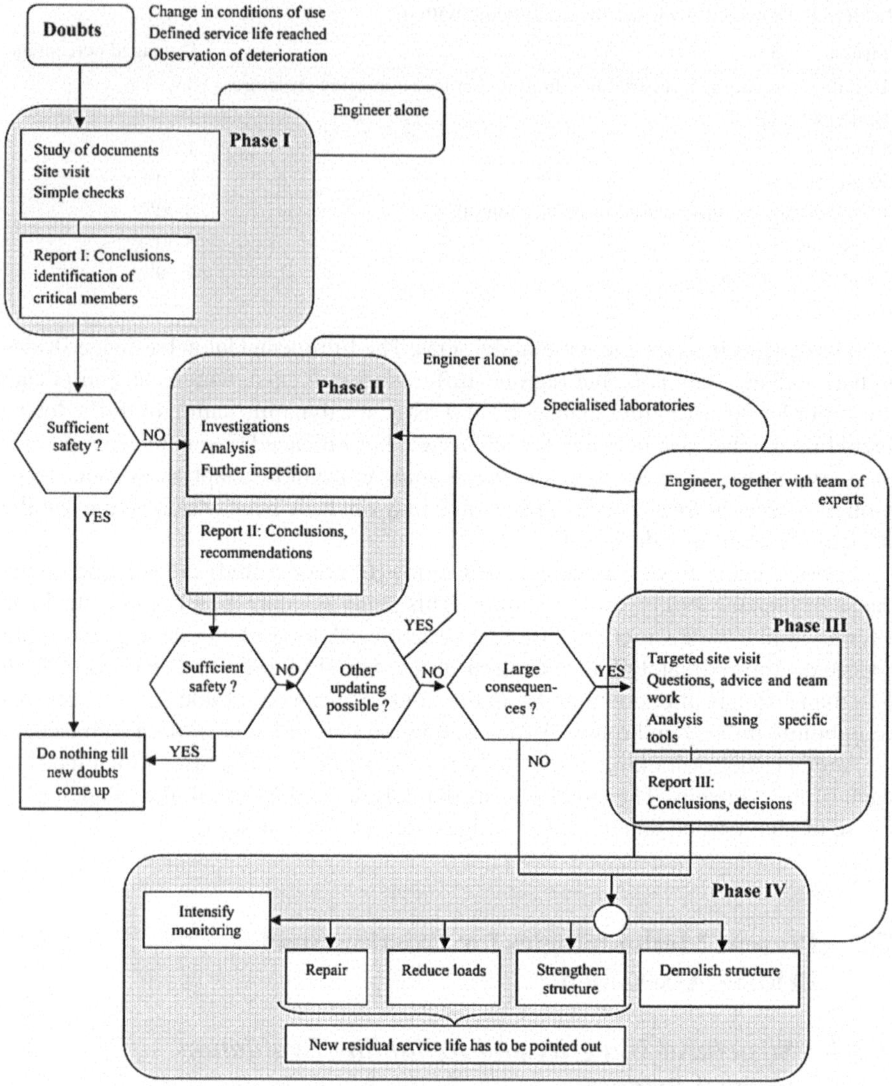

Fig. 7.1 Evaluation procedure according with [7]

risk and costs, are assessed, phase in which the engineer will use also probabilistic methods and fracture mechanics based assessment; and a *Remedial Measures* phase in order to have a fit for service structure with sufficient safety, phase in which several measures can be taken as intensification of monitoring, reduction of loads, strengthening or rehabilitation. These phases encompass risk acceptance criteria (RAC) for existing structures and further developing of the assessment is done following a fatigue evaluation. As a general level, RAC refers to redundancy of the structure, considering component behavior and their type of failure (e.g. ductile or brittle failure), the importance of the structure and the consequences of failure, criteria which take into account the high socio-economical impact and the loss of life and, the last criteria, referring to the inspection level from the inspectability point of view (e.g. structural components that are not inspectable, hidden members, etc.) and the inspection interval (as the inspection interval is higher, higher is the associated risk).

The structural safety evaluation process can be initiated for different reasons and, following the results, having various measures. The motivation for evaluation of an existing structure usually is the observation of defects caused by occurrence of corrosion following bad maintenance or appearance of cracklike type defects following increased traffic loads and speeds (e.g. railway steel bridges that, following the increasing of the trains speeds, changes the vibration characteristics).

The identification of the critical details is done following a structure assessment of the bridge and a qualitative inspection. Depending on the results, different strategies can be proposed to the owner. It includes more updating, or remedial measures (Phase IV) such as monitoring, rehabilitation, or strengthening.

7.2.2 *Steel Bridges Qualitative Evaluation Method Used in Romania*

The Romanian Highway Administration adopted a qualitative verification methodology based on the appreciation by the expert of the technical condition of the structures (Romanian Technical Instruction AND 522-2002) using quality indexes (relation 7.1), finally the result being interpreted by the expert (Table 7.2).

$$I_{\mathrm{ST}} = \sum_{i=1}^{i=5} C_i + \sum_{i=1}^{i=5} F_i \tag{7.1}$$

Table 7.2 Qualitative evaluation of the existing steel bridges

The quality index C is a sum of five aspects:	For the functional requirements also five aspects are relevant:
C_1 -is referring to the main girder	*F_1 -expresses the traffic condition on the bridge*
C_2 -is referring to the deck elements	*F_2 -expresses the loading class of the highway*
C_3 -is referring to the infrastructure and bearings	*F_3 -takes into consideration the year of construction and structure type*
C_4 -is referring to the bed river	*F_4 -is referring to the quality of fabrication, erection and operation conditions*
C_5 -is referring to the quality of the deck surface.	*F_5 -is referring to the maintenance of the structure.*
C_1 -is referring to the main girder	*F_1 -expresses the traffic condition on the bridge*
C_2 -is referring to the deck elements	*F_2 -expresses the loading class of the highway*

For every index marks (from 1–10) are given. Finally, the technical condition of the structure results from the total sum and can be ranged in one of the following categories: Very good technical condition; Good technical condition, Satisfactory technical condition; Unsatisfactory technical condition; The present technical condition cannot assure the safety of the structure.

On these conclusions, the administration can take a decision regarding the time planning of maintenance works.

This method is appropriate for concrete bridges and relatively new steel bridges (up to 20 years old). However, for older steel bridges, it serves primarily as an informational tool. A more advanced methodology should be implemented in those cases.

Given the significance of safety assessments for existing steel bridges, the standards have undergone considerable evolution. Initially, the guidelines were based on simple criteria related to damage accumulation caused by traffic loads on the bridge. Subsequently, additional criteria emerged that consider the formation and propagation of cracks due to material fatigue.

The traditional concept of fatigue assumes that a structural element is free from defects or cracks. However, the presence of discontinuities and cracks in structural components is inevitable, primarily due to material fabrication and construction processes.

The existence of cracks in structural elements fundamentally alters their fracture behavior. In this context, fracture is understood as the process of crack growth under external loads, which is significantly affected by the material's toughness. When defects are present in structural elements, even small stress ranges can facilitate the progression of cracks from initial sizes to critical dimensions (increasing failure probability). The phenomenon of crack propagation under varying stress range levels can be analyzed using the principles of fracture mechanics.

7.2.3 Fatigue Life Evaluation Procedure for Existing Steel Bridges

Fatigue failure in steel bridges has become a growing concern for infrastructure authorities across Europe, as aging structures face increased traffic loads and environmental stresses. The traditional approach of periodic visual inspections, while effective for some maintenance purposes, often fails to provide a comprehensive understanding of the complex fatigue mechanisms at work within these aging structures [8].

To address this challenge, researchers have developed advanced bridge health assessment systems that combine structural monitoring, fatigue analysis, and predictive modeling to better evaluate the condition of existing steel bridges.

In steel bridges built during the late 19th and early 20th centuries, riveting was the predominant connection method. This was later succeeded by bolted connections, which remain common today. Welding techniques began to be adopted in the mid-1930s, and currently, welding is employed in nearly all steel bridges. Steel bridges frequently exhibit fatigue cracks at areas where stress concentrations are increased.

The procedure for fatigue life assessment of the existing steel bridges has four phases (Fig. 7.2): 1) a general checking for fatigue; 2) a fatigue checking under service loads; 3) a further checking using fracture mechanics principles and 4) an

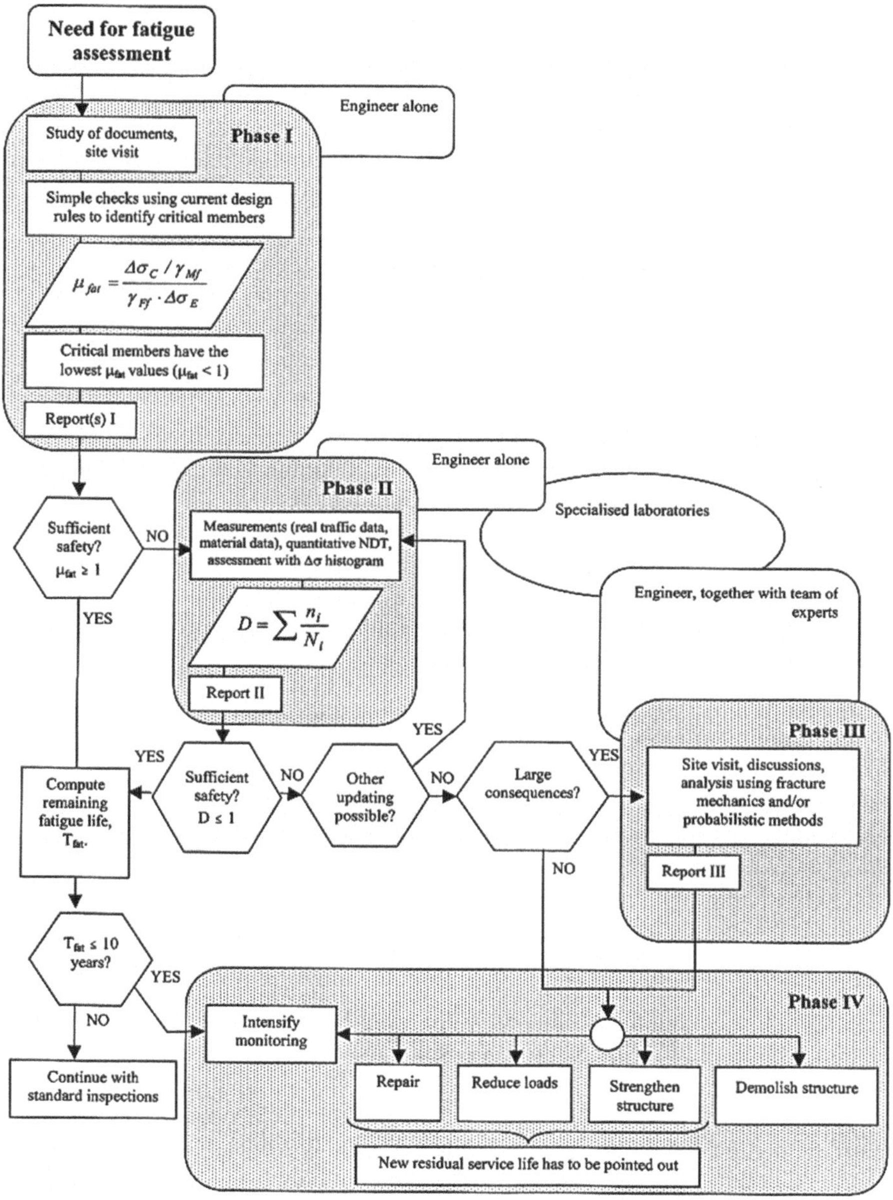

Fig. 7.2 Fatigue assessment procedure for existing steel bridges according with [7]

in-depth checking—applying fracture mechanics based on values obtained on site [7, 9, 10].

Level 1—Fatigue checking

$$\gamma_{Ff} \cdot max\Delta\sigma \leq \frac{\Delta\sigma_D}{\gamma_{Mf}} \tag{7.2}$$

where: γ_{Ff} is the safety factor for the loads/action forces (usually taken 1.00), γ_{Mf} is the safety factor of the material (according with EN1993-1-9 [11] equals between 1.00 and 1.35) and $max\Delta\sigma$ is the maximum stress block induced by the convoy.

This checking is overconservative, considering that the $max\Delta\sigma$ rarely occurs in real working conditions. Usually, level 1 is considered together with level 2 checking.

Level 2—Fatigue checking under service loads

$$D = \sum \frac{n_i}{N_i} \leq 1 \tag{7.3}$$

where n_i is the number of the $\Delta\sigma_i$ stress blocks in the in-service time period (stress spectrum) and N_i is the maximum number of cycles of intensity $\gamma_{Ff}\Delta\sigma_i$ for the considered detail, corresponding to Wöhler curves reduced by γ_{Mf}.

If condition (7.3) is not met because the history of in-service loads could not be obtained, or if D > 1.00 (indicating an overly conservative D > 0.80), it is advisable to conduct a fracture mechanics assessment [11]. Additionally, if a visual inspection of the structure uncovers crack-like flaws or other defects, a fracture mechanics assessment (level 3 or 4) must be carried out.

Level 3—using fracture mechanics

The assessment method of structural elements with cracks was developed based on the modelling possibility with known laws for crack propagation under fatigue loading. This method is based on the BS 7910/2013 [12], adapted for the case of steel bridges. In order to have crack propagation and extension, conditions given by eqns. (4–5) must be fulfilled:

$$K_{max} \leq \frac{K_c}{\gamma_{Mf}} \tag{7.4}$$

$$(\Delta K)_{max} \leq \frac{(\Delta K)_c}{\gamma_{Mf}} \tag{7.5}$$

where stress intensity factor K_{max} is calculated using the following relation:

$$K_{max} = \sqrt{\pi \cdot a} \cdot \sigma_{max} \cdot y(a) \tag{7.6}$$

where a is the length of the crack-like flaw.

The crack propagation rate can be calculated following Paris law for crack growth.

$$\frac{da}{dN} = C\left(\Delta K^m - \Delta K_{th}^m\right) \tag{7.7}$$

In relation (7.7) C and m are the material constants (determined by experiment) and ΔK depends on the flaw geometry and stress level:

$$\Delta K = \sqrt{\pi \cdot a} \cdot \Delta\sigma \cdot Y(a) \tag{7.8}$$

In relation (7.8), ΔK_{th} represents the threshold value beneath which the crack will not advance (Paris curve—zone I [11]) where K_c is the fracture toughness of the material and ΔK_c is the critical stress intensity block. If the condition (7.3) is not fulfilled, level 4 should be adopted.

Level 4 of checking—Applying fracture mechanics based on values obtained on-site

Measurements pertain to the material property parameters, specifically the stress intensity factor and crack growth rate parameters, as well as the actual stress values of the structure based on various loading scenarios. The methodology presented is essential for evaluating the acceptability of identified flaws (cracks) in steel structures or steel bridges, and it is followed by a fatigue assessment of the elements that contain these cracks. Defects (cracks) that were initially considered acceptable may grow sufficiently to lead to component failure, and without redundancy, could compromise the entire structure. In such scenarios, it is crucial to understand the remaining lifespan of the structure during which it can operate safely before any potential failure or collapse occurs.

For new designed structures, EN 1993-1-9 [11] outlines the design methods for assessing the fatigue resistance of elements and joints subjected to cyclic loads, based on the number of stress range cycles. Fatigue is categorized into two types: low cycle fatigue (LCF), which occurs at a reduced number of cycles, and high cycle fatigue (HCF), which occurs at a high number of cycles. LCF is characterized by cyclic plastic deformations and applies to cycles numbering up to 10,000. In contrast, the HCF mechanism operates within the elastic range. Key design considerations include stress concentration factors (due to construction details) and stress ranges. The transition between LCF and HCF is influenced by the stress level, marking the shift from plastic to elastic deformations, and is dependent on the material's ductility.

The methodology for calculating fatigue crack growth focuses on the increase in dimensions of a specific crack type under cyclic loading. It involves determining the number of cycles, N, required for the initial crack size (a_0) to reach the critical crack size (a_{cr}). This analysis utilizes fracture assessment data, including the geometry of the flaw or crack, parameters from fracture mechanics, and the critical flow dimension, as presented in Ref. [13].

In order to assess the implication of a flaw in the general behavior of an element is needed first to determine the load range following the load history. In case of railway steel bridges, the history of loading can be determined easily but in case of road steel bridges, only by using stochastic approaches can be done [13]. Also the material mechanical properties must be known, including the fracture toughness. Usually the tests are done on existing material samples, following Charpy tests for determining the toughness of the steel and, using [12], fracture toughness value can be calculated.

7.3 Proposal of an Algorithm for Evaluation of Existing Steel Bridges

The methodology is designed as an advanced and comprehensive analysis of structural elements with fatigue defects. It is based on the principles of fracture mechanics and consists of two main steps. The first step involves assessing the acceptability of defects using Failure Assessment Diagrams (FAD) and determining the final acceptable dimensions of the defects. The second step entails a fatigue evaluation of the structural elements in question, relying on the recorded stress history of the structure. This evaluation considers both the initial (a_0) and the critical defect sizes (a_{cr}), as well as the fracture mechanics parameters, specifically the material characteristics C and m from the Paris equation [10] (which relates to crack growth under real traffic stress). Ultimately, this process aims to accurately determine the number of cycles N required for fracture to occur and to estimate the remaining service life of the structural elements in terms of years, months, and days.

7.3.1 Failure Assessing Diagrams (FAD), a Tool for Determine the Flaw Acceptability

The assessment is typically conducted using a Failure Assessment Diagram (FAD) grounded in the principles of fracture mechanics. The vertical axis of the FAD represents the ratio of the actual conditions to those necessary for fracture, measured in fracture mechanics terms. Meanwhile, the horizontal axis indicates the ratio of the applied load to the load needed to induce plastic collapse. An assessment line is drawn on the diagram. For a given flaw, the calculations yield either the coordinates of a specific assessment point or a locus of points [9].

Planar flaws should be characterized by the height and length of their containment rectangles following the flaw type: through thickness flaws, surface flaws, embedded flaws (Fig. 7.3).

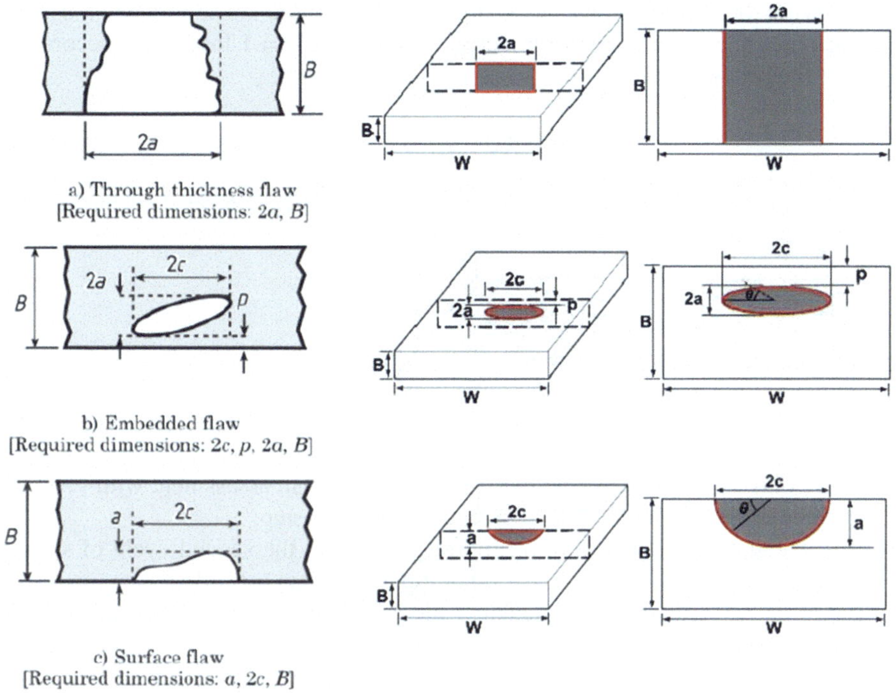

Fig. 7.3 Characterization and idealization of the flaws [9]

The first step in the evaluation of a flaw by means of FAD is the calculation of the stresses in the element with the flaw and in the area of it. Then, a fracture ratio must be determined $K_r = K_I/K_{mat}$, where $K_I = (Y\cdot\sigma)\cdot(\pi\cdot a)^{1/2}$ with $Y\cdot\sigma = (Y\cdot\sigma)P + (Y\cdot\sigma) S$. Also the ratio of stress is required $L_r = \sigma_{ref}/\sigma_y$, with σ_{ref} obtained according with a relation specific with the flaw type and σ_y being the yielding value of the material. The point/points of assessment are represented graphically in (K_r, Lr) coordinates on the FAD level 2. If the points are out of the assessment line, the flaw can cause the failure of the element due to fracture.

7.3.2 Determining the Service Lifetime

The algorithm first step (Fig. 7.3) is to fit the discovered flaw in one of the preset types and to define the initial flaw dimensions. Using FAD it can be determined the crack growth limit i.e. the critical flaw dimension.

The fatigue assessment uses the applied nominal stress ranges. These are acting on the cross section with flaw and results from the variable loading. The fatigue crack growth for a given flaw type has the following steps:

- Knowing the material constants C and m and the initial crack dimension a_0 it is calculated the crack growth da_1 corresponding to $dN = 1$ load cycle according with the following (7.9):

$$da_1 = C \cdot \Delta K^m \tag{7.9}$$

- Based on da_1 growth, it is calculated the crack dimension resulted in the first loading cycle

$$a_1 = a_0 + da_1 \tag{7.10}$$

- It is done the following checking:

$$a_1 < a_f \tag{7.11}$$

where a_f is the final crack length resulted in base of an assessment with FAD. If the relation is fulfilled, it can be proceed to the next step;
- It is calculated the growth of da_2 corresponding to the second cycle of stress, based on a_1 dimension and on the other input data: $da_2 = C \cdot \Delta K^m$
- Based on da_1 growth, it is calculated the crack dimension resulted in the first loading cycle:

$$a_2 = a_1 + da_2 \tag{7.12}$$

- It is done the following checking:

$$a_2 < a_f \tag{7.13}$$

- This procedure is done until:

$$a_i = a_f \tag{7.14}$$

The number of stress cycles N, for which it is obtained relation (7.14), represents the remaining life of the structural element (Fig. 7.4).

Considering the limits of the crack growth, in the assessment of fatigue crack growth for planar flaws, it is essential to establish a threshold for allowable crack extension that can occur without leading to failure during the service life.

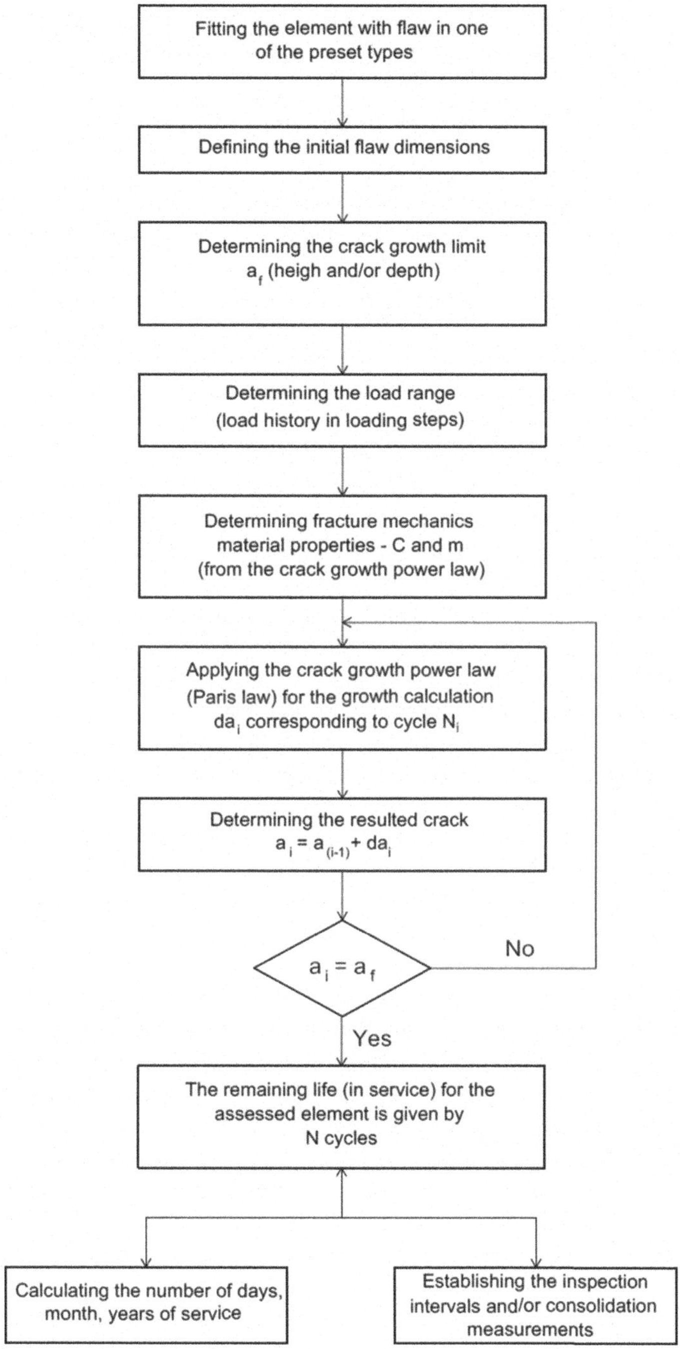

Fig. 7.4 Algorithm for evaluation of existing steel bridges

7.4 Structural Integrity Evaluation of Two Existing Steel Bridges by Means of the Proposed Algorithm

7.4.1 Historical Steel Bridge in Transilvania Region

The in case bridge is a riveted structure constructed around 1925 in Transylvania. From a geometric perspective, it features a parabolic truss main beam with a span of L = 27.86 m and a width of 6.25 m (with a roadway width of only 5.25 m). The design is similar to other bridges built during the same era, featuring a heavy deck (with pavers) composed of Zores profiles arranged on a network of beams, including stringers and cross girders, along with a bracing system that ensures the structural stability. The components are made of composite cross sections consisting of L-shaped profiles with additional steel plates.

Following the visual inspection, the bridge condition was revealed as being poor. Several flaws were discovered in the truss elements and in the deck structure, the primary cause being poor maintenance. Considering the existing structure age, it was needed to do an assessment to reveal the possible implication of the future possible discovered flaws.

The evaluation procedure was applied—staring from the assessment proposed in Sect. 7.2.3 and then the in more depth evaluation—the implication of the discovered flaws in the whole behaviour of the bridge structure following the increased traffic and taken into account the fatigue loading. First step was the calculation of the stresses in the elements. A Finite Element Method analysis was performed [9], and the main bending stresses were determined in the area of the discovered flaws. The material properties required for fracture mechanics, the necessary data were determined through experiments, including a toughness value of $K_{mat} = 71.8$ MPa·m$^{1/2}$.

A FAD assessment was conducted to identify the critical dimensions of the flaws required for fatigue evaluation. The results illustrate the acceptability level for each type of flaw through comparative graphs, and they also establish the critical dimensions of the flaws. Following the applied procedure (Sect. 7.3.1), sensibility of the deck longitudinal beam to the through thickness flaw in the element was revealed (FP-TTF-5). In case of an only 7.5 mm flaw, the element fails (Fig. 7.5 and Fig. 7.6).

Having the values of the critical flaws for different types, a further assessment was done using the algorithm presented at Sect. 7.3.2. It resulted the number of cycles until reaching the critical flaw dimension (length or height). A comparison was made between groups of flaws, in order to detect and underline the most critical flaws in term of fast crack growth and number of cycles until reaching the critical dimensions—remaining in service life time (Fig. 7.7). As one can see, the assessed through thickness flaw (FP-TTF) in the main deck beam can reduce the lifetime of the structural element, the service lifetime being only 15 years.

After assessing the service lifetime, considering the increased traffic loads and the need to widen the bridge, a solution was proposed to keep the bridge operational after retrofitting and to prevent traffic restrictions, ensuring that the future bridge can support the loads of LM1 type convoys as specified by Eurocode EN 1991–2:2004.

Fig. 7.5 Photos of the existing steel bridge (general view and joints)

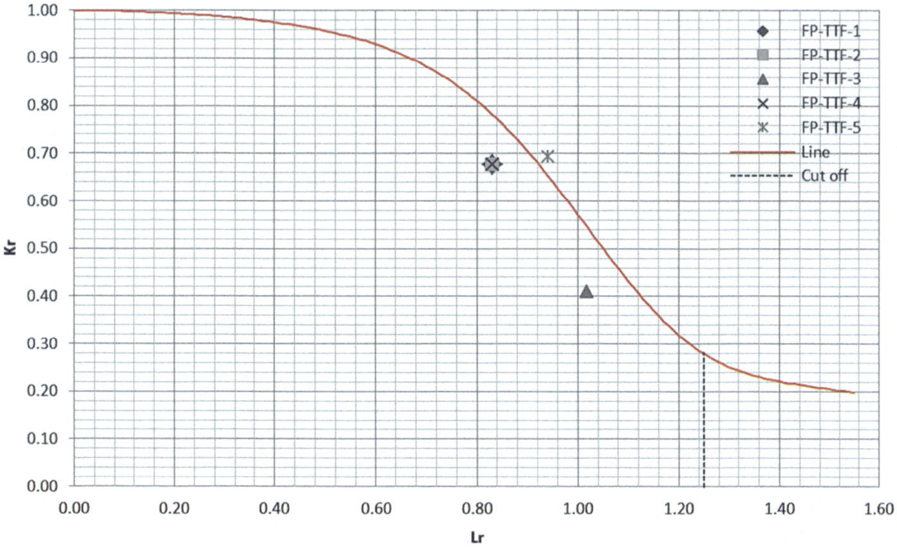

Fig. 7.6 FAD for determine the implication of different types of flaws (through thickness flaws in longitudinal beam) [10]

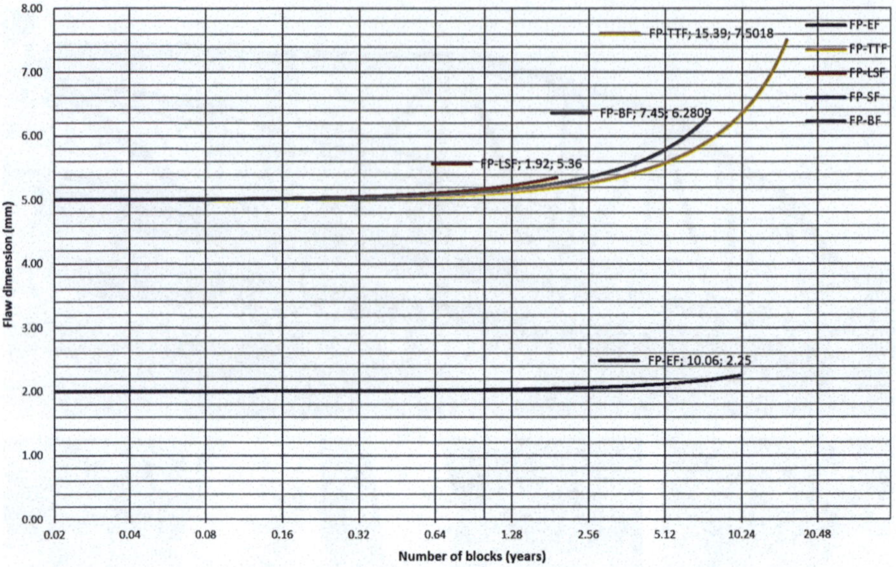

Fig. 7.7 Fatigue assessment of the different flaw cases—main deck beam [10]

This involves maintaining the existing deck structure while introducing a new deck structure in the center. This new structure features a main beam box girder that is 600 mm wide and 1800 mm high, with thickness varying from 10 mm to 15 mm, made from S355 steel. Additionally, new transverse beams will be connected to the existing beams using bolted end plate connections (Fig. 7.8). This design allows for two lanes of traffic (approximately 4.20 m wide in each direction) without any restrictions.

The results of the FEM analysis indicated the stress values for each structural element. In the existing truss beam structure, the maximum stress is observed in the first diagonal (as anticipated) and in the deck cross girders. In comparison, the stresses in the retrofitted structure are approximately 15% lower than those in the existing structure. Notably, the new carrier beam (box girder type) exhibits a high stress ratio of around 82% of its yielding strength. These findings confirm that the proposed solution is effective.

7.4.2 *Railway Steel Bridge Crossing Prahova River*

The bridge is a railway bridge which makes the connection from Câmpina to Predeal. The structure has four spans, one of 13.90 m and three of 14.10 m each and is skew (skew to the right 48°37′) and its superstructure is made out of independent plate girder decks (Fig. 7.9) and is erected in 1940. The used material is a mild steel similar with OL37 type.

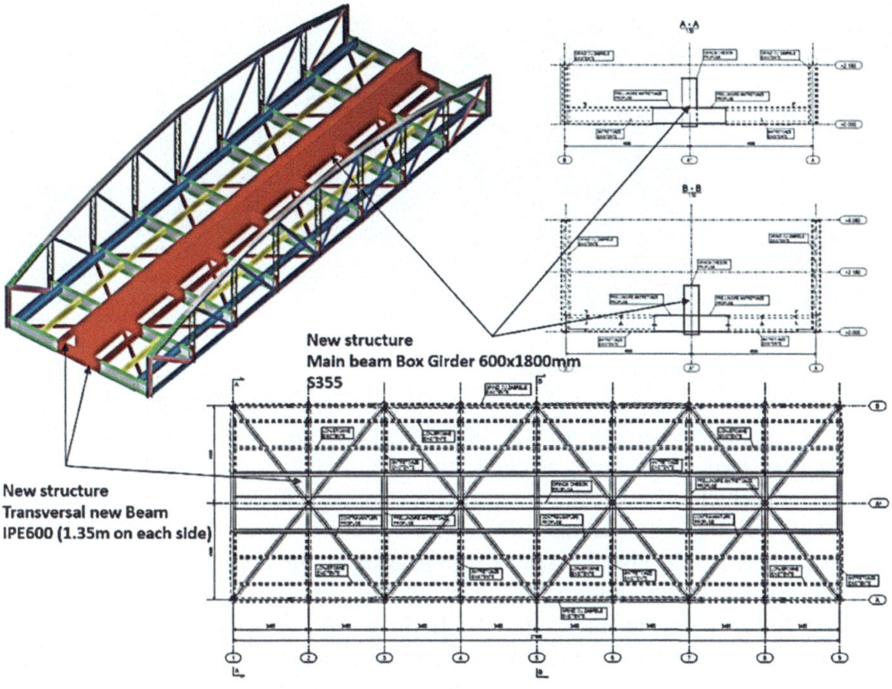

Fig. 7.8 Proposed structure—consolidation and retrofitting

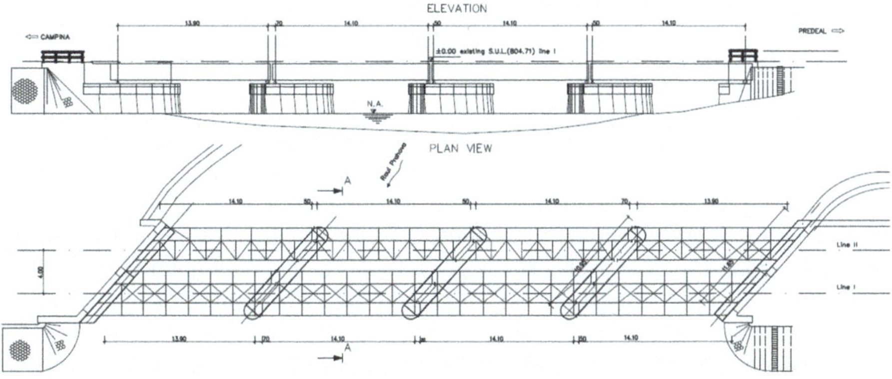

Fig. 7.9 Structural views of the bridge

The damage accumulated values (relation 7.3) were calculated following a loading spectrum (loading ranges occurrence) determined from a loading history (following Railway National Company traffic information). A stress history was created using the mixed traffic scenarios from EC 1. This represents the anticipated future traffic that will be typical for the newly rehabilitated railway corridor. All these trains were run over the structure to develop the multi-block stress history [14].

Fig. 7.10 Flaws in the existing structure

Regarding the remaining service lifetime, following several discovered flaws (Fig. 7.10), Two cases were studied: through thickness flaw with initial size $a_0 = 2.0$ mm which is undetectable because it is situated under rivet head in the web steel plate, and through thickness flaw with initial size $a_0 = 2.0$ mm which is also undetectable during a visual inspection appearing in the corner (lower flange). In the first case, the remaining fatigue life is estimated at a value of 8.31 years (corresponding to several cycles of 1,566,000). In the second case the administrator must know that propagation of the crack to the critical value will occur in 8.89 years (N = 1,676,000 cycles). In this case, the safe inspection intervals are situated between 1.2 and 2.0 years.

7.5 Conclusions

Assessing existing steel bridges is essential for maintaining public safety and ensuring that transportation infrastructure continues to function efficiently. By implementing comprehensive evaluation practices, embracing new technologies, and prioritizing funding, stakeholders can address the critical issues surrounding bridge safety and reliability effectively. The paper presents assessment methods for investigating the structural integrity of existing steel bridges with two cases for rehabilitation strategies which consider also the impacts of increased traffic loads. The proposed algorithm has a practical value in the fatigue assessment of steel bridges. With this method, historical bridge structures can remain in service, provided that a thorough individual evaluation—utilizing both experimental and numerical approaches—demonstrates their integrity. Additionally, it offers an accessible way to analyze potentially critical defects in various locations of the assessed structure or element.

References

1. Spuler T, Moor G, Meng N (2013, January 1) The role of SHM systems in planning bridge renovation works. https://doi.org/10.2749/222137813808627055
2. Omar T, Nehdi ML (2018, September 14) Condition assessment of reinforced concrete bridges: current practice and research challenges. Multidisciplinary Digital Publishing Institute 3(3):36–36. https://doi.org/10.3390/infrastructures3030036
3. Gkoumas K, Balen M van, Grosso M, Pekár F, Marques Dos Santos FL, Haq G, Ortega Hortelano A, Tsakalidis A Research and innovation in bridge maintenance, inspection and monitoring. Joint Research Centre (European Commission). https://doi.org/10.2760/719505
4. Bien J, Elfgren L, Olofsson J (2007) Sustainable bridges: assessment for future traffic demands and longer lives. Wydawnictwo Edukacyjne, Dolnosląskie
5. European Commission (2009) Commission working document—consultation on the future EU2020 strategy. COM647, Brussels
6. Oehme P (1989, November) Schäden an Stahltragwerkeneine analyse (damage analysis of steel structures). IABSE proceedings P-139/89
7. Kühn B, Lukić M, Nussbaumer A, Günther H-P, Helmerich R, Herion S, Kolstein MH, Walbridge S, Androic B, Dijkstra O, Bucak Ö (2008, February) Assessment of existing steel structures: recommendations for estimation of remaining fatigue life, prepared under the JRC—ECCS cooperation agreement for the evolution of Eurocode 3, TC 250
8. Wang X, Wang ML, Zhao Y (2005, May 17) Bridge health assessment system with fatigue analysis algorithm. SPIE. https://doi.org/10.1117/12.598852
9. Radu D, Băncilă R, Sedmak S, Aranđelović M (2022) Residual life of a historic riveted steel bridge—engineering critical assessment approach. Procedia Structural Integrity 42:1106–1112. https://doi.org/10.1016/j.prostr.2022.12.141
10. Sedmak A, Radu D, Aranđelovic M, Sedmak S, Bancilă R (2024) Residual life assessment of the historical road riveted bridges. Eng Fract Mech 299:109960. https://doi.org/10.1016/j.engfracmech.2024.109960
11. EN1993-1-9, Eurocode 3: Design of Steel Structures — Part 1–9: Fatigue, European Standard, Brussels, CEN, 2003
12. BS 7910:2013, Guide to methods for assessing the acceptability of flaws in metallic structures (British standard)
13. Radu D, Băncilă R, Bolduş D, Sedmak S, Aranđelović M (2022) Solution for consolidation and retrofitting a historical steel bridge. Procedia Structural Integrity 37:771–778
14. Petzek E, Kosteas D, Bancila R (2005) Sicherheitsbestimmung bestehender Stahlbrücken in Rumänien. Stahlbau Nr. 8, 9, ISSN 0038-9145, Ernst & Sohn

Chapter 8
Temporary Fiber-Polymer Composite Bridges, a Solution for Disaster Recovery and Military Operations

Radu Băncilă, Lucian-Attila Blaga, Dorin Radu, Ciprian Bratu, and Mihaela Malița

Abstract This manuscript is intended as an argument for the use of glass-fiber-reinforced polymeric composites, known as GFRP, in the rehabilitation of existing steel bridges. GFRPs have been used as structural materials in bridge construction since the early 1980's and today are increasingly used either in full-FRP structures or as deck elements, cables, reinforcements, beams, and trusses. The main features speaking for their beneficial use are their high specific strength and their corrosion resistance alongside their behavior under severe environments. With a density at around 20% of that of steel but at a significantly lower elastic modulus, modular GFRP bridge elements are easy to manufacture, transport, and install, thus leading to reduced installation costs, fast construction times, and structures with minimal maintenance (Blaga et al., Construct Build Mater 80: 167–179, 2015). Temporary fiber-polymer composite bridges have a wide range of applications. In disaster-prone areas, these bridges can be deployed swiftly to restore connectivity after natural disasters such as floods, earthquakes, or landslides. The military also utilizes these structures for operations requiring rapid movement of troops and supplies over difficult terrain or temporary crossings. As infrastructure needs become increasingly complex and urgent, temporary fiber-polymer composite bridges represent a vital innovation in civil engineering. With their rapid deployment capabili-

R. Băncilă · M. Malița
Politehnica University of Timişoara, Timişoara, Romania

L.-A. Blaga
LightJoin Solutions GmbH, Geesthacht, Germany

D. Radu (✉)
Transilvania University of Braşov, Braşov, Romania
e-mail: dorin.radu@unitbv.ro

C. Bratu
Transilvania University of Braşov, Braşov, Romania

Doctoral School, Politehnica University of Timişoara, Transilvania University of Braşov, Braşov, Romania

D. Radu et al. (eds.), *Countering Hybrid Threats Against Critical Infrastructures*, NATO Science for Peace and Security Series B: Physics and Biophysics, https://doi.org/10.1007/978-94-024-2304-4_8

ties, lightweight construction, and resistance to environmental deterioration, they provide effective solutions for both temporary and emergency applications. As technology continues to advance, these bridges are poised to become a standard feature in modern infrastructure, combining resilience, efficiency, and sustainability to meet the challenges of today's world.

Keywords Fiber-polymer composites · Temporary bridges

8.1 Introduction

In the ever-evolving landscape of civil engineering and construction, the demand for innovative and efficient solutions is at an all-time high. One area that has seen remarkable advancements is the development of bridges. Among various designs and materials, temporary fiber-polymer composite bridges have emerged as a revolutionary solution, meeting the needs for quick deployment, durability, and sustainability. These structures are particularly useful in scenarios where rapid access is paramount, such as disaster recovery, military operations, and remote area connectivity.

Fiber-reinforced polymer composites are a class of materials that combine high-strength fibers, such as carbon or glass, with a polymer matrix [2]. This unique combination results in a lightweight yet robust material that offers high resistance to corrosion, making it suitable for various environmental conditions. Temporary bridges constructed from fiber-polymer composites are designed for short-term usage, often deployed quickly to fill gaps in existing transportation networks or to support emergency response efforts. One of the most striking advantages of these composite bridges is their ease of installation. Unlike traditional steel or concrete bridges, which require extensive labor and heavy equipment, fiber-polymer composite bridges can be prefabricated off-site and quickly assembled on location. This is especially critical in emergency situations where time is of the essence.

The inherent properties of composite materials make these bridges significantly lighter than their traditional counterparts. This feature not only eases transportation logistics but also reduces the load on underlying structures and substrates during installation. One of the most significant challenges with traditional bridges is their susceptibility to environmental damage, particularly from water, chemicals, and other corrosive elements. Fiber-polymer composites resist such deterioration, greatly extending the lifespan of temporary bridges and reducing maintenance costs.

Unlike traditional bridge construction methods, which often require extensive time and resources, temporary fiber-polymer composite bridges can be rapidly assembled and deployed. This is particularly valuable in the aftermath of natural disasters, where infrastructure damage can impede emergency response and recovery efforts. Additionally, the use of these composite materials can provide a more cost-effective and tailored solution compared to traditional bridge construction, as they can be designed to meet specific load and environmental requirements [3].

In military contexts, the rapid establishment of bridging solutions is equally imperative. Troops on maneuvers often must cross rivers, ravines, or other obstacles with little to no advance warning. Temporary fiber-polymer composite bridges can be pre-manufactured and transported to forward operating bases, allowing military engineers to set up vital crossings with minimal delay.

The modular and lightweight characteristics make these bridges ideal for various military applications. They can be transported in standard military trucks and assembled quickly by troops with basic engineering skills. This agility allows military commanders to facilitate troop movements, logistics, and supply chains effectively, ensuring that operations can continue unhindered.

Additionally, these bridges can be designed for rapid disassembly, aligning with military needs for mobility. Once a mission is complete or a new location is required, these temporary bridges can be taken down and transported to a new site, making them a versatile solution in dynamic environments.

8.2 FRP in Bridge Construction

Most of the structural GFRP profiles are produced by the process of pultrusion and underlie the EN 13706 European Standard [4]. The European Community initiated the procedures for a Eurocode on fiber-reinforced polymeric composite (FRP) structures, which is preceded by the already existing Technical Specification CEN / TS 19101:2022, covering composite laminates, profiles, and panels for the design of civil engineering structures. Moreover, this TS applies to full-FRP as well as hybrid structures and covers several composite reinforcement types, such as aramid or carbon fibers, in addition to glass [5]. The final Eurocode for composite is expected by 2026 [6]. In the following, two application cases for fiber-polymer composites are presented.

8.2.1 Fiber-Polymer Bridge Decks

FRP bridge deck panels can be an ingenious solution for the rehabilitation of old steel bridges, given their combined lightweight and corrosion properties. This hybrid approach can significantly reduce the weight of the structure without a loss in resistance or capacity. Furthermore, service costs would be reduced, due to the minimal maintenance required. Ali et al. investigated the application of FRPs in the bridge industry in a comprehensive review, quantifying the use of FRP in bridge decks at 34% in pro-portions in bridge components around the world [7]. Moreover, FRP decks have three times the lifetime of concrete decks, making them ideal for repairing and rehabilitation from this perspective [7].

Knippers compared the conventional steel girder–concrete deck structure with a presumptive hybrid steel girder—FRP deck (connected by steel studs to the main girders) solutions for the 70-year-old structure of the Probststrasse Road Bridge in Stuttgart (15 m span), estimating an almost 60% reduction in structural weight for the hybrid solution versus the original conventional structure [8]. While the estimated costs for such a hybrid solution were significantly higher than for the conventional structure, it is worth noting that the market situation today is completely different, with high increases in production means and capacities and a multitude of suppliers on the FRP market by 2024, therefore a new valuation for shall be considered for a similar comparison.

The Prince Street Bridge in Bristol (span 27 m) is a more recent example of applying FRP in rehabilitation and thus maintaining European engineering heritage. The Grade 2-listed ("particularly important buildings of more than special interest", as stated by [9]) wrought iron swing bridge was built in 1879 and by 2015 was closed due to severe corrosion damage. The structure was finally rehabilitated and reopened for pedestrian, cyclist, and vehicular traffic in 2017 with a cost of 1.3 m £ (including the full refurbishment of the iron structure, replacement/reinforcement of corroded elements, galvanization, and FRP deck). The new FRP deck is 30% lighter than the previous, has improved durability (while reducing service life costs), and reduces the dead load by 20 t, influencing positively the overall mechanical operational performance associated with the movement of the swing bridge [10].

8.2.2 Emergency Bridges

The increasing frequency of both natural and human-made disasters in recent decades has emphasized the need for robust and for robust and quickly deployable bridge structures for extreme situations. Emergency bridges must be designed for rapid assembly, ensuring they can be erected with ease in crisis situations such as floods, earthquakes, accidents, or terrorist attacks. Traditional emergency bridges like the metallic-structured Bailey Bridge and the D-Bridge probed their effectiveness having been widely used due to their versatility, load-bearing capacities and relatively simple installation procedures. However, the development of corrosion-resistant materials, such as FRPs described in Sect. 8.2.1. could offer significant advantages. Polymer-composites not only meet the mechanical and environmental resistance requirements but also facilitate transportation and assembly of prefabricated bridge elements due to their lower weight. This can be crucial for maintaining vital transport connections during emergency situations, enabling efficient rescue and humanitarian missions [1].

Such a fiber-composite structure, namely all-GFRP, was developed at the RWTH Aachen University within a project for the German Bundeswehr. The first 20 m-span GFRP lightweight bridge for one MLC-12 vehicles was built and tested in real conditions in 2001, followed by another prototype of a 30 m-span for MLC-40 in 2004, weighing a total of only 16 tons. The proposed Warren-truss consisted of pultruded GFRP profiles joined via a hybrid bolting-adhesive solution using metallic shoes for the strengthening of the truss joint areas [11].

Fig. 8.1 Prototype of the "HUGO" auxiliary bridge (image courtesy of FiberCore Europe [12])

The Dutch companies KWS, FiberCore Europe together with the Dutch Ministry of Infrastructure and Water Management recently proposed an ingenious polymer-composite auxiliary bridge designed to reduce traffic disruption during renovation and major maintenance works on roads, viaducts, and bridges. The structure, named "HUGO", could also serve as a temporary ecoduct or as an emergency bridge on disrupted communication lines. The principle consists of a lightweight, dismountable bridge that is placed on top of an existing carriageway or structure, whereby the traf-fic is conducted on the FRP spans via metallic ramps, while the repair works can take place below the temporary structure, as shown in fig. 8.1 [12].

8.3 Innovative Proposal of a Friction-Riveted FRP Temporary Bridge Solution

Following the example of the emergency structure in [11], Blaga et al. proposed a pol-ymeric composite solution for temporary bridges with an innovative approach in joining the bridge elements [1]. The structural GFRP elements are joined via the Friction Riveting technique, whereby metallic rivets are inserted into the composites in the truss girder node area, later to be tightened together via nuts and washers using aluminum gussets. The principle and feasibility were proven on thermoplastic composite materials and titanium, at a development stage of Friction Riveting in which thermo-setting composites were not studied yet for the named technique. It must be noted that most of the pultruded GFRP structural profiles are based on thermosetting polymers such as polyester, thus limiting the use of welding or other high-temperature processes. Friction Riveting was initially developed for thermoplastic composites applied in transportation, rather than civil engineering. Within Friction Riveting, a cylindric metallic rivet is inserted via high rotational speeds into a polymeric joining partner leading to its deformation and anchoring, without full perforation of the composite. Figure 8.2. presents the static model of the proposed 25 m span bridge de-signed for the load of a single A 30 truck, consisting of only 3 different types of GFRP profiles (deck, chords and diagonals) and truss girder nodes assembled by Friction Riveting [1, 4].

The friction-riveted emergency bridge concept was taken further within the "FricRivBridge" technology transfer project of the Helmholtz Association at the Helmholtz-Zentrum Hereon in Germany. Pultruded structural GFRP profiles were used, for which the feasibility of applying Friction Riveting was successfully determined. Figure 8.3. displays a friction-riveted joint of a stainless-steel rivet within a pultruded GFRP profile alongside the potential assembly of a friction-riveted truss girder node for a truss girder structure [13].

The aspects regarding thermosetting material behavior during joining, microstructural characterizations, as well as mechanical performance under quasi-static and cyclic loading are not within the scope of this paper and will be published elsewhere. Figure 8.4. shows one of the constructed full-scale friction-riveted truss girder bridge node prototypes, whereby the assembly is gusset-less, with the rivets friction-riveted in the pultruded GFRP C-shaped chord. The truss diagonals were pre-perforated to accommodate the outer part of the rivets from the chord and consequently positioned and tightened via nuts and washers from the same stainless-steel material as the rivets.

The FricRivBridge project provided the basis for the construction of a future friction-riveted full-scale GFRP emergency bridge prototype. Furthermore, beyond the actual construction of the nodes, important knowledge was generated regarding the material behavior following the high-speed (and high-temperature) joining process as well as an understanding of the joint mechanisms and performance.

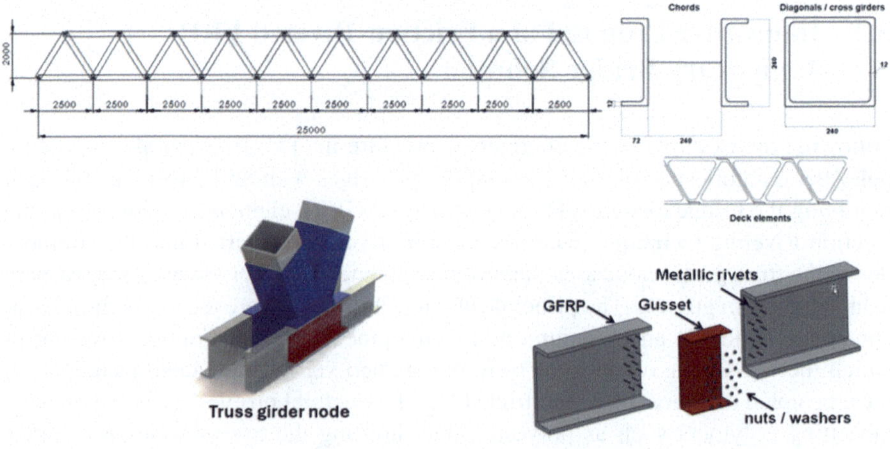

Fig. 8.2 Initial concept of a friction-riveted GFRP emergency bridge (adapted from [1, 4])

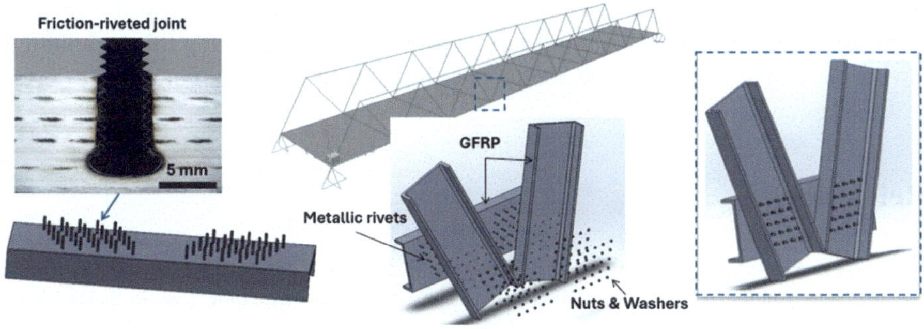

Fig. 8.3 Exemplary friction-riveted joint (stainless steel rivet within pultruded GFRP) and presumptive friction-riveted truss girder node with multiple rivets ([13])

Fig. 8.4 Friction-riveted truss girder GFRP bridge node prototype constructed within the FricRivBridge technology transfer project

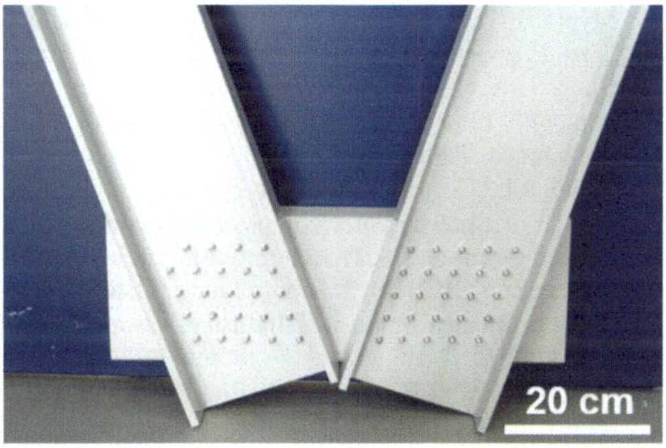

8.4 Conclusions

In today's ever-evolving world, the need for rapid and efficient infrastructure solutions is becoming increasingly critical. This is particularly evident in the contexts of disaster recovery and military operations, where the ability to quickly establish reliable transport routes can mean the difference between life and death, as well as mission success or failure. One innovative solution that has gained significant traction is the use of temporary fiber-polymer composite bridges. These modern structures offer numerous advantages over traditional bridging methods, making them ideal for various challenging scenarios.

Temporary fiber-polymer composite bridges represent a groundbreaking advancement in bridging technology, offering an effective and resilient solution to the critical challenges faced in disaster recovery and military operations. Their lightweight nature, corrosion resistance, rapid deployment capabilities, and modular design positions them as a preferred option when time and reliability are essential. As we continue to face an increasing frequency of natural disasters and complex military engagements, the ability to quickly restore connectivity through these innovative structures is not just beneficial—it is essential for the safety and success of disaster response efforts and military missions alike. Embracing this technology is vital for enhancing resilience and preparedness in an ever-changing world.

References

1. Blaga L-A, dos Santos JF, Băncilă R, Amancio-Filho ST (2015) Friction riveting (FricRiveting) as a new joining technique in GFRP lightweight bridge construction. Constr Build Mater 80(2015):167–179
2. Bakis CE, Bank LC, Brown VL, Cosenza E, Davalos JF, Lesko JJ, Machida A, Rizkalla S, Triantafillou T (2002, May 1) Fiber-reinforced polymer composites for construction—state-of-the-art review. Am Soc Civil Eng 6(2):73–87. https://doi.org/10.1061/(asce)1090-0268(2002)6:2(73)
3. Myers JJ (2007, January 1) Durability of external fiber-reinforced polymer strengthening systems. Elsevier BV, pp 247–283. https://doi.org/10.1533/9781845693565.2.247
4. Blaga L-A (2012) Innovating materials in bridge construction. Contribution to construction with fiber-reinforced materials, PhD thesis Universitatea Politehnica Timişoara, Ed. Politehnica, Timişoara
5. Ascione L, Correia JR, Keller T, Knippers J, Mottram T, Paulotto C, Sensa-Cruz J (2023) Design of fibre-polymer composite structures (CEN/TS 19101): overview, commentary and worked examples. In: Proceedings of 11th international conference on FRP composites in civil engineering CICE 2023, Rio de Janeiro, BR
6. The European Composites Industry Association Homepage, https://eucia.eu, last accessed 2024/09/07
7. Ali HT, Akrami R, Fotouhi S, Boghadi M, Saeedifar M, Yusuf M, Fotouhi M (2021) Fiber reinforced polymer composites in bridge industry. Structure 30(2021):774–785
8. Knippers J (2005) Innovative design concepts for composite bridges in Germany. In: Proceedings of COBRAE conference 2005 at the EMPA, Dübendorf, CH
9. The National Heritage List of England (NHLE) Homepage, https://historicengland.org.uk, last accessed 2024/09/07
10. Mann P (2017) Prince street swing bridge - report, place scrutiny commission 24/07/2017. Bristol City Council
11. Sedlacek G, Oppe M, Trumpf H (2005) Design and testing of an inventive GFRP-truss-bridge for 40 t trucks and 30 m span. In: Proceedings of COBRAE conference 2005 at the EMPA, Dübendorf, CH
12. FiberCore Europe Homepage, https://www.fibercore-europe.com, last accessed 2024/09/08
13. Maliţa M, Blaga L-A, dos Santos JF, Petzek E, Klusemann B (2019) Development of an innovative joint design for concept for emergency fiber-reinforced composite bridges. In: 10th international conference on bridges in the Danube Basin. Danube Bridges, Vienna, AT

Chapter 9
Future of Electrospun Nanofibrous Systems for Decontamination, Protection, and Detection Against Chemical and Biological Threats

Nimet Bölgen, Didem Demir, Sedef Salel, and Ashok Vaseashta

Abstract Biological and chemical threats pose significant security risks to all societies. While biological threats can cause health crises through the spread of pathogenic agents, chemical threats can have detrimental effects on the environment and human health through the spread of toxic substances. They can target critical infrastructure and lead to emergency management, health crises, or economic disruptions in societies. The deployment of protective equipment and sensor systems is extremely important in establishing an effective protection and detection system against these threats. While protective equipment provides a physical barrier, sensor systems provide rapid and accurate detection of threats. Electrospun nanofibers are an important material used in protective equipment and sensor systems in combating chemical and biological threats. Due to their high surface area and thin structure, these nanofibers provide effective filtration. In addition, their lightness and comfort facilitate long-term use, while their water resistance and durability extend the life of protective equipment. In sensor systems, nanofibers provide high sensitivity and fast response time, helping to rapidly and accurately detect chemical and biological agents. Surface modifications also allow the development of sensors and protective equipment specific to specific threats. This book chapter presents current approaches and future directions of electrospun nanofibers for protection, decontamination, and detection against various chemical and biological threats, based on a synthesis of recent studies.

Keywords Electrospinning · Protective · Chem.-bio threats

N. Bölgen (✉) · S. Salel
Chemical Engineering Department, Engineering Faculty, Mersin University, Mersin, Türkiye

D. Demir
Department of Chemistry and Chemical Process Technologies, Mersin Tarsus Organized Industrial Zone Technical Sciences Vocational School, Tarsus University, Mersin, Türkiye

A. Vaseashta
International Clean Water Institute, Manassas, VA, USA

D. Radu et al. (eds.), *Countering Hybrid Threats Against Critical Infrastructures*, NATO Science for Peace and Security Series B: Physics and Biophysics, https://doi.org/10.1007/978-94-024-2304-4_9

9.1 Introduction

Chemical and biological threats, which are elements of chemical, biological, radiological, and nuclear (CBRN) security, represent one of the fundamental elements of state security and have a high priority for protecting the critical infrastructure of countries. The ascending chemical and biological threats pose significant challenges to global security and, the economy, requiring advanced protective measures for people in various domains such as medical personnel, soldiers, law enforcement officers, and civilians [1]. These threats, comprising a wide range of toxic substances, require advanced detection, decontamination, and protective strategies due to their complex nature and the potential for crucial impacts on human health and safety [2]. On the other hand, when evaluated in terms of critical infrastructure, even if large-scale terrorist attacks with significant cascading effects have not yet been carried out, the existence of such a scenario requires countries to implement adequate prevention and response measures against existing threats.

The protection against threats is critical, as chemical warfare agents (CWAs), toxic industrial chemicals (TICs), and disease-causing pathogens, can severely damage first responders, leading to potentially serious outcomes which include death after exposure [3]. Traditional protective clothing, such as hazardous materials suits and toxicological agents' protective suits, provide a full barrier against contaminants but often lack the comfort and functionality needed for prolonged use in diverse environments. For instance, while these garments are impermeable to CWAs, they can be heavy, restrict movement, and complicate the decontamination process [4]. The limitations of conventional protective textiles underline the urgent need for innovative materials that can offer enhanced protection without compromising user comfort and operational efficiency.

Electrospun nanofibrous materials have emerged as a promising solution to these challenges due to their unique properties, such as high surface area, tunable porosity, lightness, and functionalization features [5]. These materials can be engineered to combine catalytic activity, selective gas permeability, and sensor functionalities, making them ideal candidates for protective textiles [6]. Electrospinning is a versatile process that creates nanofibers with diameters ranging from tens to hundreds of nanometers, producing fabrics that are lightweight, porous, and capable of integrating various functional parts [5–7]. For instance, multilayer electrospun nanofiber mats with pH-sensitive dyes have indicated great performance in detecting chemical agents, unlocking new possibilities for smart protective clothing that can provide real-time monitoring and instant visual alerts. The integration of these dyes into nanofibers allows for quick colorimetric detection of hazardous substances, raising the wearer's capability to identify and respond to threats urgently [5]. Moreover, the improvement of composite nanofibers integrating metal-organic frameworks (MOFs) has displayed significant potential in neutralizing nerve agents and TICs, further augmenting the protective capabilities of these advanced materials. MOFs are immensely porous materials including inorganic metal nodes connected by organic linkers to create crystalline structures. These constructions exhibit high surface areas and can be adapted to demonstrate specific chemical reactivities that make them effective sorbents and catalysts for neutralizing CWAs. The integration of MOFs into electrospun fibers can improve the decontamination properties of protective textiles by providing a reactive surface that can adsorb and break down toxic chemicals. For example, electrospun nanofibers integrating the MOF UiO-66-NH_2 exhibited high efficacy in degrading nerve agent simulants and TICs, offering an effective solution for next-generation protective clothing [7, 8].

Besides chemical detection and decontamination, electrospun nanofibrous systems can be designed to offer protection against biological threats. The high surface area and changeable porosity of electrospun nanofibers make them efficient filters for catching pathogens, while their ability to be functionalized with antimicrobial agents can further enhance their protective properties [4]. This multifunctionality makes electrospun nanofibers an attractive option for developing protective textiles that can address a wide range of threats.

Furthermore, current research and development in the field of electrospun nanofibers continue to explore new materials and structures that can further improve the performance and functionality of these systems. Advances in electrospinning technology, such as the development of coaxial electrospinning and the incorporation of advanced materials like graphene and carbon nanotubes, are paving the way for the next generation of protective textiles [9, 10]. These innovations hold the potential to create fabrics that are not only protective but also lightweight, breathable, and comfortable, emphasizing the key restrictions of traditional protective clothing. These unique properties of electrospun nanofibrous materials position them as a promising solution to the challenges posed by chemical and biological threats. Their ability to integrate detection, decontamination, and protective functionalities into lightweight and comfortable fabrics makes them ideal candidates for future protective textiles. Continued research and development in this field are essential to fully realize the potential of these advanced materials and to provide active protection for those at risk of exposure to chemical and biological hazards.

9.2 Electrospinning

Electrospinning has emerged as a powerful technique in material science, allowing the production of ultrafine fibers with diameters ranging from nanometers to micrometers. This method is a very significant advancement in fabricating nanostructured materials, offering unique advantages such as high surface area-to-volume ratio, lightness, flexibility, mechanical strength, tunable porosity, and the ability to integrate various functional materials into the fibers. These features make electrospun nanofibers highly desirable and suitable for a wide range of applications, such as biomedical engineering [11, 12], environmental protection [13], energy storage [14], filtration systems [15] and food safety [16]. The process is not only efficient but also remarkably adaptable, allowing for the production of fibers from a variety of polymers and composites, which is key to its widespread usage across different industries. As research in this field continues to expand, so does the potential for discovering new applications and developing existing ones. The ability to adjust the properties of electrospun fibers through the changes in the electrospinning parameters leads to countless possibilities for innovation. This part explains the basic principles and processes of electrospinning, examines the distinctive properties of electrospun nanofibers, and explores the diverse application areas of these remarkable materials.

9.2.1 Basic Principles and Process of Electrospinning

Electrospinning is a technique that uses an electric field to produce fibers from a polymer solution or melt. The electrospinning setup typically consists of a syringe pump that controls the flow rate of the polymer solution, a high-voltage power supply to generate the electric field, and a collector, which is often a flat plate or a rotating drum. The collector can affect the configuration and uniformity of the fibers. A rotating drum, for example, can be used to produce aligned fibers, which are effective in applications requesting specific mechanical properties [17]. The electrospinning process begins by loading a polymer solution into the syringe equipped with a metal needle, which is connected to the high-voltage power supply. When a sufficient voltage is applied, the polymer solution at the needle tip forms a shape known as a Taylor cone. If the electrostatic forces exceed the surface tension of the solution, a charged jet is ejected from the tip. As the jet travels through the air, the solvent evaporates, resulting in the deposition of solid fibers onto a collector [18, 19]. Several crucial parameters influence the process, including the properties of the polymer solution such as viscosity, conductivity, and surface tension, together with the operational parameters such as applied voltage, the distance between the needle and the collector, and the flow rate of the polymer solution [20, 21]. Regulating these parameters empowers control over the fiber diameter, morphology, and mechanical properties. For example, increasing the polymer concentration or reducing the distance between the needle and collector generally results in thicker fibers [6]. Moreover, the diameter of the fibers produced can range from nanometers to micrometers, depending on the specific settings used during the electrospinning process. This versatility allows for the creation of materials with high surface area-to-volume ratios and porosity. These characteristics make electrospinning a powerful tool for fabricating advanced materials tailored for specific applications. Optimizing these parameters is crucial for producing high-quality fibers with the desired properties for various industrial and research applications.

9.2.2 Properties of Electrospun Nanofibers

Electrospun nanofibers display several characteristic properties that make them highly attractive for many advanced applications. One of the significant properties is their high surface area, which is mainly because of their ultrafine diameters ranging from tens to hundreds of nanometers. This property enhances the interaction between the fibers and surrounding materials, making them particularly efficient in applications such as filtration, where capturing fine particles is essential. The small diameter also serves to create highly porous structures with small pore sizes, which further enhances their filtration efficiency and functionality in barrier applications. Along with their structural properties, electrospun nanofibers are known for their mechanical flexibility. Despite their nanoscale dimensions, these fibers can be engineered to show significant tensile strength and elasticity [22]. This combination of strength and flexibility is particularly useful in applications that require durable but lightweight materials, such as wearable protective textiles and biomedical scaffolds. The mechanical properties of these fibers can be regulated by adjusting different parameters during the electrospinning process, such as polymer concentration, solution viscosity, and applied voltage. For example, higher polymer concentrations generally produce thicker, stronger fibers, while the alignment of fibers can be controlled to improve mechanical performance [23]. Another important property of electrospun nanofibers is their ability to be easily functionalized. These fibers can be adjusted during or after the electrospinning process to integrate different functional materials such as nanoparticles, drugs, or enzymes. This capability allows for the creation of composite fibers with increased functionalities, such as enhanced chemical reactivity, electrical conductivity, or specific biological activity [24]. For example, incorporating metallic nanoparticles into electrospun fibers can enhance their performance in catalytic applications or the development of sensors [25].

Overall, the unique properties of electrospun nanofibers such as high surface area, tunable porosity, mechanical flexibility, and the ability to be functionalized make them highly versatile and practical for a wide range of applications. These fibers are particularly effective in fields that require advanced material properties, such as filtration, protective textiles, and biomedical devices. As research in this field moves forward, the potential to create more specialized and high-performing nanofibers grows, offering new opportunities in different industries.

9.2.3 Applications of Electrospun Nanofibers

Electrospun nanofibers have various applications in different industries. One of the significant applications is in the biomedical field, where their high surface area, porosity, and ability to be functionalized enable them to be used in drug delivery systems, tissue engineering scaffolds, and wound dressings. For instance, electrospun nanofibers can be engineered for controlled drug release, which makes them highly effective in targeted treatments [11]. Also, the fibers can mimic the native extracellular matrix, which provides an appropriate environment for cell growth and regeneration [26]. Another significant application area of electrospun nanofibers is filtration systems [27]. The high porosity and small pore size of the fibers make them ideal for air and liquid filtration. They are especially effective in extracting nanoparticles, bacteria, and viruses from air or water, which is crucial in healthcare services, environmental protection, and industrial processes. Furthermore, their lightweight structure and flexibility provide easy integration into existing filtration systems without significantly increasing the weight of the equipment. Moreover, in energy applications, electrospun nanofibers are used in the development of batteries [28], supercapacitors [29], and fuel cells [30]. Their high surface area enhances the electrochemical reactions, leading to enhanced energy storage and conversion efficiency. For example, nanofibers made from conductive polymers or composites are operated as electrodes in batteries and supercapacitors, leading to higher energy densities and longer cycle lives. Another application of electrospun nanofibers is in sensors [31]. The nanofibers can be functionalized with many different materials to detect various chemical and biological agents, making them important in environmental monitoring and medical diagnostics. Their large surface area and tunability allow to creation of sensors that are highly sensitive and selective and can operate under diverse conditions. Lastly, one of the most critical applications of electrospun nanofibers is in protective textiles, particularly for decontamination and protection against chemical and biological threats. The ability to functionalize these fibers with functional agents such as MOFs, metallic nanoparticles, graphene quantum dots (GQDs), etc., allows them to not only filter harmful agents but also neutralize them, providing a dual function of protection and decontamination. This multifunctionality, combined with their lightweight and breathable nature, makes electrospun nanofibers a promising solution for future protective textiles, connecting advanced material science with practical applications in safety and defense [32].

9.3 Electrospun Nanofibers for Protection, Decontamination and Detection

The preceding section of this chapter introduced the concept of nanofibers obtained by electrospinning, outlining their properties and potential applications. This part of the chapter continues with a review of the latest studies on the use of electrospun nanofibers in the field of decontamination, protection, and detection. The term "decontamination" is defined as the process of removing or neutralizing chemical or biological agents so that they no longer pose a danger. The term "protection" is a set of protective measures taken in situations where chemical, biological, radiological, or nuclear (including terrorism) dangers may exist. Finally, the concept of detection refers to the ability of a sensor to detect the presence of a specified component of any threat. The hazardous substances including chemical and biological agents that need to be removed or protected are classified in the section below. Next, the current status of protective and sensing applications using electrospun membranes to eliminate/detect these hazards is summarized.

9.3.1 Chemical and Biological Agents/Substances

It is evident that chemicals serve a plethora of purposes and are utilized as indispensable components in a multitude of manufacturing processes, laboratory procedures, and other operations. It is important to note that individuals engaged in such applications may not always have direct contact with chemicals, yet exposure can still occur as a result of the release of these substances into the environment. The potential for direct or indirect chemical exposure to reach dangerous dimensions for the ecosystem and life depends on the composition of the chemical substance in question. The hazardous chemicals are classified into three categories based on their potential consequences: Chemical Warfare Agents (CWAs), Toxic Industrial Chemicals (TICs), and New and Emerging Unconventional Agents (NTAs) (Fig. 9.1).

Traditionally known as "chemical weapons" or "chemical weapons agents", CWAs were widely used during World War I, killing millions of soldiers and civilians [33]. Since that time, over 99% of CWAs stockpiles have been destroyed following the provisions of the International Chemical Weapons Convention. Currently, 98% of the global population resides in countries that have pledged to eradicate the use of CWs. Nevertheless, instances of CWAs use persist, including in terrorist attacks involving Sarin gas, politically motivated poisoning with liquid Novichok, and battlefield conflicts involving chlorine gas. The CWAs used in warfare are classified as nerve agents, vesicants (blistering agents), blood agents (cyanogenic agents, choking agents (pulmonary agents), riot-control agents (tear gases), psychomimetic agents, and toxins [33, 34]. TICs are of industrial origin that have the

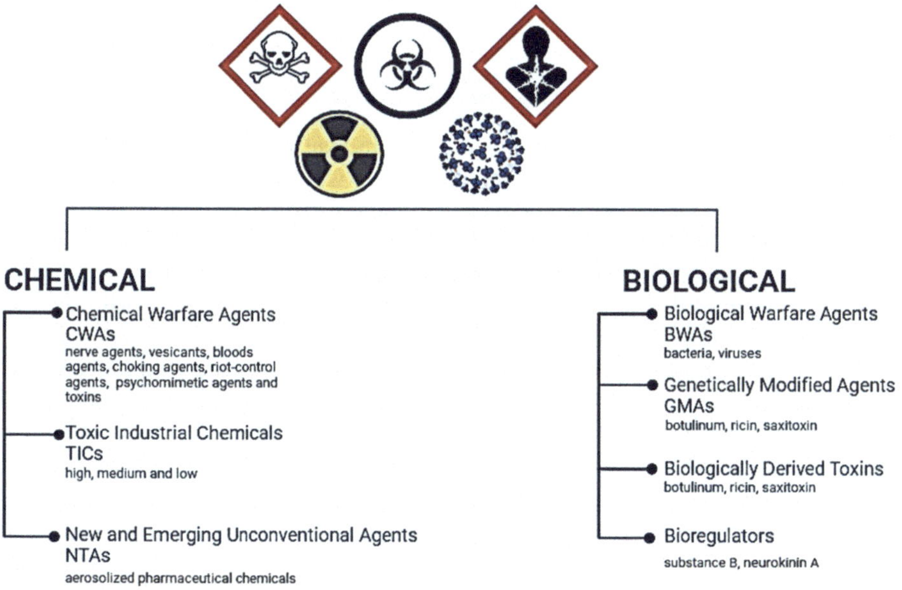

Fig. 9.1 Classification of chemical and biological threats

potential to cause harm when exposed to [35]. According to the TICs guide prepared by The Occupational Safety and Health Act (OSH Act), the number of listed chemicals is approximately 100 and they are categorized as high, medium, and low according to their hazard index [36]. Some of these are ammonia, ethylene oxide, phosgene, carbon disulfide, formaldehyde, sulfuric acid, acetone cyanohydrin, carbon monoxide, chloroacetone, arsenic trichloride, ethyl chloroformate, nitric oxide, etc. [36]. Countless chemicals can harm human life, and new chemical threat agents or toxicants not specified in other classes are considered in NTAs. An example of a threat in the NTAs class is aerosolized pharmaceutical chemicals such as fentanyl and other opioids. These chemicals, in aerosol or gas form, can rapidly fill enclosed spaces and incapacitate crowds.

When it comes to biological hazards, people working in healthcare, food processing, sewage works, laboratories, and agriculture are in direct contact with biological agents of animal, plant, or microbial origin or products containing biological agents. These microorganisms have the potential to pose significant biological threats by causing infectious diseases that have the potential to spread and cause emergencies [37]. There is also the existence of devices or agents used or intended to be used in a deliberate attempt to spread disease-producing organisms or toxins, referred to as biological weapons, by aerosol, food, water, or insect vectors. The mechanisms of action here tend to be through infection or poisoning. This type of threat, which involves the deliberate release of biological weapons to cause death or disease in humans, animals, or plants, is known as bioterrorism and the biological agents used may include bacteria, viruses, fungi, toxins, bioregulators and genetically modified agents (Fig. 9.1.) [38, 39].

9.3.2 Protection

Protective clothing and masks are essential equipment used to protect individuals against biological and chemical threats in various work environments such as the healthcare sector, laboratories, chemical production facilities, mining operations, fire departments, military and security services. The necessity for further advancement in the development of protective clothing has re-emerged in recent years, particularly in light of the global pandemic caused by the SARS-CoV-2 virus and the deliberate utilization of CWAs [40].

The current generation of protective textiles comprises two distinct layers, the outer and inner, which collectively serve to create a barrier against harmful agents. The outer layer is designed to repel the liquid or solid phases of agents, while the inner layer is intended to adsorb and separate any agents that have been absorbed or passed through. The inner layer of protective clothing typically comprises activated carbon powder, beads, or fibers that serve to adsorb toxic substances. The fiber structures utilized in the formation of these layers can be obtained through the electrospinning method. By modifying the process parameters, it is feasible to generate air-permeable, resilient, and gas-separating protective textile materials with electrospun membranes that can be manufactured with varying nanofiber diameters and porosities [41, 42].

The most commonly used textile materials such as protective clothing are masks, gloves, and clothes. Especially with the COVID-19 pandemic that has recently been faced all over the world, the number of studies in this field has increased rapidly. The development of advanced antibacterial nanomaterials has gained great importance in preventing pathogenic microbial contamination and providing biological protection. In a recent study in this area, polyimide electrospun nanofibrous membranes (PI-ENMs) produced by electrospinning were optimized in terms of mechanical strength, filtration efficiency, air permeability, and reusability. NaCl particles with different sizes were used as virus models. They showed that PI-ENMs exhibited 99.42% filtration efficiency for 20 nm particles and had the potential to provide high filtration performance against viruses of various sizes, including COVID-19, which is approximately 120 nm in size. In addition, by coating a polydimethylsiloxane layer on the surface of thermally stable PI-ENMs via chemical vapor deposition, fire resistance properties were achieved without compromising filtration efficiency and resistance [43].

In the studies conducted to date, it has been reported that electrospun nanofibers produced using different polymer compositions generally achieve more effective decontamination when used by imparting functional properties. In this context, especially in recent years, MOFs are seen as the most dominant functional agents for the elimination of CWAs and pathogens. Among these, zirconium-based MOFs such as UiO-66, UiO-66-NH_2, and MOF-808 have attracted significant attention due to their Lewis acidic nodes that serve as active sites for a wide range of catalytic reactions [44–46]. Cheng et al. studied polyimide-based electrospun nanofibers doped with MOF-808 nanoparticles and reported that nanofiber membranes prepared from the hybrid fibers exhibited 90.9% efficacy on disinfection of CEES, a mustard gas simulant [45]. Jiang et al. developed a protective material by in situ growth of UiO-66-NH_2 onto aramid nanofibers (ANFs) and assembly of this hybrid structure into a porous aerogel matrix. The high surface area, the open and interconnected cellular structure developed together with the high MOF doping and porous structure enhanced the catalytic degradation of CWAs with a removal rate of 98.9% in a short half-life of 8.15 min [44].

Apart from MOFs, organic N-halamine groups can also be used for functionalization. In a study, polyurethane (PU) nanofibers were functionalized with an organic N-halamine, a potential functional group with broad inhibitory activities among known decontaminable moieties. PU as electrospun nanofibers was chosen to better meet the properties of the actual protective clothing material such as clothes, masks, and gloves due to its high modulus and tensile strength. To evaluate the decontamination ability, 2-chloroethyl ethyl sulfide, and demeton-S-methyl chemicals were used to simulate HD-type blister agent and V-type nerve gas, respectively. The decontamination efficiency of the azido-PU electrospun nanofiber mat was determined as 69% and 16% for 2-chloroethyl ethyl sulfide and demeton-S-methyl, respectively, with a 1/1 molar ratio for 2 h at ambient conditions [47].

Table 9.1 Recent studies about electrospun nanofiber mats designed for protective textiles against chemical and biological threats

Polymer	Additive	Chemical/Biological agent	Ref
Poly(ε-caprolactone)	n-phosphonium chitosan	MHV-3 coronavirus	[49]
Poly(vinyl alcohol) Chitosan Poly-ε-caprolactone	Silver nanoparticles	Mouse hepatitis virus (MHV-3)	[50]
Polyurethane	MOF (ZIF-8)	*E. coli* *S. aureus*	[51]
Polyimide	Zirconium-MOFs	2-chloroethyl ethyl sulfide (mustard gas simulant)	[45]
Polyacrylonitrile	$Zr(OH)_4$ MOF-808	Dimethyl 4-nitrophenyl phosphate (nerve agent simulant) 2-chloroethyl ethyl sulfide (blistering agent simulant)	[52]
Polyacrylonitrile	Mono-amidoxime	Toxic organophosphates	[53]
Poly(4-vinyl pyridine)	Oxime	Dimethyl-4-nitrophenyl phosphate	[48]
Chitin	UiO-66	Methyl paraoxon (nerve agent simulant)	[54]
Polyacrylonitrile	Polyimide	N-methyl pyrrolidone N, N-dimethylformamide	[55]
Polyvinylidene fluoride Polymethylmethacrylate Polystyrene	$Ti(OH)_x$ $Zr(OH)_4$ $Ce(OH)_4$	Toxic organophosphates	[56]
Polycaprolactone chitosan	MOF graphene quantum dot	*E. coli* *S. aureus*	[57]

In the context of the functionalization of electrospun nanofibers, oxime doping also provides super nucleophilic materials that rapidly and irreversibly inactivate nerve agents and their derivatives [48]. Table 9.1 presents recent studies on electrospun nanofiber mats prepared as protective textiles against chemical/biological threats.

9.3.3 Detection

Traditional detection methods are based on instruments such as gas chromatography-mass spectrometry, quantum cascade laser, infrared laser spectroscopy, and fluorescence spectrometry. Although these instruments are considered the gold standard, they have significant drawbacks such as being expensive, time-consuming, and bulky, requiring extensive sample preparation before use, and requiring trained personnel to use them. With developments in nanotechnology and materials science, sensor systems embedded with nano-engineered materials have attracted the attention of scientists due to their size reduction, ease of use, high sensitivity, acceptable selectivity, ease of transport, low cost, and rapid response [58]. Among these nano-engineered matrices, nanofibers produced by electrospinning have high usage potential according to their high surface area-to-volume ratio, porosity, lightness, flexibility, high strength, adaptability, easy producibility, and suitability for surface modification.

The mechanism that takes place in nanofibers as sensors can be summarized as follows: target gas molecules come into contact with the nanofiber surface, and the surface undergoes physical or chemical reactions such as adsorption and desorption which lead to a change in resistance or capacitance after charge transfer. The resistance or capacitance changes in the sensor used can be used as signals to easily determine the types and concentrations of gases [59].

The use of electrospun nanofibers as sensor systems is one of the ideal remedies for monitoring and detecting all types of chemical threats (volatile organic compounds (VOCs), TICs, and CWAs) in and around critical infrastructures. Currently, nanofibers are used especially for the detection of VOCs [31]. In a recent study, Yang et al. synthesized tungsten trioxide (WO_3) based nanofibers by electrospinning for the detection of methyl salicylate (MeSa), known as a volatile plant pathogen biomarker and chemical warfare agent simulator. To increase the MeSa sensitivity of WO_3 nanofibers, doping was done with Au and Pt metals [58]. You et al. developed a wearable and eye-readable sensor for VOCs (trans-2-hexenal, ethanol, ethyl acetate, benzaldehyde, 3-carene, and terpinolene) detection by depositing the nanoscale gas-sensitive composite (dye/MOF) onto electrospinning polyacrylonitrile nanofiber membrane by vacuum filtration. Here, the electrospun membrane served as a carrier matrix, while the porous structure and unsaturated metal regions of the MOF nanoparticles provided the opportunity for sensitive VOCs detection [60]. In another study, Yagmurcukardes et al. used quartz crystal microbalances (QCM) to activate electrospun polyacrylonitrile/polypyrrole nanofibers and investigated their sensing properties against volatile organic chloroform, dichloromethane, carbon tetrachloride, benzene, toluene, and xylene [61]. Recent studies indicate that electrospun nanofibers continue to be functionalized with different nanoparticles for VOCs sensor applications.

Another application of electrospun nanofibers is pesticide detection by electrochemical immunosensors or biosensors. Since pesticides are a broad category of substances with a range of acute and chronic toxicity [62], an efficient and sensitive detection approach is desirable. Electrochemical transducers are widely used in biosensors for pesticide detection due to their high sensitivity, and electrospun nanofibers are used in surface modification of electrodes due to their unique properties [63–65]. For example, a label-free electrochemical biosensor based on the directly immobilizing nanobody onto functionalized polyvinyl alcohol/gelatin-gold nanoparticles nanofiber membrane was designed by Hu et al. for the detection of quinalphos pesticide [66]. In another study, graphite containing poly (lactic acid)/poly (butylene adipate-co-terephthalate) electrospun nanofibers were used as electrodes in the electrochemical immunosensor to be used in pesticide detection. The thermal treatment of graphite-containing fibers increased the adhesion surface which occurs in the adsorption of the analyte on the electrode, which improved the peak current in the electrochemical tests [67].

Table 9.2 Recent studies about electrospun nanofiber mats designed for sensing/biosensing applications for the detection of chemical and biological threats

Polymer	Additive	Chemical/Biological Agent	Ref
Polyacrylonitrile	MOF-199	Methanol	[69]
Polyacrylonitrile	ZIF-67	Methanol, ethanol, acetone	[70]
Polyacrylonitrile	Adenosine triphosphate with metal terbium	Norfloxacin	[71]
Polyvinyl alcohol Citric acid	–	Parathion	[72]
Polyacrylonitrile	Ti_3C_2 MXene/TiO_2	N, N-dimethylformamide	[73]
Polyvinylpyrrolidone $La(NO_3)_3{\cdot}6H_2O$ $Fe(NO_3)_3.9H_2O$	$Ti_3C_2T_x$ MXene	Sulfur dioxide	[74]
Polycaprolactone Polylactic acid	Anthocyanin	Ammonia	[75]
Polyvinylpyrrolidone $SnCl_2{\cdot}2\,H_2O$	Pd@ZIF-67	Triethylamine	[76]
Chitosan	Tetrakis (4-carboxyphenyl) porphyrin	Mercury	[77]

Another sensor application realized with electrospun nanofibers is colorimetric sensors. These sensors can detect external stimuli through color changes and can be used for various functions including body fluid property identification, temperature detection, disease diagnosis, and harmful gas detection as wearable sensors with a flexible material design such as nanofibers in the substrate. Maeng et al. designed an electrospun mat sensor by taking advantage of the flexibility of polyurethane and the colorimetric characteristics of curcumin for the colorimetric detection of ammonia. It was reported that the fiber mat showed a clear color change from orange to red when exposed to ammonia gas, and the addition of curcumin facilitated electrospinning, contributing to the design of a more flexible and reusable stretchable sensor material [68]. Current studies on electrospun nanofiber sensor applications for the detection of chemical/biological threats are also presented in Table 9.2.

9.4 Conclusion and Future Perspective

The existence of threats posed by terrorists to the international community, collective well-being, security, and safety is undoubtedly reminded of itself every day around the world. The threats of these entities become a much more important issue when they target critical infrastructure, which refers to the infrastructure elements that are vital for the functioning and security of countries. In addition, due to industrial-scale pollution, the existence of threats that disrupt individual or social security, harm the environment, affect health, and cause economic problems is inevitable. For this reason, there is an increasing need every day to develop approaches to eliminate/prevent/protect the existence of these threats that negatively affect society both on a large scale and on an individual scale.

Academic studies play an important role in protecting critical infrastructure and protecting people, the environment, and the ecosystem from current threats. Research in this field provides scientific knowledge, innovative solutions, and strategies to ensure the security of these infrastructures and reduce risks. As part of this, in this book chapter, we talked about electrospun nanofibers, a nanotechnological material developed to target chemical and biological agents that are among the threats that can be encountered in daily life around the world and that are likely to disrupt critical infrastructure due to possible terrorist attacks.

Thin-layer fibrous mats formed by stacking nano-sized fibers obtained by electrospinning have many unique features such as high surface area, porosity, mechanical resistance, flexibility, lightness, easy integration into textile materials, and fast and large-scale production. In addition, the surface of these fibers prepared on a polymer basis is suitable for easy modification with different functional groups. They also provide a suitable environment for the production of composite fibers that can carry these materials with nanoparticles/biological agents/active molecules that can be added to polymer solutions. On the other hand, fibers with different morphological structures can be produced depending on the modification of the traditional electrospinning technique and allow the formation of hybrid structures by combining them with newer production technologies such as 3D printing.

The use of electrospun nanofibers developed as a result of experimental approaches and evaluation of the obtained results has shown great impact against chemical and biological threats. Electrospun nanofibers used for protection, detection, and decontamination against these threats mostly serve as support material. They can be integrated into existing systems due to their lightness and thinness and offer the opportunity to be carried without adding weight. Thanks to their porous structure and large surface areas, they can provide a suitable environment for the retention and permeability of the specified components. They can show high sensitivity to certain components with the addition of functional groups or nanoparticles.

When recent studies on the use of electrospun nanofibrous structures in this field are examined, it is seen that they are combined with new generation nanotechnological tools. One of the most recent materials that attract attention in this field is MOFs. MOFs are defined as highly porous structures formed by the combination of metal ions or metal clusters with organic ligands, and their combination with electrospun nanofibers is an important development area in the field of materials science and nanotechnology. The combination of MOFs and electrospun fibers has enabled significant advances in the development of sensors and protective textiles against chemical and biological agents. MOFs are characterized by their high selectivity and adsorption capacity, while electrospun fibers provide high surface areas and fast response times. The combination of these two technologies enables the development of more effective and functional sensors and protective clothing, making them valuable for protecting critical infrastructure, including environmental and industrial safety, health, and defense applications.

Other current studies that draw attention in the scientific field are on the development of electrospun nanofiber-based defense methods against biological threats, the importance of which has been understood even more after the COVID-19 pandemic. Here, protective textiles are being developed against different pathogens that have the potential to cause epidemics and fatal diseases. The most important features that these textiles are desired to have been the development of breathable, self-cleaning, lightweight, thermally comfortable, and nonflammable materials for easy use.

The direction of studies on the protection and detection applications of electrospun fibers against chemical and biological threats continues toward the development of new hybrid materials. Potential studies to produce these new generation materials can be listed as follows: faster and more sensitive response to environmental changes, providing self-healing properties to protective textiles, composites capable of providing both chemical detection and biological protection, innovative designs that will enable the detection of chemical and biological agents at lower concentrations, lightweight, flexible and breathable materials that aim to increase user comfort, environmentally friendly production methods, etc. As a result, the future direction of studies on the use of electrospun fibers will be shaped in line with technological innovations, material science developments, and environmental needs. Advances in this field have the potential to provide more effective, functional, and sustainable solutions in both critical infrastructure and industrial and daily life. With developing technologies, the application areas of these materials will expand and the capacity to provide innovative solutions will increase.

References

1. Schreuder-Gibson HL, Truong Q, Walker JE, Owens JR, Wander JD, Jones WE (2003) Chemical and biological protection and detection in fabrics for protective clothing. MRS Bull [Internet] 28:574–578. https://link.springer.com/article/10.1557/mrs2003.168
2. Tin D, Cheng L, Shin H, Hata R, Granholm F, Braitberg G et al (2023) A descriptive analysis of the use of chemical, biological, radiological, and nuclear weapons by violent non-state actors and the modern-day environment of threat. Prehosp Disaster Med [Internet] 38:395–400. https://pubmed.ncbi.nlm.nih.gov/37185132/
3. McElroy CS, Day BJ (2016) Antioxidants as potential medical countermeasures for chemical warfare agents and toxic industrial chemicals. Biochem Pharmacol [Internet] 100:1. /pmc/articles/PMC4744107/
4. Ramaseshan R, Sundarrajan S, Liu Y, Barhate RS, Lala NL, Ramakrishna S (2006) Functionalized polymer nanofibre membranes for protection from chemical warfare stimulants. Nanotechnology [Internet] 17:2947. https://iopscience.iop.org/article/10.1088/0957-4484/17/12/021
5. Golmohammadi Rostami S, Sorayani Bafqi MS, Bagherzadeh R, Latifi M, Gorji M (2015) Multi-layer electrospun nanofiber mats with chemical agent sensor function. J Ind Text [Internet] 45:467–480. https://www.researchgate.net/publication/281175259_Multi-layer_Electrospun_Nanofiber_Mats_with_Chemical_Agent_Sensor_Function

6. Haider A, Haider S, Kang IK (2018) A comprehensive review summarizing the effect of electrospinning parameters and potential applications of nanofibers in biomedical and biotechnology. Arab J Chem 11:1165–1188
7. Lu AX, McEntee M, Browe MA, Hall MG, Decoste JB, Peterson GW (2017) MOFabric: electrospun nanofiber mats from PVDF/UiO-66-NH2 for chemical protection and decontamination. ACS Appl Mater Interfaces [Internet]. 9:13632–13636. https://pubmed.ncbi.nlm.nih.gov/28355051/
8. Islamoglu T, Chen Z, Wasson MC, Buru CT, Kirlikovali KO, Afrin U et al (2020) Metal-organic frameworks against toxic chemicals. Chem Rev [Internet] 120:8130–8160. https://pubmed.ncbi.nlm.nih.gov/32207607/
9. Anusiya G, Jaiganesh R (2022) A review on fabrication methods of nanofibers and a special focus on the application of cellulose nanofibers. Carbohydr Polym Technol Appl 4:100262
10. Qin X (2017) Coaxial electrospinning of nanofibers. Electrospun nanofibers [Internet]. Elsevier Inc., pp 41–71. https://www.researchgate.net/publication/311996641_Coaxial_electrospinning_of_nanofibers
11. Demir D, Güreş D, Tecim T, Genç R, Bölgen N (2018) Magnetic nanoparticle-loaded electrospun poly(ε-caprolactone) nanofibers for drug delivery applications. Appl Nanosci [Internet] 8:1461–1469. https://link.springer.com/article/10.1007/s13204-018-0830-9
12. Demir D, Özdemir S, Ceylan S, Yalcin MS, Sakım B, Bölgen N (2022) Electrospun composite nanofibers based on poly (ε-caprolactone) and styrax liquidus (liquidambar orientalis miller) as a wound dressing: preparation, characterization, biological and cytocompatibility results. J Polym Environ 30:2462–2473
13. Zhang W, He Z, Han Y, Jiang Q, Zhan C, Zhang K et al (2020) Structural design and environmental applications of electrospun nanofibers. Compos Part A Appl Sci Manuf 137:106009
14. Zhu J, Yan C, Li G, Cheng H, Li Y, Liu T et al (2024) Recent developments of electrospun nanofibers for electrochemical energy storage and conversion. Energy Storage Mater 65:103111
15. Yu Z, Fan T, Liu Y, Li L, Liu J, Yang B et al (2024) Efficient air filtration through advanced electrospinning techniques in nanofibrous materials: a review. Sep Purif Technol 349:127773
16. Goksen G, Demir D, Echegaray N, Bangar SP, Gomes da Cruz A, Shao P et al (2024) New insights of active and smart natural-based electrospun mats for food safety in meat and meat products. Food Biosci 59:104159
17. Robinson AJ, Pérez-Nava A, Ali SC, González-Campos JB, Holloway JL, Cosgriff-Hernandez EM (2021) Comparative analysis of fiber alignment methods in electrospinning. Matter [Internet] 4:821–844. https://pubmed.ncbi.nlm.nih.gov/35757372/
18. Doshi J, Reneker DH (1995) Electrospinning process and applications of electrospun fibers. J Electrost 35:151–160
19. Qin XH, Wang SY (2006) Filtration properties of electrospinning nanofibers. J Appl Polym Sci [Internet] 102:1285–1290. https://onlinelibrary.wiley.com/doi/full/10.1002/app.24361
20. Vaseashta A, Demir D, Sakım B, Aşık M, Bölgen N (2022) Hierarchical integration of 3D printing and electrospinning of nanofibers for rapid prototyping. Electrospun Nanofibers Princ Technol Nov Appl 2022:631–655
21. Bölgen N, Demir D, Aşık M, Sakım B, Vaseashta A (2022) Introduction and fundamentals of electrospinning. In: Vaseashta A, Bölgen N (eds) Electrospun nanofibers principles, technology and novel applications. Springer, Cham, pp 3–34
22. He JH, Wan YQ, Xu L (2007) Nano-effects, quantum-like properties in electrospun nanofibers. Chaos Solitons Fract 33:26–37
23. Dokuchaeva AA, Timchenko TP, Karpova EV, Vladimirov SV, Soynov IA, Zhuravleva IY (2022) Effects of electrospinning parameter adjustment on the mechanical behavior of poly-ε-caprolactone vascular scaffolds. Polymers (Basel) 14:349. https://www.mdpi.com/2073-4360/14/2/349/htm
24. Medeiros GB, Lima FA, de Almeida DS, Guerra VG, Aguiar ML (2022) Modification and functionalization of fibers formed by electrospinning: a review. Membranes (Basel) 12:861. https://www.mdpi.com/2077-0375/12/9/861/htm

25. Korotcenkov G (2021) Electrospun metal oxide nanofibers and their conductometric gas sensor application. Part 1: nanofibers and features of their forming. Nanomaterials [Internet] 11:1544. https://www.mdpi.com/2079-4991/11/6/1544/htm
26. Anjum S, Li T, Arya DK, Ali D, Alarifi S, Yulin W et al (2023) Biomimetic electrospun nanofibrous scaffold for tissue engineering: preparation, optimization by design of experiments (DOE), in-vitro and in-vivo characterization. Front Bioeng Biotechnol 11:1288539
27. Saleh M, Demir D, Ozay Y, Yalvac M, Bolgen N, Dizge N (2021) Fabrication of basalt embedded composite fiber membrane using electrospinning method and response surface methodology. J Appl Polym Sci [Internet]. 138:50599. https://onlinelibrary.wiley.com/doi/full/10.1002/app.50599
28. Senthilkumar SH, Ramasubramanian B, Rao RP, Chellappan V, Ramakrishna S (2023) Advances in electrospun materials and methods for Li-ion batteries. Polym Int 15:1622. https://www.mdpi.com/2073-4360/15/7/1622/htm
29. Liang J, Zhao H, Yue L, Fan G, Li T, Lu S et al (2020) Recent advances in electrospun nanofibers for supercapacitors. J Mater Chem A [Internet] 8:16747–16789. https://pubs.rsc.org/en/content/articlehtml/2020/ta/d0ta05100d
30. Liu Z, Gu Y, Bi L (2023) Applications of electrospun nanofibers in solid oxide fuel cells—a review. J Alloys Compd 937:168288
31. Bölgen N, Demir D, Vaseashta A (2015) Nanofibers for the detection of VOCs. Nanosci Adv CBRN Agents Detect Inf Energy Secur [Internet] 39:159–165. https://link.springer.com/chapter/10.1007/978-94-017-9697-2_16
32. Lu P, Ding B (2008) Applications of electrospun fibers. Recent Pat Nanotechnol [Internet] 2:169–182. https://pubmed.ncbi.nlm.nih.gov/19076051/
33. Ganesan K, Raza SK, Vijayaraghavan R (2010) Chemical warfare agents. J Pharm Bioallied Sci 2:166
34. Brennan RJ, Waeckerle JF, Sharp TW, Lillibridge SR (1999) Chemical warfare agents: emergency medical and emergency public health issues. Ann Emerg Med 34:191–204
35. Hincal F, Erkekoğlu P (2006) Toxic industrial chemicals (TICs)—chemical warfare without chemical weapons. Fabad J Pharm Sci 31:220–228
36. Toxic Industrial Chemicals (TICs) Guide | Occupational Safety and Health Administration
37. Rossodivita A, Visconti A, Saporito T, Rizzardini G (2019) Bioterrorism: toxins as potential biological weapons—an emerging global health threat. Int J Infect Dis 79:55
38. ASTM International (2016) ASTM E96 / E96M-05, Standard Test Methods for Water Vapor Transmission of Materials. ASTM Int
39. Blatny JM (2022) Biological threats. Cham, Adv Sci Technol Secur Appl. Springer, pp 47–78
40. Cheung YH, Ma K, Van Leeuwen HC, Wasson MC, Wang X, Idrees KB et al (2021) Immobilized regenerable active chlorine within a zirconium-based MOF textile composite to eliminate biological and chemical threats. J Am Chem Soc 143:16777–16785
41. Lee J, Seo J, Cho KM, Heo J, Jung H, Park S et al (2022) Ultralight and ultrathin electrospun membranes with enhanced air permeability for chemical and biological protection. ACS Appl Mater Interfaces 14:32522–32532
42. Schreuder-Gibson HL, Truong Q, Walker JE, Owens JR, Wander JD, Jones WE (2003) Chemical and biological protection and detection in fabrics for protective clothing. MRS Bull 28:574–578
43. Li Y, Zhong L-B, Zhang QJ, Guo CY, He BQ, Younas M et al (2024) Highly breathable and durable waterproof polyimide electrospun nanofibrous membrane for potential reusable protective clothing application: preparation, characterization, and performance. J Memb Sci 693:122354
44. Jiang N, Liu H, Zhao G, Li H, Yang S, Xu X et al (2023) Aramid nanofibers supported metal-organic framework aerogel for protection of chemical warfare agent. J Colloid Interface Sci 640:192–198

45. Cheng Z, Li X, Zhang L, Yuan Z, Zheng H, Guo H et al (2023) Flexible and MOF-808 uniformly assembled polyimide nanofiber membranes with excellent mechanical properties for rapid degradation of chemical warfare agents. Chem Eng J 475:145912
46. Ma K, Cheung YH, Kirlikovali KO, Wang X, Islamoglu T, Xin JH et al (2023) Protection against chemical warfare agents and biological threats using metal-organic frameworks as active layers. Accounts Mater Res 4:168–179
47. Choi J, Moon DS, Ryu SG, Lee B, Bin YW, Lee KJ (2018) N-chloro hydantoin functionalized polyurethane fibers toward protective cloth against chemical warfare agents. Polymer (Guildf) 138:146–155
48. Biswas P, Shuster DB, Sonmez Baghirzade B, Scanga RA, Harris SA, Tran CN et al (2023) Oxime-functionalized, nonwoven nanofabrics for rapid, inexpensive decontamination of a nerve agent simulant. ACS Appl Nano Mater [Internet] 6:3425–3434. https://pubs.acs.org/doi/abs/10.1021/acsanm.2c05216
49. de Moura Junior CF, Ochi D, Calais GB, Rocha Neto JBM, d'Ávila MA, Beppu MM (2024) Electrospun nonwoven fabric of poly(ε-caprolactone)/n-phosphonium chitosan for antiviral applications: fabrication, characterization, and potential efficacy. Int J Biol Macromol 278:134861
50. de Freitas CF, Souza PR, Jacinto GS, Braga TL, Ricken YS, Souza GK et al (2024) Silver nanoparticles in situ synthesized and incorporated in uniaxial and core–shell electrospun nanofibers to inhibit coronavirus. Pharmaceutics [Internet] 16:268. https://www.mdpi.com/1999-4923/16/2/268/htm
51. Cai X, Gao L, Wang J, Li D (2023) MOF-integrated hierarchical composite fiber for efficient daytime radiative cooling and antibacterial protective textiles. ACS Appl Mater Interfaces 15:8537–8545
52. Wu T, Qiu F, Xu R, Zhao Q, Guo L, Chen D et al (2023) Dual-function detoxifying nanofabrics against nerve agent and blistering agent simulants. ACS Appl Mater Interfaces 15: 1265–1275
53. Wang L, Jiang P, Niu YL, Cao JP, Guo TY (2024) Fabrication of molecularly imprinted electrospun nanofibers with mono-amidoxime functional ligand for efficient decontamination of toxic organophosphates. Chinese J Polym Sci English Ed [Internet] 42:446–456. https://link.springer.com/article/10.1007/s10118-024-3075-y
54. Seo JY, Song Y, Lee JH, Na J, Baek KY (2024) Robust and highly reactive membranes for continuous disposal of chemical warfare agents: effects of nanostructure and functionality in MOF and nanochitin aerogel composites. Carbohydr Polym 324:121489
55. Liu X, Cao J, Chen J, Gao Q, Zheng H, Gao M et al (2024) Mechanically robust superhydrophobic and oleophobic nanofiber membranes with breathability, solvent-resistance, and high-temperature-resistance. Mater Des 244:113153
56. Dwyer DB, Liu J, Gomez JC, Tovar TM, Davoodabadi A, Bernier WE et al (2019) Metal hydroxide/polymer textiles for decontamination of toxic organophosphates: an extensive study of wettability, catalytic activity, and the effects of aggregation. ACS Appl Mater Interfaces [Internet] 11:31378–31385. https://pubs.acs.org/doi/abs/10.1021/acsami.9b10440
57. Zou Y, Weng J, Qin Z, Zhang Y, Ji S, Zhang H (2023) Metal–organic framework- and graphene quantum dot-incorporated nanofibers as dual stimuli-responsive platforms for day/night antibacterial bio-protection. Chem Eng J 473:145365
58. Yang B, Thi Hanh To D, Sobolak D, Mendoza ER, Myung NV (2024) High-performance methyl salicylate gas sensor based on noble metal (Au, Pt) decorated WO_3 nanofibers. Sensors Actuators B Chem 413:135741
59. Zhang X, Zhang J, Li C, Zhang X, Yun J, Cao D (2024) A review on nanofiber-based composites for toxic and flammable gas sensing. Adv Compos Hybrid Mater [Internet] 7:1–29. https://link.springer.com/article/10.1007/s42114-024-00922-6
60. You Z, Zhao M, Lu H, Chen H, Wang Y (2024) Eye-readable and wearable colorimetric sensor arrays for in situ monitoring of volatile organic compounds. ACS Appl Mater Interfaces [Internet]. 16:19359–19368. https://pubs.acs.org/doi/full/10.1021/acsami.4c00312

61. Yagmurcukardes N, Ince Yardimci A, Yagmurcukardes M, Capan I, Erdogan M, Capan R et al (2023) Electrospun polyacrylonitrile (PAN)/polypyrrole (PPy) nanofiber-coated quartz crystal microbalance for sensing volatile organic compounds. J Mater Sci Mater Electron [Internet] 34:1–12. https://link.springer.com/article/10.1007/s10854-023-11281-1
62. Kaur R, Choudhary D, Bali S, Bandral SS, Singh V, Ahmad MA et al (2024) Pesticides: an alarming detrimental to health and environment. Sci Total Environ 915:170113
63. Iftikhar FJ, Shah A, Wali Q, Kokab T (2023) Advancements in nanofiber-based electrochemical biosensors for diagnostic applications. Biosens [Internet] 13:416. https://www.mdpi.com/2079-6374/13/4/416/htm
64. Halicka K, Cabaj J (2021) Electrospun nanofibers for sensing and biosensing applications—a review. Int J Mol Sci [Internet] 22:6357. /pmc/articles/PMC8232165/
65. Hara TO, Singh B (2021) Electrochemical biosensors for detection of pesticides and heavy metal toxicants in water: recent trends and Progress. ACS Environ Sci Technol Water [Internet] 1:462–478. https://pubs.acs.org/doi/full/10.1021/acsestwater.0c00125
66. Hu JC, Wen P, Wang Y, Yang JY, Xiao ZL, Xu ZL et al (2024) Fabrication of a label-free electrochemical immunosensor by functionalized nanofiber membrane for the ultrasensitive detection of quinalphos. Food Control 162:110423
67. Pozza Junior MC, Rosenberger AG, da Silva FF, Dragunski DC, Muniz EC, Caetano J (2024) Application of a PLA/PBAT/Graphite sensor obtained by electrospinning on determination of 2,4,6-trichlorophenol. Environ Technol [Internet] 45:2388–2401. https://pubmed.ncbi.nlm.nih.gov/36734624/
68. Maeng B, Kim S, An H, Jung D (2023) A tough and stretchable colorimetric sensor based on a curcumin-loaded polyurethane electrospun fiber mat for hazardous ammonia gas detection. Sensors Actuators B Chem 394:134420
69. Niu B, Zhai Z, Yu S, Li T, Wang J, Zhou Y et al (2023) Preparation of MOF-199/polyacrylonitrile nanofiber membrane and its application in the preparation of flexible VOC gas sensors. Microchem J 191:108815
70. Zhai Z, Sun Y, Hao X, Li C (2023) Capacitive gas sensors based on a ZIF-67/PAN nanofiber membrane to detect volatile organic compounds. Appl Surf Sci 621:156833
71. Wang J, Feng Y, Zhao X, Tian Y, Duan Y (2023) Electrospun nanofiber-based indicator paper sensing platform for fluorescence and visualization detection of norfloxacin. Biosens Bioelectron 238:115562
72. Yin WJ, Zhang JX, Wang H, Wang Y, Zeng X, Xu ZL, Yang JY, Xiao ZL, Hammock BD, Wen P (2023) A highly sensitive electrochemical immunosensor based on electrospun nanocomposite for the detection of parathion. Food Chem 404:134371
73. Wang T, Zhu L, Yue Y, Asghari MR, Samani BH, Yamamoto T et al (2024) N,N-dimethylformamide detection and refractive index sensing using an electrospun polymer/Ti3C2 MXene-TiO2 modified optical fiber sensor. Sensors Actuators B Chem 417:136143
74. Zhou L, Niu C, Hu Y, Zhang H, Shao X, Ding Z et al (2024) High-performance SO2 gas sensor based on MXene/LaFeO3 nanotubes by electrospinning technology. J Mater Sci Mater Electron [Internet]. 35:1–12. https://link.springer.com/article/10.1007/s10854-024-13030-4
75. Al-Qahtani SD, Al-Senani GM (2024) Development of toxic gas sensor from anthocyanin-embedded polycaprolactone-co-polylactic acid nanofibrous mat. Int J Biol Macromol 267:131649
76. Wang Y, Liu Z, Li Y, Liu Y, Liang X, Liu F et al (2024) Trace PdO and co-MOF derivative-modified SnO2 nanofibers for rapid triethylamine detection with little humidity disturbance. Sensors Actuators B Chem 403:135239
77. Ma Y, Liu X, Fu Z, Zhang L, Guo Y, Wang H et al (2024) Sensitive and environmentally friendly sensor on porphyrin functionalized chitosan fiber membrane for Hg2+. Nano Res [Internet] 17(11):1–6. https://link.springer.com/article/10.1007/s12274-024-6890-0

Chapter 10
Cybersecurity Maturity Models: A Systematic Literature Review

Alvaro Rocha, Fadele Ayotunde Alaba, Haliru Musa, Maria Jose Sousa, Jose Braga de Vasconcelos, and Rúben Pereira

Abstract Cybersecurity maturity models (CMMs) are essential for organizations to assess and improve their cybersecurity posture. This systematic literature review provides a comprehensive overview of existing CMMs, their applications, and implications for research and practice. The methodology involves a systematic search of academic databases, yielding a corpus of relevant studies that were analyzed and synthesized. Key findings highlight the diversity of CMMs available, their common components and frameworks, and the challenges and opportunities associated with their implementation. The review underscores the importance of CMMs in enhancing organizational cybersecurity resilience and offers insights for future research and practical applications. The systematic literature review revealed a diverse landscape of cybersecurity maturity models, ranging from generic frameworks to industry-specific standards. Common components identified across these models include governance, risk management, compliance, and technical controls. The review also highlighted the importance of contextual factors, such as organizational culture, size, and sector, in shaping CMM implementation strategies. Challenges encountered in practice include resource constraints, stakeholder engagement, and the dynamic nature of cyber threats. Despite these challenges, CMMs offer significant benefits for organizations seeking to enhance their cybersecurity resilience, including improved risk visibility, decision-making, and resource allocation. Future research should focus on developing tailored maturity models for

A. Rocha (✉)
ISEG, University of Lisbon, Lisbon, Portugal

F. A. Alaba · H. Musa
Federal University of Education, Zaria, Nigeria

M. J. Sousa · R. Pereira
ISCTE-IUL – Instituto Universitário de Lisboa, Lisbon, Portugal

J. B. de Vasconcelos
Lusofona University, Porto, Portugal

D. Radu et al. (eds.), *Countering Hybrid Threats Against Critical Infrastructures*, NATO Science for Peace and Security Series B: Physics and Biophysics, https://doi.org/10.1007/978-94-024-2304-4_10

emerging technologies and sectors and evaluating CMM effectiveness and scalability in diverse organizational contexts. Overall, the SLR's findings emphasize the importance of CMMs in advancing cybersecurity maturity and resilience across organizations.

Keywords Cybersecurity maturity models · Frameworks · Organizational resilience · Risk management · Implementation strategies · Cyber threats · Effectiveness evaluation

10.1 Introduction

Cyberattacks have increased in the digital age, including in Singapore. Ransomware, phishing, and susceptible online and mobile app attacks greatly cost organizations, and they need cyber security to respond to risks quickly [1]. SMEs and multinationals are susceptible to cyberattacks because they lack the funds to invest in cybersecurity. Due to their budgetary constraints, startups, especially those under 200 people, are more susceptible to these attacks. Cybersecurity is increasingly becoming a commercial risk; thus, entrepreneurs must invest in security [2]. Cybersecurity maturity models, or CMMs, are essential instruments businesses may use to evaluate, monitor, and enhance their cybersecurity capabilities throughout certain periods. While improving their cybersecurity capabilities, companies advance through a succession of maturity levels or processes these models define [3]. A separate stage of cybersecurity maturity is represented by each maturity level, which is characterized by increasing degrees of complexity, efficacy, and resilience in the face of cyber attacks. It allows businesses to build focused plans and efforts to move to higher maturity levels and enhance their cybersecurity defenses. This is accomplished by analyzing their present degree of maturity and finding holes in their cybersecurity defenses [1].

The CMMI, established by the Software Engineering Institute at Carnegie Mellon University, is considered one of the most well-known methodologies for assessing cybersecurity's maturity level. The CMMI model defines five distinct maturity levels. These levels range from initial (Level 1) to optimal (Level 5), and each level represents a different state of maturity regarding cybersecurity standards and procedures. Another important cybersecurity maturity model is the NIST Cybersecurity Framework (CSF), established by the National Institute of Standards and Technology (NIST) in response to the growing number of cybersecurity threats businesses are experiencing. The CSF provides a risk-based approach to cybersecurity management [4]. It enables businesses to identify, evaluate, and prioritize cybersecurity risks based on their business goals and risk tolerance. The capacity of cybersecurity maturity models to offer businesses a formal framework for evaluating existing cybersecurity capabilities, determining areas in which they may develop, and prioritizing cybersecurity investments and efforts is the primary reason for the relevance of these models [5]. Adopting a maturity-based approach to cybersecurity management enables firms to evaluate the risks associated with cybersecurity, prioritize investments, monitor progress, show compliance, and improve cyber resilience [6].

Thus, cybersecurity maturity models play an essential part in cybersecurity management by supplying businesses with a systematic framework that enables them to evaluate, measure, and improve their cybersecurity capabilities [4]. A maturity-based strategy allows firms to identify risks, prioritize investments, monitor progress, show compliance, and improve cyber resilience. It is all accomplished via the adoption of the method. Organizations looking to navigate the complex and ever-changing cybersecurity risk will continue to find that cybersecurity maturity models are critical tools [7]. It is because cyber threats continue to change and intensify. The systematic literature review (SLR) on cybersecurity maturity models aims to identify current models, examine their features and components, assess their relevance and efficacy in real-world contexts, and identify gaps and challenges. The models are gathered from many sources, including academic literature and industry reports, to provide a thorough inventory for future research. In addition, the SLR highlights shortcomings in current models, such as scalability, flexibility to future risks, and usability in various organizational situations. The SLR's goal in tackling these goals is to help enhance cybersecurity practices and establish more robust maturity models.

10.1.1 Definition of Key Terms

Maturity Model: A maturity model is a structured framework that describes a set of maturity levels or stages organizations go through as they improve their capabilities in a certain area. Maturity models are used to describe the development of organizations. When businesses use maturity models, they can see where they are, what needs fixing, and how to go to the next level of maturity [8].

Levels of Maturity: In a maturity model, levels represent different phases of developing capabilities or reaching maturity. Different maturity levels indicate varying degrees of competence, efficiency, and insight into the evaluated topic. Organizations advance through these levels in a logical progression, with each level expanding upon the skills attained at the level before it [9].

Evaluation Criteria: Evaluation criteria are the particular measures, factors, or signs that measure how well an organization can do compared to the norms or requirements in a maturity model. Organizational structure, performance indicators, technological infrastructure, rules and procedures, and budget allocation are all potential considerations. Organizations may more accurately gauge their current level of maturity and pinpoint where they might progress using assessment criteria [10].

Cybersecurity Maturity Model: A cybersecurity maturity model is a specific framework that is meant to evaluate and develop the capabilities of an organization in terms of cybersecurity. Cybersecurity areas like governance, risk management, compliance, incident response, and technological infrastructure have unique maturity levels and evaluation criteria defined by these models. Organizations may improve their cyber resilience and reduce cyber risks with maturity models, which provide a systematic way to manage cybersecurity [11].

Baseline: A baseline is an initial or reference point against which subsequent improvements or modifications are evaluated. The baseline in a maturity model is the point at which an organization is considered to have reached a certain level of development within the area under evaluation. Organizations may monitor their development over time and assess the efficacy of their endeavors to improve maturity and capabilities by establishing a baseline [12].

Gap Analysis: Gap analysis identifies inconsistencies or gaps between an organization's present condition and its intended state or goals. Gap analysis, as it pertains to maturity models, compares an organization's current maturity with the amount of maturity the model specifies as ideal. By doing this research, businesses can identify their weak spots, set priorities for development, and ultimately reach a greater degree of maturity.

10.1.2 Definition of Key Terms

For the SLR of It study, the following research questions were definitively proposed:

1. What key cybersecurity maturity models are identified in academic literature and industry reports?
2. What are the common characteristics, components, and assessment criteria used in cybersecurity maturity models?
3. How effective are cybersecurity maturity models in assisting organizations in improving their cybersecurity capabilities?
4. What are the main challenges and limitations of current cybersecurity maturity models?

10.2 Overview of Cybersecurity Maturity Models

The increasing need for enterprises to determine and enhance their cybersecurity capabilities has contributed to developing cybersecurity maturity models. Successive models built on earlier ones, like Carnegie Mellon University's Capability Maturity Model (CMM), established criteria for evaluating an organization's degree of development in areas including technical controls, risk management, compliance, and governance [13]. The proliferation of sector-specific models coincided with the increasing industry-specificity of cybersecurity threats. For example, healthcare and financial services businesses adopted models such as the HIPAA Security Rule and the Federal Financial Institutions Examination Council (FFIEC) Cybersecurity

Assessment Tool. These models provide thorough frameworks for integrating cybersecurity standards with industry best practices, aligning cybersecurity procedures with widely recognized norms and standards [2]. Current cybersecurity maturity models prioritize resilience and risk management in light of the ever-changing nature of cyber threats. As a result, businesses can better assess cybersecurity risks, set priorities, and devise preventative measures to deal with them [3]. Companies may plan for and react to cybersecurity crises and disruptions using resilience-focused maturity models to ensure business continuity and minimize the effect of cyber attacks. Strategic planning and decision-making, meeting regulatory requirements, reducing risk and preventing incidents, and continuously improving and adapting are the four main areas where cybersecurity maturity models shine [14]. Developing comprehensive cybersecurity strategies and prioritizing investments in projects based on risk exposure and business goals may be achieved by measuring companies' maturity levels across key areas [15].

Organizations may evaluate their cybersecurity posture regularly, spot new threats and trends, and change their controls and policies in response with the help of these models, which encourage continuous development and adaptation in cybersecurity procedures. Organizations may strengthen their cybersecurity resilience and remain one step ahead of developing risks by maintaining agility and responsiveness to shifting cyber threats and business needs [8]. Consequently, cybersecurity maturity models have experienced immense growth over time, reflecting the shifting environment of cyber threats, technology breakthroughs, and regulatory requirements. In today's increasingly digital environment, organizations may proactively manage cybersecurity risks, achieve regulatory compliance, and develop resilience by adopting several strategies.

10.2.1 National Institute of Standards and Technology (NIST) Cybersecurity Framework

The cybersecurity framework developed by the NIST is widely acknowledged and extensively utilized as a framework for managing cybersecurity risk. This framework comprises rules, standards, and best practices intended to assist companies in managing and mitigating the risks associated with cybersecurity [16]. The framework is built on the foundation of five fundamental functions: identifying, protecting, detecting, responding, and recovering. Within cybersecurity risk management, these functions provide a comprehensive strategy, directing businesses through identifying, defending against, detecting, reacting to, and recovering from cybersecurity events and threats [2, 17]. The identify function is primarily concerned with gaining knowledge of the cybersecurity risks, assets, and vulnerabilities that apply to a company. Asset management, risk assessment, governance, and risk management procedures are fundamental components that must be included. For organizations to prioritize and solve cybersecurity concerns, it is strongly recommended that they identify their key assets, evaluate the risks associated with cybersecurity, build governance structures, and implement risk management systems [18].

Protecting entails putting protections and controls in place to protect against potential cybersecurity threats. Some components include data security, defensive technology, awareness and training, and access control. Putting in place access restrictions, educating staff about cybersecurity and providing them with training, encrypting sensitive data, and deploying defensive technologies like firewalls, antivirus software, and intrusion detection systems are all recommendations for organizations [14]. Discovering cybersecurity issues and threats promptly is the primary emphasis of detection. Some components include Anomaly finding, continuous monitoring, and incident response planning [19]. To address cybersecurity incidents quickly and efficiently, organizations should build incident response strategies, install methods for identifying abnormal activity, and continually monitor networks and systems for cybersecurity events [20]. Coordinated and effective responses to cybersecurity issues are what we mean when we talk about reacting. Response planning, communication, mitigation, and recovery are all components that make up the incident response system. To limit the effect of events and begin recovery efforts to restore operations and systems, organizations should build and execute incident response plans, and establish communication protocols for reporting and coordinating responses to cybersecurity issues, and establish communication protocols [7]. Some important advantages enterprises may get from the NIST cybersecurity framework include flexibility, a risk-based strategy, standard alignment, and a better cybersecurity posture. The framework is still useful for companies that want to fortify their cybersecurity and safeguard their data, systems, and assets, even if cybersecurity threats are always changing [21].

10.2.2 Capability Maturity Model Integration (CMMI) Framework

The Software Engineering Institute (SEI) at Carnegie Mellon University is responsible for developing a framework for process improvement known as the CMMI. In areas such as software development, project management, and service delivery, it aids businesses in improving their processes, increasing their performance, and reaching greater degrees of maturity. Five separate maturity levels make up the framework [21]. Each level represents a particular degree of process maturity and organizational competence. These maturity stages provide a road map for businesses to gradually enhance their processes, transitioning from ad hoc and chaotic practices to disciplined and optimized procedures, contributing to the company's success [4]. Capability Areas, Process Management Areas, and Support Areas are the three categories that make up the CMMI framework itself, which consists of a total of 25 process areas. Capability Areas concentrate on certain technical tasks that are associated with the production of products and services. At the same time, process management areas are concerned with the management of organizational and project administration procedures. Support Areas provide supplementary functions and resources to enhance processes [22].

The term "generic practices" refers to common techniques applied to various process areas and maturity levels within the CMMI framework. These practices are considered to be the best practices for the improvement of processes and the development of organizational capabilities [23]. The CMMI framework offers organizations implementation instructions and tools to assist them in efficiently using it. It includes providing direction on conducting assessments, creating development targets, and choosing suitable process improvement solutions. Additionally, it includes extensive definitions of each maturity level, process area, and generic practice. To assist enterprises in accomplishing their process improvement objectives and implementing the CMMI framework, the SEI provides training, consulting services, and other tools [24].

The CMMI framework provides businesses with several major benefits [18]:

1. Process Improvement: The framework offers a systematic approach to process improvement, which enables businesses to identify areas of weakness, put into practice best practices, and improve their processes to achieve better results.
2. Performance Measurement: The framework incorporates measurement and analytical procedures that assist businesses in evaluating the efficiency of their operations, monitoring performance over a while, and determining areas in which they may make further improvements.
3. Organizational Learning: By using the CMMI framework, businesses can foster a culture of continuous learning and development, allowing workers to work together, innovate, and create good change.
4. Competitive Advantage: Organizations with higher CMMI maturity levels show their dedication to quality and excellence, giving them an edge in the market and drawing in clients who appreciate consistency, dependability, and top-notch performance.

With maturity models, businesses can better comprehend and implement processes, capabilities, and overall performance changes. Three fundamental ideas are the foundation for them [2]: maturity levels, domains, and evaluation methodologies. From Level 1, which represents early or ad hoc practices, to Level 5, which represents optimal or completely mature practices, maturity levels indicate distinct degrees of organizational competence and process maturity. The implementation of process improvements and the adoption of best practices in various domains are how these levels are accomplished. Software engineering, project management, cybersecurity, quality management, and organizational management are domains to which a maturity model might apply [25]. The maturity levels, process areas, and evaluation criteria that are associated with each domain may be different from one another. Assessment methods are approaches used to evaluate an organization's maturity in a certain area. Some examples of assessment methods include self-evaluation, external assessment by third-party auditors or consultants, and benchmarking against industry standards or best practices [26]. Some models include quantitative measurements and performance indicators to provide a more objective measurement of maturity stages. Companies can evaluate their performance compared to industry norms or established benchmarks via benchmarking, an essential component of maturity models. In today's highly competitive business world, businesses can improve their processes, skills, and overall performance by gaining a knowledge of maturity models and successfully implementing them [27].

10.3 Literature Review Findings

This section summarizes and synthesizes findings from selected studies on cybersecurity maturity models. [1] reviewed cybersecurity maturity models from numerous industries and emphasized the necessity for modification models to accommodate particular organizational circumstances. The authors emphasized that maturity models should be adaptable and extensible to meet the needs of organizations with varying organizational structures and cybersecurity needs. A study by [28] examined how well cybersecurity maturity models enabled organizations to improve their cybersecurity. Using maturity models, they determined that threat detection capabilities, incident response times, and overall cybersecurity resilience were all improved for organizations. The research highlighted the need to regularly evaluate and improve maturity models to tackle changing cyber threats. The suitability of several cybersecurity maturity models for SMEs was examined by [15]. Results showed that SMEs have difficulty using maturity models established for bigger companies. To meet the specific requirements of SMEs, the authors suggested creating maturity models that are both easy to understand and efficient with resources. The function of maturity models in enhancing cloud security posture was studied [29].

According to their research, organizations can tackle cloud security issues using maturity models designed for cloud-based environments, including data protection, identity management, and compliance. The research emphasized that cloud-specific maturity models are crucial for protecting the resiliency of cloud-based systems [30]. evaluated how company culture affects the validity of the maturity model. According to their findings, cultivating a security-conscious culture inside enterprises is crucial to successfully applying cybersecurity maturity models. To foster a cybersecurity knowledge and best practices culture, the research highlighted the importance of leadership's commitment, staff's awareness, and training programs. An analysis of the historical development of cybersecurity maturity models was presented by [31]. According to the authors, mature models have changed from basic frameworks to more detailed and situation-specific models developed for various business and company needs. According to the research, developing maturity models is an iterative process that requires constant updating and refining to allow for new cyber risks and technological developments [28]. looked at the pros and cons of maturity model adoption in government agencies.

According to their study, factors impacting the effective application of cybersecurity maturity models in government organizations include organizational culture, financial restrictions, and regulatory compliance. The research stressed the need to coordinate maturity model aims with the firm and its stakeholders to achieve optimal results [14]. investigated the function of maturity models in strengthening cybersecurity in the supply chain. The researchers concluded that coordinated methods across supply chain participants are crucial to successfully using cybersecurity maturity models across linked networks. The research highlighted the need for standardized frameworks, information exchange methods, and risk management procedures to reduce cybersecurity threats in the supply chain ecosystem. Examining the function, efficacy, and generalizability of cybersecurity maturity models in different company environments is a key takeaway from this research. Maturity models provide a framework for strengthening cybersecurity, but whether or not they work relies on company culture, available resources, and alignment with goals. The effectiveness of cybersecurity maturity modeling in addressing growing cyber threats requires more study into emerging trends and best practices [32].

There is a lack of standardization, complexity, scalability, subjective judgment in assessments, coverage of emerging threats, resource intensity, static nature, stakeholder engagement, ease of effectiveness measurement, and limited tailoring to organizational context regarding cybersecurity maturity models. Due to these characteristics, organizations may find it difficult to compare their cybersecurity maturity to industry norms, so more standardized models are needed. Furthermore, not all models can handle new threats well; some may be too complicated, necessitating time, energy, and knowledge. Furthermore, resistance and ineffectual cybersecurity improvements may result from approaches that put compliance ahead of good risk management and resilience. Cybersecurity experts, industry stakeholders, and lawmakers must work together continuously to develop maturity models that are more standardized, flexible, and effective to conquer these challenges.

10.4 Methodology

The methodological processes adopted in It SLR are provided in this section.

10.4.1 Search Strategy and Selection Criteria

The study gathered research articles from the following digital databases: IEEE Explore (ieeeexplore.ieee.org); Scopus (www.scopus.com); Springer (www.springer.com); Web of Science (http://apps.webofknowledge.com). This study targeted research articles from January 2011 to March 2024 using the following keywords: ("cybersecurity maturity model" OR "cybersecurity maturity framework") AND ("review" OR "survey" OR "assessment"); ("cybersecurity maturity model"

OR "cybersecurity maturity framework") AND ("review" OR "survey" OR "assessment"); ("cybersecurity maturity model" OR "cybersecurity maturity framework") AND ("review" OR "survey" OR "assessment"); ("cybersecurity maturity model" OR "cybersecurity maturity framework") AND ("review" OR "survey" OR "assessment").

10.4.2 Inclusion and Exclusion Criteria

Table 10.1 presents the criteria for including and excluding criteria in It study. These criteria were established to conduct a focused search that aligns with the research goals. A reference management system that manages academic publications, Mendeley (www.mendeley.com) is where all the chosen research papers have been saved.

Using the search strings stated in the identification stage, this study identified 1772 research papers published in IEEE Explore, Scopus, Springer, and Web of Science. These articles included duplicates. The keywords utilized in the systematic literature review were employed to assist in the identification process. As can be seen in Fig. 10.1, there were 1472 publications retrieved from Springer, 51 items acquired from IEEE Explore, 175 articles from Scopus, and 74 articles from Web of Science.

The first screening kept 620 research papers after removing duplicates. Table 10.1 shows that 550 research papers were omitted from consideration throughout the screening process based on the inclusion and exclusion criteria. After that, we reviewed the synopses of the remaining 70 studies and excluded 27. Despite including pertinent keywords in the titles, abstracts, and body of the omitted research publications, their emphasis was on cyber risk rather than cyber security maturity, and the reverse was also true. Additionally, the research issues posed by this study

Table 10.1 Inclusion and Exclusion Criteria

Criteria	Inclusion	Exclusion
Publication year	Studies published between 2011 and 2024.	Studies published before 2011 or after 2024.
Research focus	Studies focused on cybersecurity maturity models.	Studies not related to cybersecurity maturity models.
Study type	Peer-reviewed articles, conference papers, and scholarly publications.	Grey literature includes blog posts, opinion pieces, and news articles.
Language	Studies available in English.	Non-English studies or studies lacking full-text availability.
Relevance	Relevance to the research topic and objectives.	Studies lacking relevance to the research topic or objectives.
Methodology	Studies employing sound research methodologies.	Studies with flawed methodologies or insufficient methodological detail.

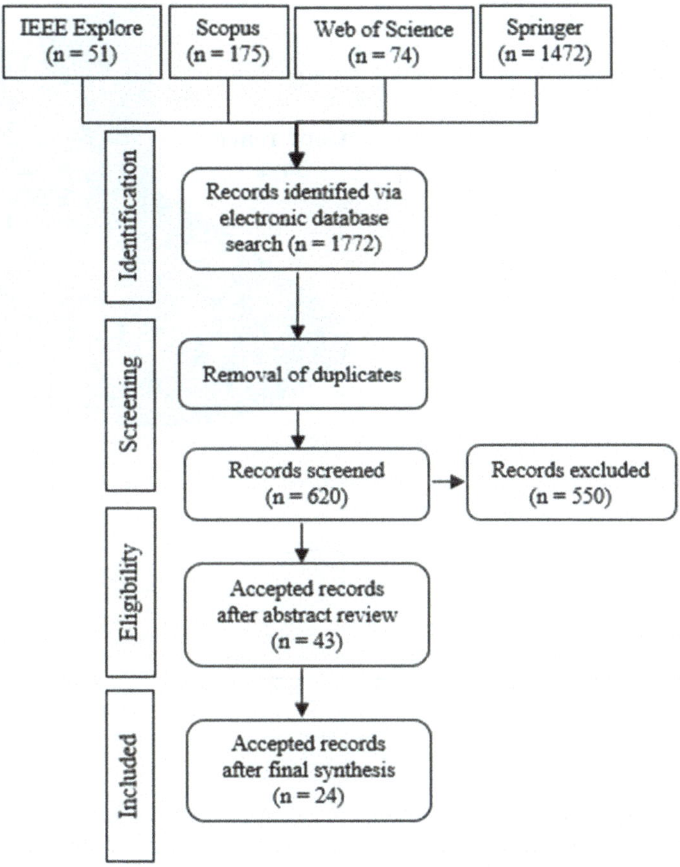

Fig. 10.1 PRISMA flowchart

were not addressed in these journals. Twenty-four (24) research publications were chosen for a thorough evaluation after the last synthesis.

10.4.3 Overview of Selected Studies

This study was based on the selection of 24 papers. 15 papers were published in journals, and 9 were included in conference proceedings. In Fig. 10.2, the percentages are shown. The 24 publications that made it into this comprehensive literature evaluation are shown in Fig. 10.3, which demonstrates the number of papers by year of publication. Starting in 2019, the graph shows that the number of publications is increasing significantly. In 2024, as of March 31st, the paper distribution was minimal.

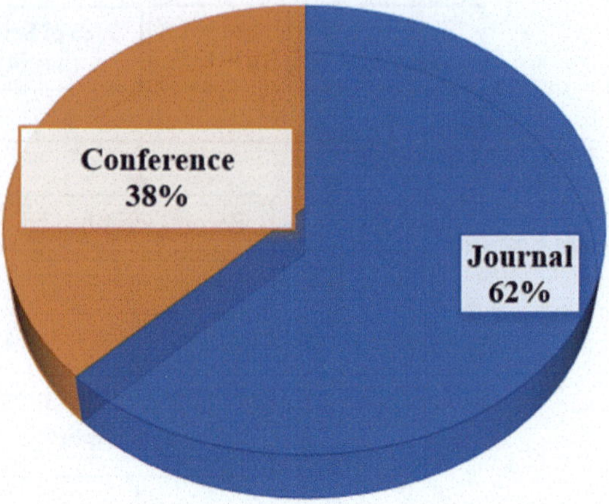

Fig. 10.2 Percentage of collected studies

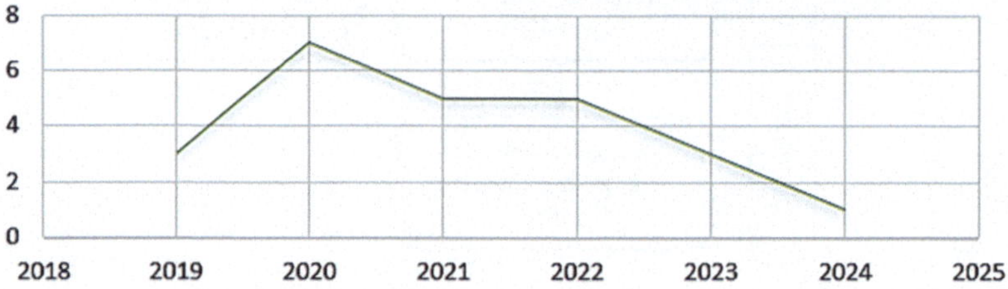

Fig. 10.3 Number of papers by year of publication

The systematic literature evaluation relies on 24 publications, the distribution shown in Fig. 10.3 according to the publication year. From 2019 onwards, the graph shows a rise in publishing. The low paper distribution rate in 2024 was as of March 31st.

Imperfect data extraction and the possibility of bias might limit our results and undermine the validity of our SLR. Not all relevant articles may be returned from the considered databases when first searching. Each paper's reference section was manually reviewed to hone down on the articles relevant to the study field. Each of the 43 articles was reviewed independently to ensure they related to the study's objectives and questions. We used the PRISMA standards while selecting these 24 journal articles to make sure we did not overlook any relevant ones, that they could answer our research objectives, and that they met our inclusion and exclusion criteria. The relevant publications were not inadvertently excluded by using several permutations of the search keywords. The articles chosen and evaluated according to the PRISMA standards are among the best in their fields, relevant to the identified research objectives, and conducted unbiased and impartial.

10.5 Discussion and Findings

The VOSviewer program was used for this purpose. The VOSviewer assisted in determining the keywords that were used the most often in the articles and the connections between the authors. The larger bubbles showed keywords that occurred more often. This analysis requires gathering study findings to meet the research questions (RQs) selected in this paper. A data extraction process was carried out on the chosen research articles (nD24), and the outcomes have been addressed concerning the RQs of this study:

RQ1: What key cybersecurity maturity models are identified in academic literature and industry reports?

Several notable models have been discovered that address RQ1, relating to the importance of cybersecurity maturity models identified in academic literature and industry publications. Academic literature and industry studies have recognized many significant cybersecurity maturity models, including the CMMI cybermaturity platform, NIST Cybersecurity Framework (CSF), ISO/IEC 27001, CERT Resilience Management Model (CERT-RMM), and C2M2 [2]. These models attempt to help organizations analyze, improve, and maintain their cybersecurity posture in an ever-changing threat scenario. The CMMI Cybermaturity Platform offers a systematic way of assessing an organization's cybersecurity processes across several domains, such as governance, risk management, and incident response.

The NIST CSF is a widely used framework that outlines principles, standards, and best practices for addressing cybersecurity risks. It has five key functions: identify, protect, detect, respond, and recover [33]. ISO/IEC 27001 is a comprehensive framework for developing, implementing, maintaining, and upgrading an information security management system (ISMS). Organizations may reach various degrees of maturity depending on their compliance with the standards standards and the efficacy of their security policies. CERT-RMM emphasizes incorporating cybersecurity practices into broader risk management and business operations. It comprises numerous process areas with specified maturity levels, allowing organizations to analyze and improve their cybersecurity capabilities methodically [34]. The C2M2 maturity model was created by the United States Department of Energy (DOE) to evaluate and enhance the cybersecurity capabilities of energy sector organizations. An analysis of scholarly articles and business feedback revealed many important cybersecurity maturity frameworks, of which these models are a subset. Organizations often use different models in tandem to handle cybersecurity concerns and improve resilience against cyber attacks [35]. Thus, Table 10.2 shows the publications that were chosen for data extraction.

RQ2: What are the common characteristics, components, and assessment criteria used in cybersecurity maturity models?

RQ2 examines cybersecurity maturity model features, components, and evaluation criteria. These models follow a systematic framework that divides

Table 10.2 Selected papers addressing research question 1

Paper title	Authors/ year	Publication/conference
"Aeronautic maturity cybersecurity: a framework"	[36]	Sandia National Laboratories
"Systematic Literature Review of Threat Modeling Concepts"	[32]	International Conference on Information Systems Security and Privacy
"Industry 4.0 data security: A cybersecurity frameworks review"	[37]	Journal of Industrial Information Integration
"An Analysis of Assessment Approaches and Maturity Scales Used for Evaluation of Information Security and Cybersecurity User Awareness and Training Programs: A Scoping Review"	[38]	Organizational Cybersecurity Journal: Practice, Process and People
"Information Security Maturity Model for NIST Cyber Security Framework"	[35]	NIST Special Publication
"Cybersecurity capability maturity models review and application domain"	[39]	Advances in Science, Technology and Engineering Systems
"An Explanatory Review on Cybersecurity Capability Maturity Models"	[5]	NIST Special Publication
Cybersecurity Capability Maturity Model (C2M2) for Electricity Sector	[2]	Computer Networks
Cybersecurity Maturity Model: An Introduction	[3]	SANS Institute
Cybersecurity Maturity Model: Putting It All Together	[15]	SANS Institute
Developing a Cybersecurity Maturity Model for the Automotive Industry	[31]	Automotive Engineering Magazine
Evaluating Organizational Cybersecurity Maturity: A Model and an Application	[40]	NIST Special Publication 800-55
Government Cybersecurity Maturity Model (GCMM)	[41]	NISTIR 8011
Healthcare Sector Cybersecurity Framework Implementation Guide	[16]	Computers & Security
How Cybersecurity Maturity Models Can Help	[42]	ISACA Journal
Hybrid Cybersecurity Maturity Model for Small and Medium-sized Enterprises	[43]	Computers & Security
Introducing a Cybersecurity Maturity Model for Organizations	[1]	IEEE
Investigating the Applicability of Cybersecurity Maturity Models in Small and Medium-sized Enterprises	[38]	Journal of Computer Information Systems
An International Standard for Enhancing Cybersecurity Maturity	[32]	IEEE Security & Privacy
Mapping Cybersecurity Maturity Models to the NIST Cybersecurity Framework	[44]	Proceedings of the 50th Hawaii International Conference on System Sciences
Measuring Cybersecurity Maturity: The CMMI Institute's Cyber Maturity Platform	[25]	CMMI Institute

cybersecurity into various categories or domains. They have numerous maturity levels, each indicating a different state of cybersecurity maturity [39]. These models cover various topics, including governance, risk management, compliance, incident response, and technological controls. They are intended to be scalable and adaptable to diverse organizational sizes, sectors, and situations, allowing for organizations of varied maturity levels. Cybersecurity maturity models encompass governance and leadership, risk management, security controls, incident response and recovery, and continuous improvement [45]. Governance and leadership are concerned with developing cybersecurity policies, processes, and organizational structures. Risk management includes discovering, evaluating, managing, and monitoring cybersecurity threats. Security controls include implementing technological, administrative, and physical safeguards to protect assets, data, and systems against attacks. Incident response and recovery define the methods for identifying, reacting to, and recovering from cybersecurity events.

Continuous improvement is also emphasized in maturity models, which include procedures for monitoring, measuring, assessing, and improving cybersecurity capabilities over time [46]. Criteria for evaluation contain predetermined criteria or indicators for evaluating cybersecurity competence across all domains and stages of development. Benchmarking criteria compare an organization's cybersecurity maturity to industry standards, best practices, or comparable organizations, allowing businesses to assess their performance and discover areas for improvement. Evidence-based evaluation standards guarantee the consistency and validity of maturity assessments. Cybersecurity maturity models provide a structured approach to assessing, improving, and benchmarking cybersecurity capabilities, increasing cybersecurity resilience, and aligning efforts with organizational goals and objectives by incorporating these common characteristics, components, and assessment criteria [47].

Table 10.3 summarizes major studies covering typical characteristics, components, and evaluation criteria used in cybersecurity maturity models and their publication years.

RQ3: How effective are cybersecurity maturity models in assisting organizations in improving their cybersecurity capabilities?

The RQ3 assesses the effectiveness of cybersecurity maturity models in enhancing organizations' cybersecurity capabilities. It highlights the benefits of maturity models, including a structured approach to improvement, gap identification and prioritization, a roadmap for progression, alignment with industry best practices and standards, a continuous improvement culture, enhanced risk management, and resource optimization. Maturity models provide a structured framework for assessing an organization's cybersecurity posture and identifying areas for improvement. They help organizations prioritize their efforts and focus on enhancing critical cybersecurity capabilities. They also help identify gaps between an organization's current state and desired maturity levels, allowing organizations to prioritize actions to address these gaps [60]. Maturity models also offer a roadmap for progression, outlining incremental steps to advance cybersecurity maturity. It helps

Table 10.3 Selected papers addressing research question 2

Paper title	Authors	Publication/Conference
Common Criteria for Cybersecurity Maturity Models	[48]	ACM Digital Library
A Systematic Review of Cybersecurity Maturity Models	[49]	Proceedings of the 15th International Conference on Evaluation of Novel Approaches to Software Engineering
Common Components in Cybersecurity Maturity Models	[50]	Journal of Cybersecurity Education, Research and Practice
A Comparative Analysis of Cybersecurity Maturity Models	[24]	International Journal of Cybersecurity Intelligence and Cybercrime
Assessing the Effectiveness of Cybersecurity Maturity Models	[26]	IEEE International Conference on Cybersecurity and Privacy
Components and Characteristics of Cybersecurity Maturity Models	[51]	Journal of Information Security and Cybersecurity
Evaluation Frameworks for Cybersecurity Maturity Models	[52]	IEEE Symposium on Security and Privacy
Key Components of Cybersecurity Maturity Models	[53]	International Conference on Information Security
The Role of Assessment Criteria in Cybersecurity Maturity Models	[54]	ACM Conference on Computer and Communications Security
Understanding Common Characteristics in Cybersecurity Maturity Models	[55]	Journal of Information Assurance and Security
Analysis of Assessment Criteria in Cybersecurity Maturity Models	[56]	IEEE Transactions on Dependable and Secure Computing
A Framework for Assessing Cybersecurity Maturity Models	[57]	ACM Transactions on Information and System Security
Common Components and Characteristics of Cybersecurity Maturity Models	[58]	International Journal of Cybersecurity Research
The Evolution of Assessment Criteria in Cybersecurity Maturity Models	[59]	ACM Symposium on Access Control Models and Technologies
Best Practices for Developing Cybersecurity Maturity Models	[57]	Journal of Cybersecurity Best Practices

organizations make informed decisions about resource allocation and risk management. They also help organizations understand and manage cybersecurity risks more effectively. In conclusion, maturity models are valuable tools for organizations to enhance their security posture, mitigate risks, and protect their assets and data from cyber threats [5]. However, their effectiveness depends on their implementation, integration into organizational processes, and adaptation to each organization's specific needs and context.

Therefore, Table 10.4 shows the primary publications and the years of publication that explain cybersecurity maturity models that help organizations improve their cybersecurity abilities.

Table 10.4 Selected papers addressing research question 3

Paper Title	Authors	Publication/Conference
A Comprehensive Cybersecurity Maturity Model	[2]	Computers
A Cybersecurity Maturity Model for Cloud Services	[61]	Computers and Security
A Review of Cybersecurity Maturity Models	[62]	IEEE Access
An Assessment of Cybersecurity Maturity Model Development and Implementation	[9]	Computers and Security
Applying the Cybersecurity Maturity Model to Enhance Critical Infrastructure Protection	[10]	Computers and Security
Assessing Cybersecurity Maturity in Industrial Control Systems	[11]	Computers and Security
Building a Cybersecurity Maturity Model for Critical Infrastructure	[63]	IEEE Access
Cybersecurity Capability Maturity Model (C2M2) for Electricity Sector	[64]	IEEE Access
Cybersecurity Maturity Model: An Introduction	[65]	IEEE Access
Cybersecurity Maturity Model: Putting It All Together	[66]	IEEE Access
Developing a Cybersecurity Maturity Model for the Automotive Industry	[67]	IEEE Access
Evaluating Organizational Cybersecurity Maturity: A Model and an Application	[63]	IEEE Access
Investigating the Applicability of Cybersecurity Maturity Models in Small and Medium-sized Enterprises	[68]	Journal of Computer Information Systems

RQ4: What are the main challenges and limitations of the current cybersecurity maturity models?

RQ4 investigates the primary issues and limits of existing cybersecurity maturity frameworks. These issues include overemphasis on compliance rather than security, limited acceptance and awareness, difficulty quantifying return on investment (ROI), complexity, resource intensity, subjective evaluation, and lack of customization to organizational context. Other issues include static models, a lack of attention to changing risks and technologies, and a lack of customization. Companies with fewer resources and less experienced staff may struggle with complexity and overload. When maturity models are implemented, time, money, and competent people are generally largely needed [28]. Because of the inherent subjectivity in the evaluation process, outcomes from maturity evaluations may be inconsistent and unreliable.

Organizations may struggle to adapt to new cybersecurity threats and enhance their cybersecurity capabilities due to the static nature of models. Instead of fixing real security threats and vulnerabilities, certain maturity models may prioritize compliance with standards and laws [69]. Due to a lack of understanding of maturity models' advantages and value propositions and the difficulties associated with implementing them, adoption rates may be low among smaller organizations or those operating in certain sectors. It may be difficult for organizations to establish the ROI of maturity model implementations as the results of maturity assessments

Table 10.5 Selected papers addressing research question 4

Paper title	Authors	Publication/Conference
Challenges in assessing cybersecurity maturity	[71]	IEEE international conference on cybersecurity
Limitations of current cybersecurity maturity models	[72]	Journal of cybersecurity best practices
Addressing the challenges of cybersecurity maturity models	[73]	ACM symposium on access control models and technologies
The complexity of cybersecurity maturity models	[74]	IEEE transactions on dependable and secure computing
Inadequate scope of current cybersecurity maturity models	[75]	ACM conference on computer and communications Security
Lack of standardization in cybersecurity maturity models	[76]	International journal of cybersecurity research
Challenges in tailoring cybersecurity maturity models	[77]	Journal of information assurance and Security
Overreliance on self-assessment in cybersecurity maturity models	[78]	IEEE symposium on Security and privacy
Difficulty in measuring effectiveness of cybersecurity maturity models	[42]	ACM conference on data and application Security and privacy
Theoretical underpinnings of cybersecurity maturity models	[64]	Journal of cybersecurity education, research and practice
Lack of integration with emerging threats in cybersecurity maturity models	[79]	International conference on cybersecurity
Challenges in adapting cybersecurity maturity Models to different industries	[80]	ACM symposium on applied computing

and improvement efforts are not always easy to quantify. Cybersecurity maturity models require constant work to make them more accurate, more applicable, and more widely used and understood by businesses so that they can overcome these constraints and difficulties. Organizations could combine maturity models with other cybersecurity frameworks, tools, and practices to address unique demands and overcome restrictions [70].

Table 10.5 contains a list of significant studies highlighting the key issues and limits of existing cybersecurity maturity models and the years of publication. These publications give insight into the challenges and limitations faced in existing cybersecurity maturity models in various areas, giving helpful recommendations for development.

10.5.1 Critical Analysis of Cybersecurity Maturity Models

Organizations serious about improving their cybersecurity posture now cannot do without cybersecurity maturity models. Organizations may be guided through several stages of cybersecurity maturity using these models, which systematically measure and upgrade skills [2]. Although there are several advantages to using cybersecurity maturity models, it is important to examine their limits critically. To better understand cybersecurity maturity models [50], this article comprehensively analyses their benefits and drawbacks.

10.5.2 Benefits of Cybersecurity Maturity Models

1. Structured Framework: The structured framework that cybersecurity maturity models provide is one of the key advantages of these models. Its framework offers organizations a systematic approach to the enhancement of their cybersecurity. These models provide organizations with a clear path by defining maturity levels, domains, and evaluation criteria [38].
2. Capabilities for Benchmarking: Cybersecurity maturity models make it possible for organizations to compare their cybersecurity procedures to the standards and best practices followed by the industry. A company may find its development areas and reasonable improvement targets by comparing its maturity level to its peers or recognized benchmarks [39].
3. Adaptable Solutions: Depending on a company's size, industry, and unique cybersecurity requirements, many cybersecurity maturity models provide individualized recommendations. Due to its personalization, businesses can zero in on the specific cybersecurity concerns that pertain to their operations and risk profile [28].
4. Continuous Improvement: Cybersecurity maturity models encourage organizations to advance through maturity stages over time, which fosters a culture of continuous improvement to be implemented. Organizations may stay ahead of the curve in cybersecurity by consistently evaluating their capabilities and making necessary upgrades [15].

10.5.3 Limitations of Cybersecurity Maturity Models

1. Assessing cybersecurity maturity is not without its inherent subjectivity, which is one of the main obstacles to using these models. There could be discrepancies in how various organizations determine their maturity levels if assessors have differing interpretations of maturity standards [2].
2. A cybersecurity maturity model's implementation and upkeep can be demanding on resources, necessitating substantial resources (time, energy, and money). It could be difficult for smaller companies or those with less cybersecurity resources to use maturity models fully [4].
3. Cybersecurity maturity models may be dynamic or static, depending on how fast they adapt to new threats and technology. Consequently, businesses can struggle

to deal with emerging cybersecurity threats that the model does not account for [38].

4. A lack of adaptability means certain cybersecurity maturity models provide a cookie-cutter solution that would not work for every company. Because of its rigidity, the model may be unable to adapt to different settings [1].
5. The focus of certain maturity models is on compliance requirements rather than proactive cybersecurity practices, which is a mistake. While organizations prioritize compliance, they should push themselves above and beyond to build cybersecurity resilience [3].

Finally, cybersecurity maturity models provide helpful structures for businesses to evaluate and improve their cybersecurity capacities. They do have certain limits. Organizations may improve the efficacy of their cybersecurity efforts, fill up any gaps, and make knowledgeable decisions on cybersecurity tactics by critically evaluating maturity models. Attaining optimum cybersecurity resilience requires a well-rounded strategy considering cybersecurity maturity models'benefits and drawbacks [28].

10.5.4 Gap Analysis

The literature on cybersecurity maturity models is limited, lacking research on dynamic models that adapt to changing threats, organizational contexts, and technological advancements. There is a need for real-time maturity assessment and continuous improvement methodologies to address evolving cyber threats. There is also a lack of research on sector-specific models tailored to unique cybersecurity challenges and regulatory requirements [70]. The literature also lacks research on the behavioral aspects of maturity, including the role of organizational culture, human behavior, and employee awareness. Future research should explore methodologies for incorporating behavioral science principles into maturity models to promote positive cybersecurity behaviors and enhance risk awareness. There is also a gap in research on integrating emerging technologies such as AI, machine learning, blockchain, and IoT into maturity assessment and improvement [51]. Future research should focus on empirical, interdisciplinary, longitudinal case studies and best practices from organizations that have successfully implemented maturity models. In conclusion, the gap analysis highlights opportunities for future research in cybersecurity maturity assessment and improvement. By addressing research gaps, advancing methodologies, and exploring emerging trends, researchers can contribute to developing more effective, adaptive, and practical maturity models that help organizations enhance their cybersecurity posture and resilience in an ever-changing threat landscape [81].

Cybersecurity maturity models are a systematic technique for analyzing an organization's cybersecurity posture and driving long-term changes. However, their usefulness and scalability might vary greatly across different organizational situations.

Size and complexity, industry and sector, organizational culture, scalability, and integration with current processes are all important considerations. Size and complexity are critical for big organizations with substantial IT infrastructure, but smaller organizations may benefit from simpler models customized to their individual goals and available resources [82]. Regulations, threat environments, and operational goals provide distinct cybersecurity problems for each industry and sector. Maturity models should be adaptive to meet these issues and compliance requirements. Organizational culture influences attitudes toward cybersecurity and risk management. Scalability is critical for maturity models to support organizational development and changing cybersecurity threats.

They should provide flexibility regarding assessment scope, maturity levels, and development routes, enabling organizations to adapt their projects to changing conditions [57]. Resource needs must be matched to guarantee that organizations of various sizes and capabilities may successfully execute them. Maturity models have important implications for cybersecurity governance, risk management, and compliance processes. They offer a formal framework for aligning cybersecurity activities with organizational objectives and priorities, allowing for accountability and transparency by establishing explicit roles, responsibilities, and performance measures [59]. They provide systematic risk assessment and prioritization, allowing organizations to allocate resources to high-priority areas and manage hazards proactively. Maturity models encourage continuous development and provide a road map for increasing cybersecurity maturity. They assist organizations in achieving and maintaining regulatory compliance by aligning cybersecurity activities with applicable laws, regulations, and industry standards. Finally, maturity models help auditors and assurance professionals by offering a holistic picture of cybersecurity controls, processes, and procedures. In summary, maturity models are critical for improving organizations' cybersecurity governance, risk management, and compliance procedures [83].

10.5.5 Implications

Maturity models are critical to cybersecurity governance, risk management, and compliance. They offer a formal framework for aligning cybersecurity activities with organizational goals and objectives, allowing organizations to prioritize investments and resources based on strategic priorities. They provide defined roles, duties, accountability mechanisms for cybersecurity governance, and performance measures and benchmarks that allow senior management and board members to track progress and make informed choices. Maturity models encourage continuous development by promoting proactive actions such as frequent assessments, audits, and reviews to detect and resolve new risks and weaknesses. Systematic risk assessment helps organizations identify and prioritize risks based on their probability and possible effects.

Maturity models also assist organizations in implementing risk mitigation strategies and remedial steps by detecting gaps and weaknesses in cybersecurity defenses. Assessing many maturity dimensions, such as technological, procedural, and cultural characteristics, provides a holistic perspective of cybersecurity risk throughout the company. Compliance is made easier by aligning with regulatory requirements and industry standards, which provide a formal framework for showing compliance and adherence to rules. Maturity models enhance audit and assurance efforts, enabling organizations to show internal and external stakeholders how successful their cybersecurity program is. Maturity models allow continuous monitoring and reporting on cybersecurity performance and compliance status. Maturity models have important implications for cybersecurity governance, risk management, and compliance. They may help organizations improve resilience, minimize risks, and ensure compliance with regulatory requirements and industry standards.

For citations of references, we prefer the use of square brackets and consecutive numbers. Citations using labels or the author/year convention are also acceptable. The following bibliography provides a sample reference list with entries for journal articles [1], an LNCS chapter [2], a book [3], proceedings without editors [4], as well as a URL [5].

10.6 Future Research Directions

The future of cybersecurity maturity evaluation and development looks bright, with research concentrating on dynamic maturity models that respond to changing threats, organizational settings, and technical improvements. These models may be customized for many companies and sectors, increasing cybersecurity resilience and compliance [15]. Behavioral dimensions of maturity are also being investigated, emphasizing the importance of organizational culture, human behavior, and employee knowledge in encouraging beneficial cybersecurity behaviors. Emerging technologies like AI, machine learning, blockchain, and IoT are included in maturity assessments and improvements, increasing their efficacy in dealing with emerging cyber threats [15]. Recommendations for improving usability include creating user-friendly interfaces, providing guidance and best practices to organizations, providing training programmes and capacity-building initiatives, and fostering collaboration among cybersecurity practitioners, researchers, and policymakers via collaborative platforms. Predictive analytics, cybersecurity metrics and measurement, cybersecurity ecosystem maturity, and cyber resilience and adaptive maturity are all emerging trends and areas of innovation [15].

Predicted analytics may improve proactive decision-making, resource allocation, and risk mitigation techniques using historical data and predicted indications. Cybersecurity metrics and measurement frameworks help quantify cybersecurity investments' impact on organizational resilience, efficiency, and effectiveness. The scope of cybersecurity maturity assessment may be broadened beyond individual organizations to include whole ecosystems, supply chains, and digital ecosystems, analyzing participants' interconnection, interdependence, and maturity levels. Cyber resilience and adaptive maturity in cybersecurity may be developed on agility, flexibility, and responsiveness. Focusing on these prospective research areas, suggestions, and emerging trends will allow the field of cybersecurity maturity assessment and improvement to grow and meet complicated issues in a more linked and digital world [15].

10.7 Conclusion

The SLR on cybersecurity maturity models thoroughly assesses this area's present state of knowledge. It focuses on various maturity models, each with its framework, methodology, and focal areas. These models address various topics, including technological controls, organizational procedures, and human aspects. They have been utilized extensively in various fields, including cybersecurity governance, risk management, compliance, benchmarking, and strategic planning. However, the analysis highlights the limits of maturity models for increasing cybersecurity posture and organizational resilience. Scalability, complexity, and adaptation to changing threats and technology are some of them.

The study also identifies research gaps and potential for future research in maturity models, including dynamic, sector-specific, and behaviorally informed models, predictive analytics, cyber resilience, and ecosystem-level maturity evaluation. Finally, maturity models are critical tools for improving an organization's cybersecurity posture, resilience, and effectiveness. They provide a formal framework for evaluating, upgrading, and measuring cybersecurity capabilities, allowing businesses to identify gaps, prioritize investments, and track progress over time. However, they also confront issues, including scalability, applicability, and adaptation to changing threat environments. Maturity models may help policymakers influence cybersecurity policies, legislation, and standards, encouraging organizations to adopt best practices and strengthen their resistance to cyber-attacks. Researchers may help by filling research gaps, improving methodology, and devising novel ways. Finally, maturity models are effective instruments for enhancing organizations' cybersecurity posture and resilience.

References

1. Gourisetti SNG, Mylrea M, Patangia H (2020) Cybersecurity vulnerability mitigation framework through empirical paradigm: enhanced prioritized gap analysis. Futur Gener Comput Syst 105:410–431. https://doi.org/10.1016/j.future.2019.12.018
2. Chatziamanetoglou D, Rantos K (2024) Cyber threat intelligence on Blockchain: a systematic literature review. Computers 13(3):60. https://doi.org/10.3390/computers13030060
3. Dube DP, Mohanty RP (2023) Application of grounded theory in construction of factors of internal efficiency and external effectiveness of cyber security and developing impact

models. Organ Cybersecurity J Pract Process People 3(1):41–70. https://doi.org/10.1108/ocj-04-2022-0009
4. Marican MNY, Razak SA, Selamat A, Othman SH (2023) Cyber Security maturity assessment framework for technology Startups: a systematic literature review. IEEE Access 11(January):5442–5452. https://doi.org/10.1109/ACCESS.2022.3229766
5. Garba AA, Siraj MM, Othman SH (2020) An explanatory review on cybersecurity capability maturity models. Adv Sci Technol Eng Syst 5(4):762–769. https://doi.org/10.25046/AJ050490
6. Gupta Gourisetti N, Mylrea M, Patangia H (2019) Application of rank-weight methods to blockchain cybersecurity vulnerability assessment framework. In: 2019 IEEE 9th Annu. Comput. Commun. Work. Conf. CCWC 2019, pp 206–213. https://doi.org/10.1109/CCWC.2019.8666518
7. Hein-Pensel F et al (2023) Maturity assessment for industry 5.0: a review of existing maturity models. J Manuf Syst 66(December 2022):200–210. https://doi.org/10.1016/j.jmsy.2022.12.009
8. Russo N, Reis L, Silveira C, Mamede HS (2024) Towards a comprehensive framework for the multidisciplinary evaluation of organizational maturity on business continuity program management: a systematic literature review. Inf Secure J 33(1):54–72. https://doi.org/10.1080/19393555.2023.2195577
9. Patterson CM, Nurse JRC, Franqueira VNL (2023) Learning from cyber security incidents: a systematic review and future research agenda. Comput Secur 132:103309. https://doi.org/10.1016/j.cose.2023.103309
10. Iyer VR, Babu K, Guruswamy VR (2024) Cyber Security frameworks through the lens of foreign direct investment (FDI): a systematic literature review. Int J Intell Syst Appl Eng 12(4s):279–291
11. Aljowder T, Ali M, Kurnia S (2023) Development of a maturity model for assessing smart cities: a focus area maturity model. Smart Cities 6(4):2150–2175. https://doi.org/10.3390/smartcities6040099
12. Vargas VB, De Oliveira Gomes J, Fernandes PC, Vallejos RV, De Carvalho JV (2023) Influential factors for hospital management maturity models in a post-COVID-19 scenario—systematic literature review. J Inf Syst Eng Manag 8(1):1–10. https://doi.org/10.55267/iadt.07.12868
13. Rohan R, Pal D, Hautamäki J, Funilkul S, Chutimaskul W, Thapliyal H (2023) A systematic literature review of cybersecurity scales assessing information security awareness. Heliyon 9(3):e14234. https://doi.org/10.1016/j.heliyon.2023.e14234
14. Yeoh W, Liu M, Shore M, Jiang F (2023) Zero trust cybersecurity: critical success factors and a maturity assessment framework. Comput Secur 133, no. July:103412. https://doi.org/10.1016/j.cose.2023.103412
15. Muller SR (2023) An analysis of the design of the Cybersecurity Maturity Model Certification (CMMC) and its direct effect on supply chain management. Real-World Solut Divers Strategy Chang Organ Dev Perspect Heal Educ Business Technol no. September:220–243. https://doi.org/10.4018/978-1-6684-8691-7.ch014
16. Castillo-Villar FR, Castillo-Villar RG (2023) Mobile banking affordances and constraints by the elderly. Mark Intell Plan 41(1):124–137. https://doi.org/10.1108/MIP-01-2022-0045
17. Homeland Security (2015) Commercial facilities sector cybersecurity framework implementation guidance. p 36. https://www.us-cert.gov/sites/default/files/c3vp/framework_guidance/commercial-facilities-framework-implementation-guide-2015-508.pdf
18. Khalid Khan S, Shiwakoti N, Stasinopoulos P, Warren M (2023) Modelling cybersecurity regulations for automated vehicles. Accid Anal Prev 186, no. February:107054. https://doi.org/10.1016/j.aap.2023.107054
19. Weerabahu WMSK, Samaranayake P, Nakandala D, Hurriyet H (2023) Digital supply chain research trends: a systematic review and a maturity model for adoption. Benchmarking 30(9):3040–3066. https://doi.org/10.1108/BIJ-12-2021-0782

20. Georgescu TM, Iancu B, Zurini M (2019) Named-entity-recognition-based automated system for diagnosing cybersecurity situations in IoT networks. Sensors (Switzerland) 19(15):3380. https://doi.org/10.3390/s19153380
21. Kolarić S, Mandičák T, Vukomanović M, Mesároš P (2018) BIM training in construction management educational practices in Croatia and Slovakia. Proceedings of the Creative Construction Conference. Creative Construction Conference 2018, CCC 2018, 30 June–3 July 2018, Ljubljana, Slovenia. doi: https://doi.org/10.3311/ccc2018-130
22. Aras A, Büyüközkan G (2023) Digital transformation journey guidance: a holistic digital maturity model based on a systematic literature review. Systems 11(4):1–31. https://doi.org/10.3390/systems11040213
23. Palko D et al (2023) Cyber Security risk Modeling in distributed information systems. Appl Sci 13(4):2393. https://doi.org/10.3390/app13042393
24. Tokerud S, Jansen J, Niemimaa M, Järveläinen J (2023) Designing extended zero trust maturity model—from technical to socio-technical. ICIS 2023 Proc., January, 2023. [Online]. https://aisel.aisnet.org/icis2023/cyber_security/cyber_security/5
25. Li J, Lyu L, Liu X, Zhang X, Lyu X (2022) FLEAM: a federated learning empowered architecture to mitigate DDoS in industrial IoT. IEEE Trans Ind Inform 18(6):4059–4068. https://doi.org/10.1109/TII.2021.3088938
26. Al Rawajbeh M et al (2023) A new model for security analysis of network anomalies for IoT devices. Int J Data Netw Sci 7(3):1241–1248. https://doi.org/10.5267/j.ijdns.2023.5.001
27. Toussaint M, Krima S, Panetto H (2024) Industry 4.0 data security: a cybersecurity frameworks review. J Ind Inf Integr 39(March):100604. https://doi.org/10.1016/j.jii.2024.100604
28. Yigit Ozkan B, van Lingen S, Spruit M (2021) The cybersecurity focus area maturity (CYSFAM) model. J Cybersecurity Priv 1(1):119–139. https://doi.org/10.3390/jcp1010007
29. Abohatem AY, Ba-Alwi FM (2024) Cybersecurity maturity assessment of information Systems for Yemen Telecoms. Int J Intell Syst Appl Eng 12(8s):539–548
30. de Jesus C, Lima RM (2020) Literature search of key factors for the development of generic and specific maturity models for industry 4.0. Appl Sci 10(17):5825. https://doi.org/10.3390/app10175825
31. Shaheen K, Zolait AH (2023) The impacts of the cyber-trust program on the cybersecurity maturity of government entities in the Kingdom of Bahrain. Inf Comput Secur 31(5):529–544. https://doi.org/10.1108/ICS-06-2022-0108
32. Lohmann PA, Albuquerque C, Machado R (2023) Systematic literature review of threat modeling concepts. Int Conf Inf Syst Secure Priv no. March:163–173. https://doi.org/10.5220/0011783000003405
33. Yi H, Nie Z (2018) Side-channel security analysis of UOV signature for cloud-based Internet of things. Futur Gener Comput Syst 86:704–708. https://doi.org/10.1016/j.future.2018.04.083
34. Antonishyn M, Misnik O (2019) Analysis of testing approaches to android mobile application vulnerabilities. CEUR Workshop Proc 2577:270–280
35. Almuhammadi S, Alsaleh M (2017) Information Security maturity model for NIST cyber Security framework. NIST Spec. Publ, pp 51–62. https://doi.org/10.5121/csit.2017.70305
36. Brito-Acuña G (2023) Aeronautic maturity cybersecurity: a framework. DYNA 90(227):24–34. https://doi.org/10.15446/dyna.v90n227.107420
37. Abdullahi M et al (2022) Detecting cybersecurity attacks in the Internet of things using artificial intelligence methods: a systematic literature review. Electron 11(2):1–27. https://doi.org/10.3390/electronics11020198
38. Muronga K, Herselman M, Botha A, Da Veiga A (2019) An analysis of assessment approaches and maturity scales used for evaluation of information security and cybersecurity user awareness and training programs: a scoping review. 2nd Int. Conf. Next Gener. Comput. Appl. 2019, NextComp 2019—Proc., no. September, pp 1–6. doi:https://doi.org/10.1109/NEXTCOMP.2019.8883535

39. Abdullahi Garba A, Musa Bade A, Yahuza M, Nuhu Y (2020) Cybersecurity capability maturity models review and application domain. Int J Eng Technol 9(3):779. https://doi.org/10.14419/ijet.v9i3.30719
40. Nadir I, Bakhshi T (2018) Contemporary cybercrime: a taxonomy of ransomware threats & mitigation techniques. 2018 Int. Conf. Comput. Math. Eng. Technol. Inven. Innov. Integr. Socioecon. Dev. iCoMET 2018—Proc., vol. 2018-January, no. February, pp 1–7. doi:https://doi.org/10.1109/ICOMET.2018.8346329
41. Mbona I, Eloff JHP (2022) Detecting zero-day intrusion attacks using semi-supervised machine learning approaches. IEEE Access 10(July):69822–69838. https://doi.org/10.1109/ACCESS.2022.3187116
42. Sahi SK (2017) A study of Wannacry ransomware attack. Int J Eng Res Comput Sci Eng 4(9):5–7
43. Shulha O, Yanenkova I, Kuzub M, Muda I, Nazarenko V (2022) Banking information resource cybersecurity system Modeling. J Open Innov Technol Mark Complex 8(2):80. https://doi.org/10.3390/joitmc8020080
44. Neisse R, Steri G, Fovino IN, Baldini G (2015) SecKit: a model-based Security toolkit for the Internet of things. Comput Secur 54:60–76. https://doi.org/10.1016/j.cose.2015.06.002
45. Ramanayaka CDD, Venkatachalam S (2015) Reflection on BIM development practices at the pre-maturity. Proc Eng 123:462–470. https://doi.org/10.1016/j.proeng.2015.10.092
46. Abbas S (2019) An efficient sybil attack detection for the Internet of things. In: Advances in intelligent systems and computing. Springer Verlag, pp 339–349. https://doi.org/10.1007/978-3-030-16184-2_33
47. Shohel Rana M, Gudla C, Sung AH (2018) Evaluating machine learning models for android malware detection—a comparison study. ACM Int Conf Proc Ser no. December:17–21. https://doi.org/10.1145/3301326.3301390
48. Leka E, Selimi B, Lamani L (2019) Systematic literature review of blockchain applications: smart contracts. 2019 Int. Conf. Inf. Technol. InfoTech 2019—Proc., no. September, pp 19–20. doi: https://doi.org/10.1109/InfoTech.2019.8860872
49. Cetindamar D, Shdifat B, Erfani E (2022) Understanding big data analytics capability and sustainable supply chains. Inf Syst Manag 39(1):19–33. https://doi.org/10.1080/10580530.2021.1900464
50. Al-rimy BAS, Maarof MA, Shaid SZM (2018) Ransomware threat success factors, taxonomy, and countermeasures: a survey and research directions. Comput Secur 74:144–166. https://doi.org/10.1016/j.cose.2018.01.001
51. Heydari J et al (2019) Towards a circular economy for packaging waste by using new technologies: the case of large multinationals in emerging economies. J Clean Prod 11(1):438–450. https://doi.org/10.1016/j.resconrec.2017.08.017
52. Golosova J, Romanovs A (2018) The advantages and disadvantages of the blockchain technology. 2018 IEEE 6th Work. Adv. Information, Electron. Electr. Eng. AIEEE 2018—Proc., no. November. doi: https://doi.org/10.1109/AIEEE.2018.8592253
53. Williams P, Kaylan I, Daoud H, Bayoumi M (2022) A survey on security in Internet of things with a focus on the impact of emerging technologies. Internet of Things 19:100564. https://doi.org/10.1016/j.iot.2022.100564
54. Agarwal R, Thapliyal T, Shukla SK (2021) Vulnerability and transaction behavior based detection of malicious smart contracts. IEEE Commun Mag 4(2):1–13
55. Tsabary I, Manuskin A, Eyal I (2022) LedgerHedger: gas reservation for smart-contract Security. IACR Cryptol. ePrint Arch 2022:56. https://eprint.iacr.org/2022/056
56. Beillahi SM, Ciocarlie G, Emmi M, Enea C (2020) Behavioral simulation for smart contracts. Proc ACM SIGPLAN Conf Program Lang Des Implement 20(4):470–486. https://doi.org/10.1145/3385412.3386022
57. Ivanov D, Dolgui A (2019) New disruption risk management perspectives in supply chains: digital twins, the ripple effect, and resileanness. IFAC-PapersOnLine 52(13):337–342. https://doi.org/10.1016/j.ifacol.2019.11.138

58. Hu Y, et al. (2021) Security threats from bitcoin wallet smartphone applications: vulnerabilities, attacks, and countermeasures. CODASPY 2021—Proc. 11th ACM Conf. Data Appl. Secure. Priv., pp. 89–100. doi:https://doi.org/10.1145/3422337.3447832
59. Jurcut A, Niculcea T, Ranaweera P, Le-Khac N-A (2020) Security considerations for Internet of things: a survey. SN Comput Sci 1(4):1–19. https://doi.org/10.1007/s42979-020-00201-3
60. Faghihi F, Abadi M, Tajoddin A (2018) SMSBotHunter: a novel anomaly detection technique to detect SMS botnets. 2018 15th Int. ISC (Iranian Soc Cryptology) conf inf secur cryptology, ISC. 2018, pp. 1–6. doi: https://doi.org/10.1109/ISCISC.2018.8546934
61. Sánchez-García ID, Feliu Gilabert TS, Calvo-Manzano JA (2023) Countermeasures and their taxonomies for risk treatment in cybersecurity: a systematic mapping review. Comput Secur 128:103170. https://doi.org/10.1016/j.cose.2023.103170
62. Alam G, Mahmood S, Alshayeb M, Niazi M, Zafar S (2024) Maturity model for secure software testing. J Softw Evol Process 36(5):e2593. https://doi.org/10.1002/smr.2593
63. Nallakaruppan MK, Somayaji SRK, Fuladi S, Benedetto F, Ulaganathan SK, Yenduri G (2024) Enhancing security of host-based intrusion detection systems for the Internet of things. IEEE Access 12(December 2023):31788–31797. https://doi.org/10.1109/ACCESS.2024.3355794
64. Maghrabi LA et al (2024) Enhancing cybersecurity in the Internet of things environment using bald eagle search optimization with hybrid deep learning. IEEE Access 12(December 2023):8337–8345. https://doi.org/10.1109/ACCESS.2024.3352568
65. Mohammad Shah IN, Ismail ES, Samat F, Abd Rahman NN (2023) Modified generalized feistel network block cipher for the Internet of things. Symmetry (Basel) 15(4):900. https://doi.org/10.3390/sym15040900
66. Dzemydienė D, Burinskienė A, Čižiūnienė K, Miliauskas A (2023) Development of E-service provision system architecture based on IoT and WSNs for monitoring and management of freight intermodal transportation. Sensors 23(5):2831. https://doi.org/10.3390/s23052831
67. Tauseef M, Kounte MR, Nalband AH, Ahmed MR (2023) Exploring the joint potential of blockchain and AI for securing Internet of things. Int J Adv Comput Sci Appl 14(4):885–895. https://doi.org/10.14569/IJACSA.2023.0140498
68. Uchani Gutierrez OC, Xu G (2023) Blockchain and smart contracts to secure property transactions in smart cities. Appl Sci 13(1):66. https://doi.org/10.3390/app13010066
69. Ahsan M, Nygard KE, Gomes R, Chowdhury MM, Rifat N, Connolly JF (2022) Cybersecurity threats and their mitigation approaches using machine learning—a review. J Cybersecurity Priv 2(3):527–555. https://doi.org/10.3390/jcp2030027
70. Safarov F, Basak M, Nasimov R, Abdusalomov A, Cho YI (2023) Explainable lightweight block attention module framework for network-based IoT attack detection. Futur Internet 15(9):297. https://doi.org/10.3390/fi15090297
71. Centobelli P, Cerchione R, Del Vecchio P, Oropallo E, Secundo G (2022) Blockchain technology for bridging trust, traceability, and transparency in the circular supply chain. Inf Manag 59(7):103508. https://doi.org/10.1016/j.im.2021.103508
72. Khan JA, Chowdhury MM (2021) Security analysis of 5G network. IEEE Int Conf Electro Inf Technol, vol. 2021-May, pp. 1–6. doi: https://doi.org/10.1109/EIT51626.2021.9491923
73. Chen S, Pang Z, Wen H, Yu K, Zhang T, Lu Y (2021) Automated labeling and learning for physical layer authentication against clone node and sybil attacks in industrial wireless edge networks. IEEE Trans Ind Inform 17(3):2041–2051. https://doi.org/10.1109/TII.2020.2963962
74. Rathore S, Park JH (2018) Semi-supervised learning based distributed attack detection framework for IoT. Appl Soft Comput J 72:79–89. https://doi.org/10.1016/j.asoc.2018.05.049
75. Veksler VD, Buchler N, LaFleur CG, Yu MS, Lebiere C, Gonzalez C (2020) Cognitive models in cybersecurity: learning from expert analysts and predicting attacker behavior. Front Psychol 11(June):1–13. https://doi.org/10.3389/fpsyg.2020.01049
76. Moore C (2016) Detecting ransomware with honeypot techniques. Proc.—2016 cybersecurity cyberforensics conf. CCC 2016, pp. 77–81. doi:https://doi.org/10.1109/CCC.2016.14
77. Hatzivasilis G et al (2019) WARDOG: awareness detection watchdog for botnet infection on the host device. IEEE Trans Sustain Comput 6:4–18. https://doi.org/10.1109/tsusc.2019.2914917

78. Araki S, Takahashi K, Hu B, Kamiya K, Tanikawa M (2019) Subspace clustering for interpretable botnet traffic analysis. IEEE Int Conf Commun, vol. 2019-May, pp. 0–5. https://doi.org/10.1109/ICC.2019.8761218
79. Ali S, Li Q, Yousafzai A (2024) Blockchain and federated learning-based intrusion detection approaches for edge-enabled industrial IoT networks: a survey. Ad Hoc Netw 152(December):103320. https://doi.org/10.1016/j.adhoc.2023.103320
80. Hasan MZ, Mohd Hanapi Z (2023) Efficient and secured mechanisms for data link in IoT WSNs: a literature review. Electron 12(2):458. https://doi.org/10.3390/electronics12020458
81. Hu T, Liu X, Chen T, Zhang X, Huang X, Niu W, Lu J, Zhou K, Liu Y (2021) Transaction-based classification and detection approach for ethereum smart contract. Inf Process Manag 58(2):102462. https://doi.org/10.1016/j.ipm.2020.102462
82. Butun I, Osterberg P, Song H (2020) Security of the Internet of things: vulnerabilities, attacks, and countermeasures. IEEE Commun Surv Tutorials 22(1):616–644. https://doi.org/10.1109/COMST.2019.2953364
83. Wang Z, He B, Yang Y, Shen C, Peña-Mora F (2020) Building a next generation AI platform for AEC: a review and research challenges. Inform Technol Constr Conf 12(9):27–45. https://doi.org/10.46421/2706-6568.37.2020.paper003

Chapter 11
Protecting Water and Wastewater Critical Infrastructure from Hybrid Threats

Termine Khachikyan, Ashok Vaseashta, Gor Khachatryan, and Gor Gevorgyan

Abstract Modern life and human progress are supported by the operation of critical infrastructure. Although they ensure the well-being of human life, they also present risks that are difficult to manage. These risks are particularly exacerbated when critical infrastructures are interdependent and are maintained by the public and private sectors. Water supply systems are among the most important, yet vulnerable, critical infrastructures in modern society. Safe drinking water is a prerequisite for ensuring public health and human activities, and treated wastewater is essential for the prevention of waterborne diseases and environmental safety. Today, as in the past, the world is witnessing health, economic, environmental, and other problems caused by intentional anthropogenic impacts on water and wastewater facilities. Improving the safety and stability of this critical infrastructure is therefore a national strategic priority for all countries around the world. We present certain historical and current events leading to the disruption of the security of the water and wastewater facilities, especially from hybrid threats. Existing legal standards and current regulations for the protection of critical infrastructures, including the water and wastewater sector, are discussed. Based on our comprehensive analysis of the available data, methodological solutions to modernize the framework for the protection of water and wastewater critical infrastructure are proposed.

Keywords Critical infrastructure · Water · Wastewater · Risk assessment · Structural monitoring · Counterterrorism

---

T. Khachikyan · G. Khachatryan · G. Gevorgyan (✉)
Scientific Center of Zoology and Hydroecology, National Academy of Sciences of RA, Yerevan, Armenia
e-mail: gor.gevorgyan@sczhe.sci.am

A. Vaseashta
Office of Research, International Clean Water Institute, Manassas, VA, USA

Academy of Sciences of Moldova, Chisinau, Moldova

D. Radu et al. (eds.), *Countering Hybrid Threats Against Critical Infrastructures*, NATO Science for Peace and Security Series B: Physics and Biophysics, https://doi.org/10.1007/978-94-024-2304-4_11

11.1 Introduction

Water is one of the most fundamental requirements for sustaining life on our planet. Water and society are interconnected [1, 2] and since the beginning of civilization, societies have been built around water resources. Water collection, distribution, storage, and other aspects, such as wastewater treatment—were later termed as part of the critical infrastructure. Fast forward to modern times, the term critical infrastructure (CI) is defined differently by various countries, however, fundamentally, it represents systems, assets, and/or their components that are of primary importance and are essential for maintaining vital social functions, security, and economic stability (Fig. 11.1; [3]). The CI in the USA, and other countries, now includes the following sectors: chemical; defense industrial base; healthcare and public health; commercial facilities; emergency services; information technology; communications; energy; nuclear reactors, materials, and waste; advanced manufacturing; financial services; transportation systems; bridges and dams; food and agriculture; water and wastewater systems; and government facilities [3, 4].

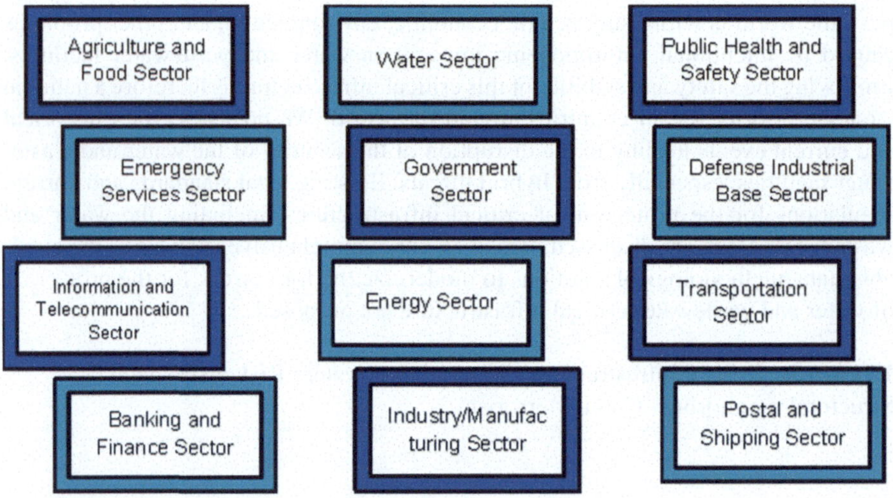

Fig. 11.1 Traditional CI sectors

11.2 Threats and Vulnerabilities

Each sector of CI has its own vulnerabilities and unique security requirements. These sectors are often exposed to various threats—natural, incidental, accidental, and intentional such as climate change, natural disasters, epidemics, terrorist attacks, and cyberattacks. In addition to natural and physical threats, cyber threats have shown a significant upward trend in recent decades [5, 6]. In the general context of threats, so-called hybrid threats are distinguished (Fig. 11.2), some of which are related to the fusion of technologies and information security systems.

In the modern era, the information factor plays a crucial role in various spheres of society. While this factor is a vital component for public development, it can also pose significant risks to CI when used in hybrid tactics (Fig. 11.3). Management of these infrastructures often relies on a range of technologies, including individual devices, operating systems, software, network packages, information and cloud technologies, and the Internet. Notably, active hacker groups have been identified in over 130 countries, continuously developing new malicious programs and executing cyberattacks that collectively amount to several 100 million annually [7, 8]. The purpose of these attacks is to inflict financial and economic damage on opponents or competing business organizations or to obtain personal profit through financial extortion or psychological victories. Recently, there has been a noticeable increase in cyber-based ransomware attacks (Fig. 11.4; [9]). These attacks involve viruses that encrypt valuable computer files, which hackers then use for various purposes. Although cyber tools are increasingly accessible, the damage they cause can be significant.

The damage caused to the global economy and CI by cybercrime and cyberattacks continues to rise each year. For instance, Africa, with over 500 million Internet users, is particularly vulnerable to cybercriminals (Fig. 11.5). In 2021 alone, cyberattacks–mostly small-scale–cost the continent nearly $4 billion and reduced its GDP by 10%. South Africa was hit hardest by ransomware attacks in the first quarter of 2021, according to Interpol's African Cyberthreat Assessment Report. Government agencies in South Africa are now prime targets for hackers. The

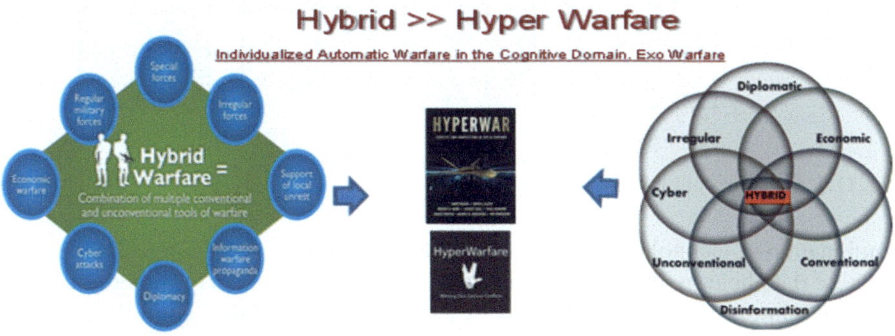

Fig. 11.2 Context of the hybrid nature of threats

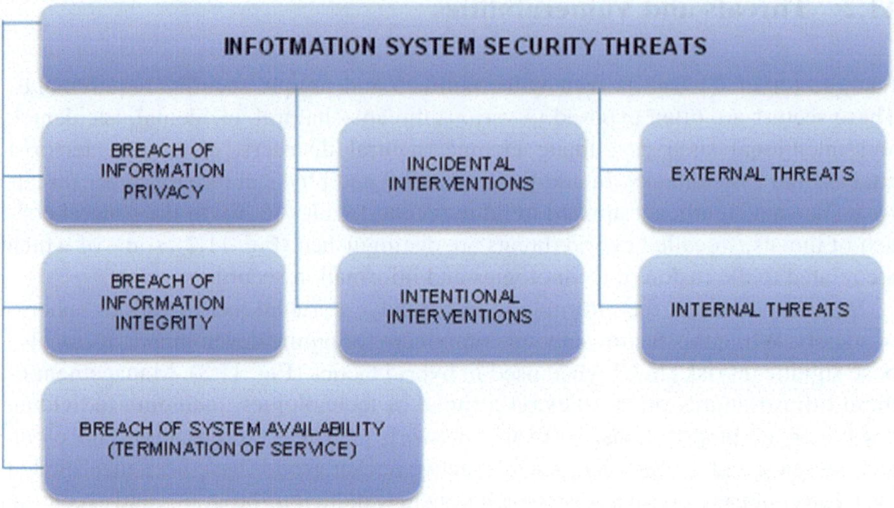

Fig. 11.3 Classification of information system security threats [10]

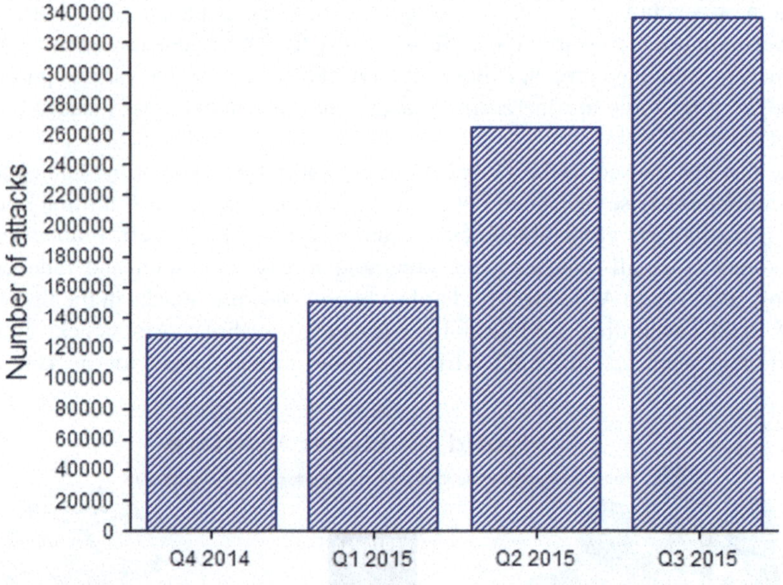

Fig. 11.4 Dynamics of ransomware attacks within one year

combination of technology gaps, insufficient funding, lack of training, and the high value of data held by these organizations makes them particularly susceptible. The economic impact on African states is expected to increase, highlighting the urgent need for enhanced cybersecurity measures and infrastructure protection [7].

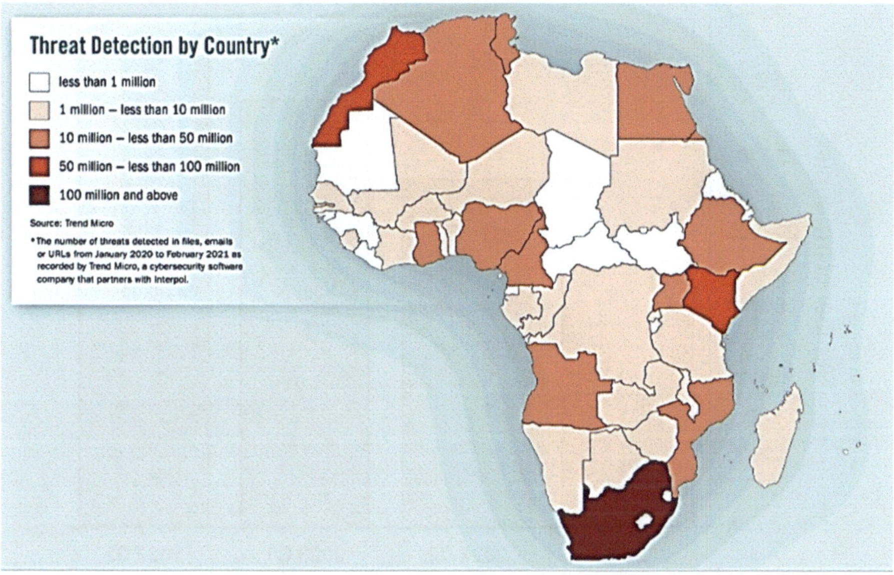

Fig. 11.5 Cyberattacks detected in Africa from January 2020 to February 2021 (ADF, [33])

The American Internet Crime Complaint Center (IC3) also analyzed data on attacks against American CI for 2021. The number of ransomware attacks alone totaled approximately 649, with the Healthcare and Public Health sector being the most affected, followed by the Financial Services and Information Technology sectors. According to IC3 reports, California, Florida, Texas, and New York were the most impacted states during this period [11].

According to the latest RTK-Solar report, 325,000 information security incidents were detected in Russia in the second quarter of 2023. This represents a 12% increase from the first quarter and a 38% rise compared to the same period last year (Fig. 11.6). A noticeable trend of increasing cyberattacks was also recorded in India from 2015 to 2022 (Fig. 11.7). The largest cyberattacks recorded in 2023 are mentioned in Fig. 11.7.

During the Russia-Ukraine war, bilateral cyberattacks targeting CI were carried out alongside conventional military operations. Ukrainian networks across various sectors, including finance, telecommunications, energy, media, military, and government, experienced extensive cyberattacks. These included denial-of-service (DoS) attacks and large-scale phishing campaigns. Conversely, Russia also faced cyber-attacks, with some federal agency websites, including the Department of Energy, being compromised by unknown attackers, disrupting supply chains. Cyber-attacks against Russia targeted state-related entities, as well as sectors such as military, banking, logistics, transport, and energy. In the context of these conflicts, the use of cyber tools often extends the boundaries of the conflict, involving other states in cyber operations (Figs. 11.8 and 11.9; [8, 26]).

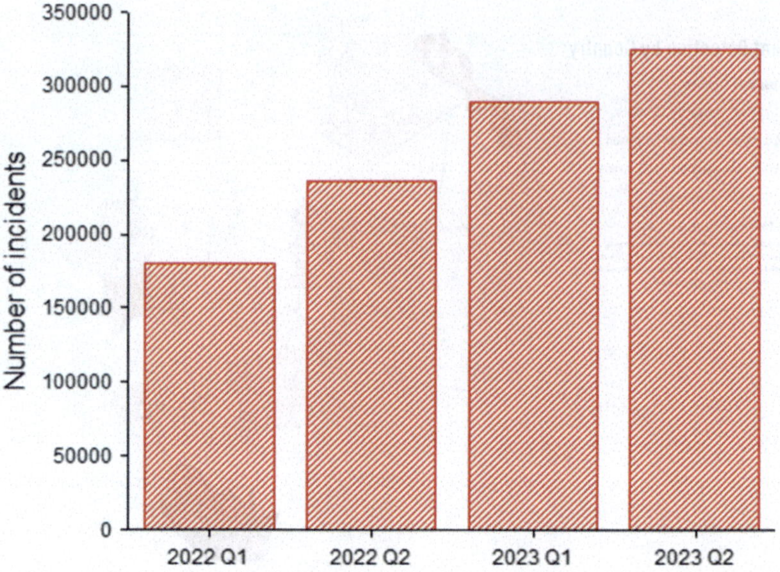

Fig. 11.6 Comparative dynamics of cyberattacks on the Russian information security system (RG, [34])

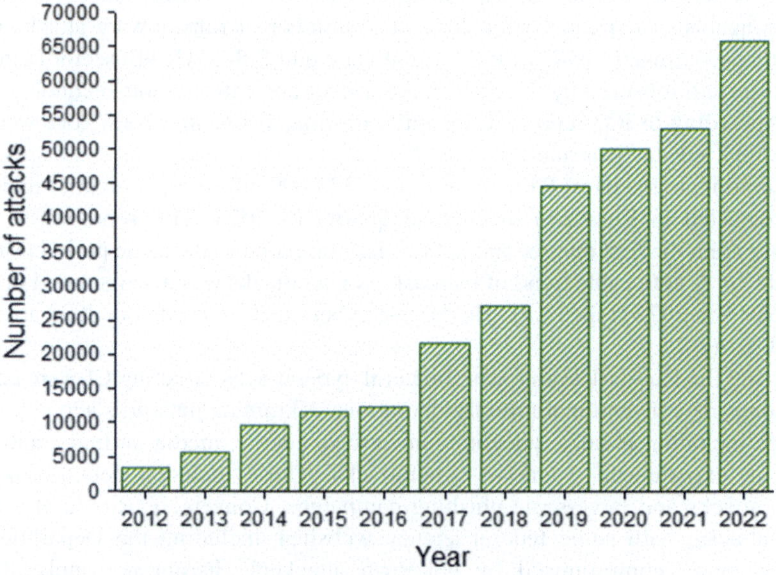

Fig. 11.7 Multi-year dynamics of reported cyberattacks in India (Statista [25])

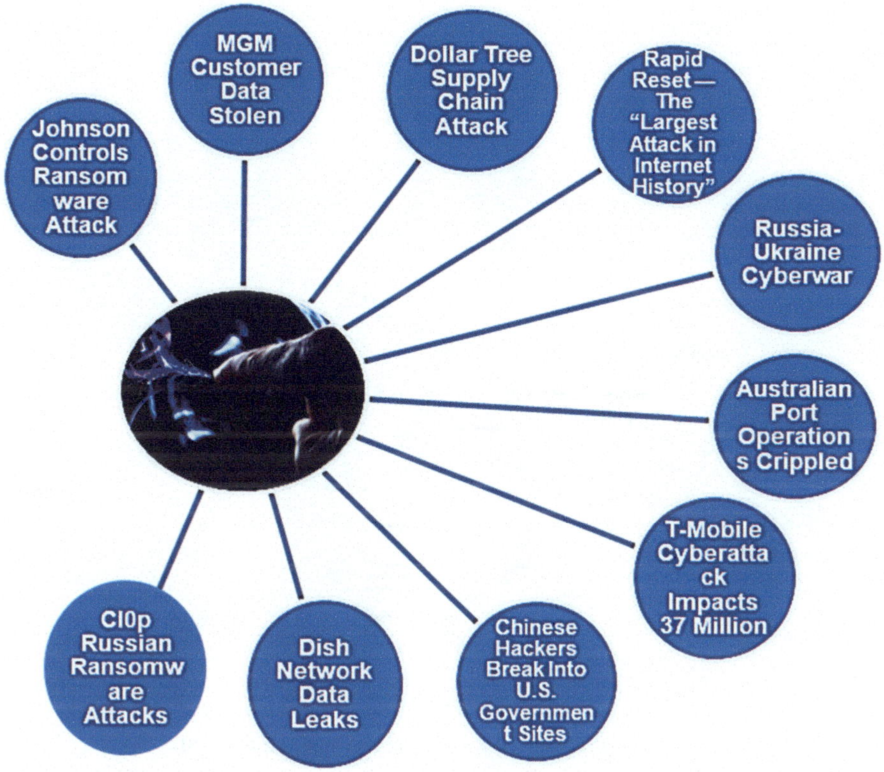

Fig. 11.8 Largest cyberattacks recorded in 2023 (MSSP Alert [26])

These cases highlight the critical need to invest in cybersecurity to protect not only digital assets but also the physical safety of citizens and the nation's CI. Among the CIs, water and wastewater infrastructure is especially important. The importance and criticality of water CI are validated, as it is generally agreed that water is (just after air) the most precious commodity for mankind to survive [12, 13]. At the same time, the trends of increasing demographic indicators, the diversification of ways of consumption of water resources due to the increase in living standards, and the increase in volumes have led to a noticeable deterioration of both quantitative and qualitative indicators of water [14, 15]. Trends in negative shifts over time are also predictable [16].

The operation of drinking water and wastewater treatment systems is interconnected with areas of vital importance for society, such as health care and uninterrupted operation of sectoral infrastructures, transport, energy, agriculture, food security, etc. (see Fig. 11.10).

This infrastructure encompasses water treatment plants, distribution systems, dams, and reservoirs. Disruptions to this infrastructure can lead to severe consequences, including water shortages, public health risks, and even loss of life. In the

Fig. 11.9 Cyberattacks associated with the Russia-Ukraine War in Europe [8]

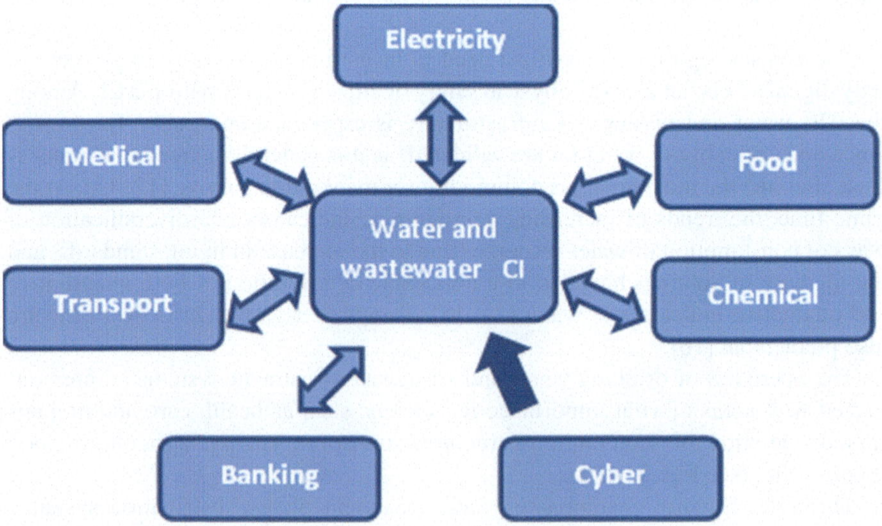

Fig. 11.10 Dependency of water CI on other CI systems (Birkett [13])

U.S., approx. 153,000 public drinking water systems serve about 80 percent of the population, and over 16,000 wastewater treatment plants are responsible for sewage treatment. Water supply and wastewater management are crucial national functions, and disruptions, inefficiencies, or corruption in these sectors can lead to significant negative impacts on public health, economic security, and national security [17].

The role and maintenance of water infrastructure are especially critical in countries facing physical water shortages or challenging climate-geographic conditions. This infrastructure is highly vulnerable to various threats and attacks, making its protection a top priority. Key dangers include global climate change, escalating ecological risks, natural disasters, and the adverse effects of ecosystem degradation caused by artificial infrastructures and other anthropogenic activities. Additionally, hybrid threats in their various forms pose significant risks to this vital infrastructure. Gleick identified 32 countries as deficient in water resources, highlighting the need for innovative water sustainability initiatives [16]. This anticipated deficiency not only increases the value of water as a critical resource but also heightens the risk of adverse human intervention targeting drinking water supplies. Furthermore, the scarcity of water raises its profile as a potential target for terrorist attacks [13]. Historical examples of attacks on water-based CI with an increasing trend underscore this vulnerability (see Fig. 11.11).

As water infrastructure systems increasingly connect to the Internet and rely on computer control and regulation, they become more vulnerable, particularly to cyberattacks (Fig. 11.12). Hackers can exploit malware and other techniques to gain unauthorized access and disrupt system performance. Various tools underpin these cyberattacks on water infrastructure, with the main forms outlined in Table 11.1. These tools can be combined with different attack strategies to achieve their intended effects (Fig. 11.13).

In 2004, 50 incidents of unauthorized intrusion into water systems were identified in the U.S. by 10 federal agencies. These incidents involved SCADA computer systems that control water supply and wastewater systems [19, 20]. In 2015, the

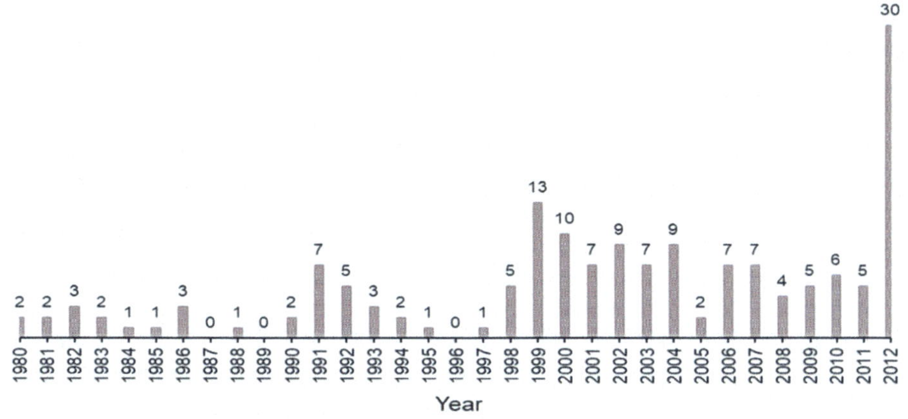

Fig. 11.11 Number of recorded attacks on water and wastewater CI from 1980 to 2012 [13]

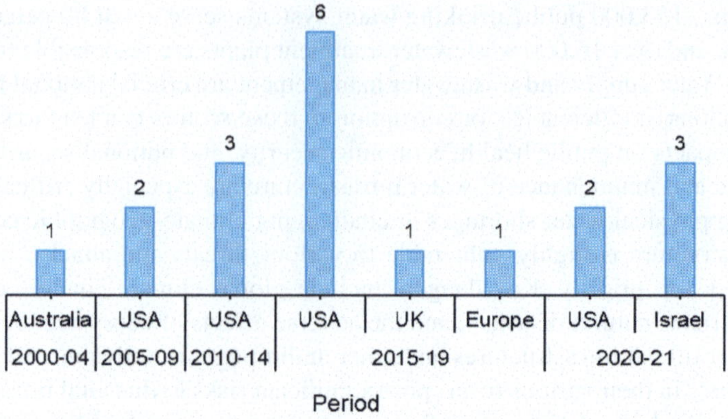

Fig. 11.12 Number of publicly known cyberattacks on water and wastewater CI from 2000 to 2021 [18]

Table 11.1 Main tools used in cyberattacks on water and wastewater CI

Ransomware	This malicious malware encrypts a system's data, rendering it inaccessible. Attackers then demand a ransom payment to restore access. Disrupting water treatment or distribution to force a ransom payment can have devastating consequences.
Malware	Malicious software, such as viruses or worms, can infiltrate water system controls, potentially disrupting operations, damaging equipment, or even altering water treatment processes.
Denial-of-Service (DoS)	This attack floods a system with excessive traffic, overwhelming it and blocking legitimate users from accessing critical controls. A denial-of-service (DoS) attack on a water treatment plant could prevent operators from monitoring or adjusting essential systems.
Man-in-the-Middle (MitM)	In this attack, hackers intercept communications between systems, potentially stealing data or issuing false commands to disrupt operations. An attacker using a man-in-the-middle (MitM) attack could alter water treatment chemical dosages or interfere with communication between control systems.
Social engineering	Social engineering manipulates individuals into disclosing information or access. This technique could be used to gain entry to control systems or steal credentials for more sophisticated cyberattacks.

U.S. Department of Homeland Security (DHS) reported 245 incidents, with the water sector being the fourth largest category for which DHS provided incident response support (Fig. 11.14). Clearly, these incidents had the potential to directly impact water supply systems [21].

In 2020, the Israeli government reported cyberattacks targeting water supply and treatment facilities, which aimed to increase chlorine levels in the water. Authorities urged these facilities to change their passwords in response [22]. In 2020, Greenville Water in the USA announced a targeted cyberattack that affected online payments for the 500,000 people it serves. That same month, a breach exposed customer-sensitive information from 300 accounts at the Detroit Water & Sewerage

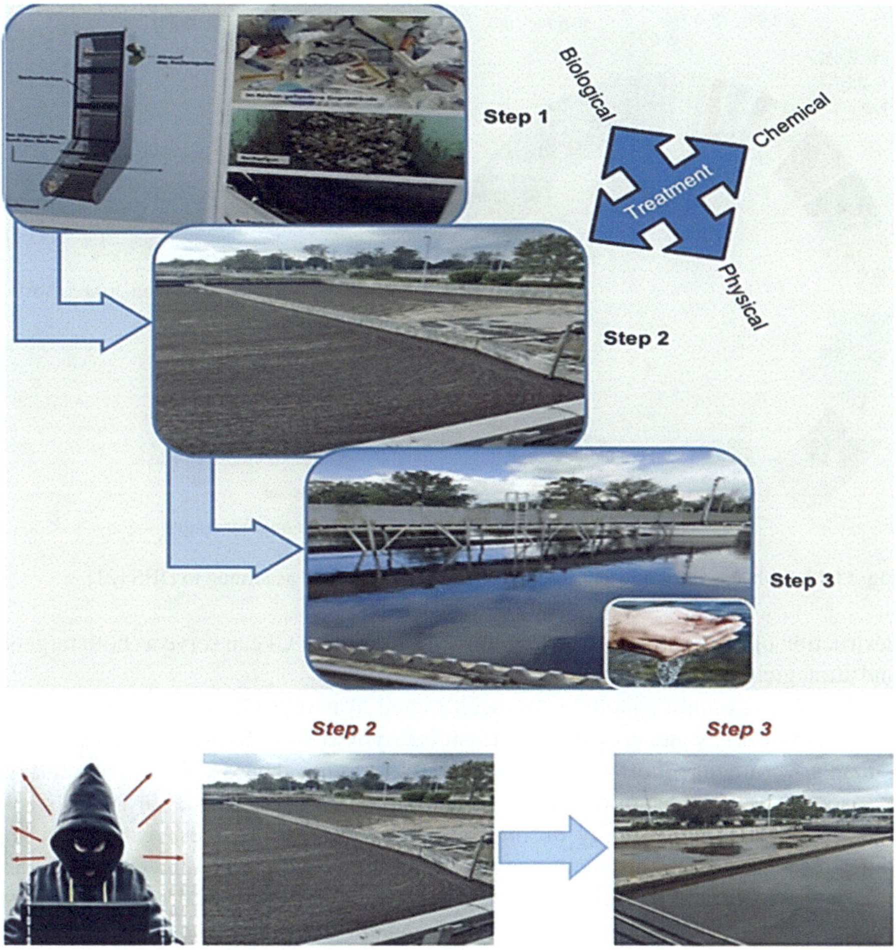

Fig. 11.13 Scheme of a cyberattack on a wastewater treatment plant

Department in Michigan [18]. In 2021, a water treatment plant in Oldsmar, Florida, was hacked, allowing the attacker to temporarily increase sodium hydroxide levels. The situation was quickly noticed and corrected by an operator before any significant harm could occur [18].

The risk of cyberattacks is a constant concern in the water and wastewater sector. However, in the context of hybrid threats, this infrastructure is also highly vulnerable to terrorist attacks and challenges arising from military-political conflicts. These threats are characterized by new developments against a backdrop of historical precedents. Throughout military operations, attacks on water resources have been frequent. According to the "Water Conflict Chronology List" prepared by the Pacific Institute, such attacks, though varied, can be categorized into four main groups: diversion or cutting off of water, flooding, poisoning of water resources, and

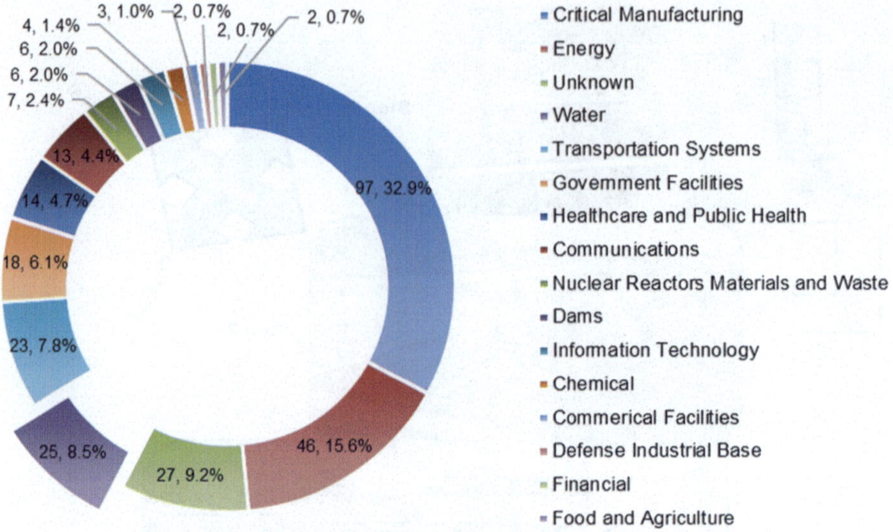

Fig. 11.14 Cybersecurity incidents recorded in the USA in 2015 according to DHS [21]

destruction of water infrastructure. In this context, water CI can serve as both targets and strategic tools.

Historical examples include the Arab Israeli War in 1948, during which Arab troops cut off the water supply from Rosh Ha'ayin to West Jerusalem. In 1964, the Cuban government cut off water supplies to the U.S. naval base at Guantanamo Bay in response to the arrest of Cuban fishermen off the coast of Florida. In 1982, Israel cut off Beirut's water supply during the city's siege. Sometimes, restrictions on water resources and infrastructure are used not merely to weaken the enemy but as a strategic lever to forcefully achieve specific goals. For example, Kazakhstan's failure to send coal to Kyrgyzstan led Kyrgyzstan to close the reservoirs supplying water to Kazakhstan until 1999. In 2000, Uzbekistan briefly cut off Kazakhstan's water supply due to non-payment of debt [23]. The approaches mentioned above raise significant concerns, as many states share transboundary water resources. Upstream countries could potentially use similar strategic tactics to exert control over downstream nations.

Major powers can leverage control over critical water management systems to gain geopolitical advantages over other states. For example, the People's Republic of China, by constructing an extensive network of dams and managing the water resources of the Tibetan Plateau, controls the flow of key rivers such as the Mekong and Brahmaputra. This gives China significant influence over countries like Vietnam, Bangladesh, India, Nepal, and Bhutan regarding water availability. The Indo-Pacific region has emerged as a potential hotspot for water-related conflicts, with China using its control over transboundary water flows to exert financial, commercial, and political leverage over its neighbors. By manipulating this asymmetric control, China aims to dominate the water resources of the Indo-Pacific region [24, 25].

Another alarming trend is the use of water control strategies by armed non-state actors. The Taliban, for example, were among the first to employ this tactic. Between 1998 and 2001, the Taliban cut off the flow of the Helmand River towards Sistan-Baluchistan in Iran multiple times. This was motivated by drought conditions in Afghanistan. Since Sistan-Baluchistan's irrigation depended heavily on the Helmand River, the Taliban's actions pressured Iran to deploy forces into Afghanistan to build canals and restore water supplies to the region. Similar strategic tactics have since been used by terrorist groups in Syria, Yemen, and Libya.

There are several instances where floods have been deliberately caused as a military tactic during conflicts. One notable historical example is from the Sino-Japanese War in 1939, when the Chinese Nationalist Army blew up the flood-control walls of the Yellow River in an attempt to halt the Japanese advance on Wuhan. Despite causing thousands of casualties and widespread destruction, this tactic ultimately failed to achieve its objective (Futehally, 2019). In November 1944, Soviet troops deliberately flooded the area south of the Istra Reservoir near Moscow as a strategy to impede the German advance on the city. Shortly thereafter, German forces employed a similar tactic, creating a water barrier to halt the Soviet 16th Army's progress [19].

Despite the unpredictability of using flooding as a military tactic, it was later adopted by non-state actors. In 2003, an unknown terrorist group in Baghdad carried out bomb attacks that damaged water pipes and caused street flooding. Since then, terrorist groups have increasingly exploited control of water CI as a strategic tactic. This has led to the deliberate flooding of villages and towns in Iraq, Syria, and other regions [23].

Currently, advancements in technology significantly increase the risks of intentional flooding, especially when combined with cyberattacks. Within the scope of attacks on water resources, there have been instances where sources of drinking water were damaged and water CI was polluted. For example, in 2007, an unknown group bombed an oil pipeline near Baiji in Iraq, contaminating the Tigris River. This river was a major source of drinking water for nearby communities.

Following the 09/11 attacks on New York, numerous false reports of terrorist activity surfaced globally. U.S. law enforcement and intelligence agencies discovered that Al-Qaeda members were gathering information on water supply and wastewater management practices both in the U.S. and internationally. In 2003, Al-Qaeda issued a threat through a London-based, Saudi-owned magazine, stating that "Al-Qaeda does not rule out using sarin gas and poisoning drinking water in the U.S. and other Western cities". Armed non-state actors, such as the Taliban and Hezbollah, have since exploited these threats by carrying out deliberate attacks on water infrastructure to achieve their various objectives [23]. In 2003, 2005, and 2007, the Taliban targeted the Kajaki Dam, a structure of strategic importance for Afghanistan. The group consistently employed threats against water CI in various situations. For instance, in 2008, when the Pakistani Army launched a major offensive against the Taliban in Swat and Bajaur, the Taliban retaliated by threatening to blow up the Warsak Dam, which serves as the primary water supply for Peshawar [23]. In the early 2000s, the Revolutionary Armed Forces of Colombia (FARC) targeted national water and wastewater infrastructure as part of their strategy to undermine the Colombian government. They used various tactics to disrupt essential public services, aiming to cause economic damage and gain political leverage. Specifically, in 2002, FARC detonated an explosive device on the Chingasa dam valve, and in 2003 and 2004, they used poisons to contaminate water facilities.

Disputes between African farmers and Arab herders over water scarcity and pastures contributed to the Darfur conflict that began in 2003. The conflict saw the involvement of government-backed Arab militias known as the Janjaweed, who were accused of deliberately contaminating wells by dumping corpses into them in Central and South Darfur. This tactic exacerbated the humanitarian crisis and served as a weapon of war, further intensifying the violence and suffering in the region [23]. The poisoning of water resources can occur through various methods. Water infrastructure might be directly targeted to damage processing or delivery systems. Alternatively, water can be contaminated by introducing poisons, disease-causing pathogens, or harmful chemicals into the distribution systems, a tactic sometimes referred to as "backflow" [13]. In 1939, the Japanese military poisoned Soviet water sources with intestinal typhoid bacteria in an area near the former Mongolian border [13]. Between 1939 and 1942, the Japanese military unit known as Unit 731 poisoned Chinese water sources, wells, and bulk water storages with typhoid and other chemical and biological pathogens [19]. Therefore, overcoming these hybrid threats is crucial for the maintenance of water CI. Flexible and integrated mechanisms must be implemented to effectively address these threats and ensure the protection and sustainability of vital water systems.

11.3 Threat Mitigation Strategies

Addressing these challenges effectively is not only a matter of national security but often involves broader concerns. This is due to the cross-border nature of some CIs, the dynamic interactions between different infrastructure systems, or their interconnections. National efforts to enhance the safety and resilience of CI frequently depend on the ability of infrastructure managers and coordinators in both the public and private sectors to manage these systems effectively in both stable and crises. In this context, risk management at national, regional, state, and local levels is a fundamental factor.

Protecting water and wastewater infrastructure from hybrid threats requires a multi-layered approach that integrates physical security, cybersecurity, and organizational resilience strategies. It also involves developing and enforcing industry regulations and ensuring compliance with relevant laws. The protection mechanism involves several key steps, which are crucial components of a strategic plan for safeguarding CI (Fig. 11.15). A primary step in this process is conducting a thorough risk assessment of water and wastewater infrastructure to identify vulnerabilities and potential threats [21, 26].

Fig. 11.15 Schematic plan for the protection of water and wastewater CI from hybrid threats

Table 11.2 Strategic goals, actions, and performance indicators for risk management

Strategic goal	Actions	KPIs
Cyber security system upgrade	Implement cyber security measures	Quantitative and qualitative indicators for evaluating the effectiveness of actions
Improved security systems: material/ technical parts	Update physical security components	
Ensuring regulatory bases	Develop national legislative acts; knowledge of international agreements and sectoral laws	

The risk assessment should be conducted by a qualified expert group, considering factors such as the geographical location and critical importance of the water infrastructure, the availability of necessary institutional and programmatic components, the analysis of staff competencies, regional security issues, and interactions with other CIs. Following this comprehensive evaluation, it will be possible to identify the main types of hybrid threats facing the infrastructure and develop an action plan to address them. Next, it is essential to develop strategic goals aligned with the identified risks and to define targeted actions for their implementation (Table 11.2). To evaluate the effectiveness of these actions, both quantitative and qualitative key performance indicators (KPI) should be established (Table 11.3). These indicators can be monitored in the short, medium, and long term to facilitate benchmarking and guide the planning of subsequent actions.

Table 11.3 Indicators of action efficiency

Indicators	Description
Quantitative indicators	
Number of attacks detected	Quantify the frequency of attempted hybrid attacks on the infrastructure over a specific period.
Downtime	Measure the total downtime experienced due to hybrid attacks, including both planned and unplanned outages.
Financial losses	Calculate the direct financial losses incurred due to hybrid attacks, including costs related to infrastructure damage, loss of revenue, and expenses for recovery and remediation efforts.
Data breach impact	Quantify the extent of data breaches resulting from hybrid attacks, including the volume of sensitive information compromised.
Response time	Measure the average time taken to detect and respond to hybrid attacks, from the initial breach to containment and resolution.
System availability	Quantify the percentage of time that critical systems and services remain available despite attempted hybrid attacks
Resource utilization	Assess the impact of hybrid attacks on resource utilization, such as bandwidth, storage, and computing resources, during peak attack periods.
False positive rate	Measure the rate of false positives generated by security systems, which could indicate inefficiencies in threat detection and response mechanisms.
Effectiveness of mitigation measures	Quantify the effectiveness of mitigation measures implemented to protect against hybrid attacks, such as the percentage reduction in successful attacks or the number of incidents prevented.
Compliance metrics	Evaluate compliance with relevant cybersecurity regulations and standards, using quantitative measures such as the number of compliance violations or the percentage of regulatory requirements met.
Incident response quality	Assess the effectiveness and efficiency of incident response procedures, including the coordination of response teams and the clarity of communication during crises.
Resilience and adaptability	Evaluate the infrastructure's resilience to hybrid attacks and its ability to adapt to evolving threats, based on qualitative assessments of system architecture, redundancy, and flexibility.
Collaboration and information sharing	Qualitatively evaluate the level of collaboration with external stakeholders, such as government agencies, industry partners, and cybersecurity experts, through feedback and anecdotal evidence
Training and awareness	Assess the effectiveness of cybersecurity training and awareness programs through qualitative feedback from employees, focusing on their ability to recognize and respond to hybrid threats.
Innovation in security measures	Qualitatively evaluate the organization's commitment to innovation in cybersecurity, based on the adoption of cutting-edge technologies and proactive measures to stay ahead of emerging threats.
Cultural attitudes towards security	Gauge the organization's culture regarding cybersecurity, including attitudes toward risk management, accountability, and continuous improvement, through qualitative assessments of employee behavior and leadership priorities.

11.3.1 Cyber Security Measures

- Implement robust cybersecurity measures, including firewalls, intrusion detection and prevention systems, network segmentation, encryption, and regular security updates. These measures are crucial for protecting against cyber threats such as hackers, malware, and ransomware attacks.
- Develop an incident response plan that details the steps to be taken in the event of a security breach or other emergencies. Regularly update the plan and conduct routine exercises to test its effectiveness.
- Provide staff members with regular training and awareness programs to educate them about potential risks, including how to recognize and respond to suspicious activities, such as phishing attempts and social engineering tactics.
- Ensure supply chain security by evaluating suppliers and contractors, implementing contractual security requirements, and controlling third-party access to critical systems and data.

11.3.2 Improvement of Material and Technical Components of the Security System

- Implement physical security measures, such as access control systems and area control systems, to prevent unauthorized access to CI facilities.
- Build infrastructure resilience by implementing appropriate measures, backup systems, and disaster recovery plans to minimize the impact of failures or physical attacks.

11.3.3 Ensuring Regulatory Bases and Compliance

This is necessary to stay informed about relevant regulations and requirements related to the safety of water and wastewater CI and ensure compliance with applicable laws and standards. Laws and regulations for protecting water CI from hybrid attacks can vary by country or region. Many countries have enacted specific national legislation to safeguard water CI. These laws typically cover areas such as physical security, cybersecurity, and emergency preparedness and response planning. For example, in the U.S., several key regulations apply, including:

- Federal Water Pollution Control Act (Clean Water Act)
- Safe Drinking Water Act [27]
- Critical Infrastructure Protection Act [28]
- National Infrastructure Protection Plan

At the same time, international agreements, standards, and directives also play a role in the legal regulation of the sector.

11.3.4 Continuous Monitoring and Improvement

By considering both quantitative and qualitative indicators, organizations can gain a comprehensive understanding of the effectiveness of their actions in protecting water and wastewater CI from hybrid attacks. This approach enables the review and updating of security tools to adapt to evolving threats and vulnerabilities. In subsequent rounds, the defense plan can be further refined to achieve better results.

11.4 Conclusion, Policy Recommendations, and Future Pathways

Disruptions to water and wastewater CI pose a complex and critical challenge with significant consequences for society. Addressing these challenges requires a multifaceted comprehensive approach to include modernization, cross-sector collaboration, robust cyber security, training, and education, effective communication, revised policy planning including local, state, and the whole of the government approach, and contingency planning. In terms of threat mitigation strategies, it is imperative to conduct a thorough risk assessment of water and wastewater infrastructure to identify potential vulnerabilities and weak areas by expert groups. This would identify targets actions implementation strategy and key indicators to assess success. Due to our dependence on SCADA systems, one of the key areas of strategic importance is cybersecurity. Implementing robust cybersecurity measures includes installing firewalls, intrusion detection and prevention systems, network segmentation, encryption, and regular security updates. These measures are crucial for protecting against cyber threats such as hackers, malware, and ransomware attacks. For other policy matters, it is necessary to stay informed about relevant regulations and requirements related to the safety of water and wastewater CI and ensure compliance with applicable laws and standards. Many countries have enacted specific national legislation to safeguard water CI and these laws typically cover areas such as physical security, cybersecurity, and emergency preparedness and response planning. New and emerging chemicals, micro/nano plastics, volatile organic compounds (VOCs), returned pharmaceuticals, and other contaminants pose [29, 30] a serious challenge. Since the laws are always slower to develop than the pace of technology, it is always prudent to be updated with the latest research, to keep ahead of the curve. The use of advanced sensor platforms [2, 31] with bioinformatics capability [30, 32] to sense, detect, and identify contaminants in real-time is always beneficial for water safety, security, and sustainability.

Acknowledgments The work was partly supported by the Higher Education and Science Committee of MESCS RA in the frame of research project no. 23AA-1F022.

References

1. Vaseashta A (2021) Introduction to water safety, security and sustainability. In: Vaseashta A, Maftei C (eds) Water safety, security and sustainability. Springer International Publishing, Cham, pp 3–22. Advanced sciences and technologies for security applications. https://doi.org/10.1007/978-3-030-76008-3_1
2. Vaseashta A, Khudaverdyan S (eds) (2013) Advanced sensors for safety and security. NATO science for peace and security series B: physics and biophysics. Springer Netherlands, Dordrecht. https://doi.org/10.1007/978-94-007-7003-4
3. Kovacevic A, Nikolic D (2015) Cyber attacks on critical infrastructure: review and challenges. In: Handbook of research on digital crime. IGI Global, Hershey, PA, pp 1–18
4. Riggs H, Tufail S, Parvez I, Tariq M, Khan MA, Amir A, Vuda KV, Sarwat AI (2023) Impact, vulnerabilities, and mitigation strategies for cyber-secure critical infrastructure. Sensors 23:4060. https://doi.org/10.3390/s23084060
5. COBALT (2023) Top cybersecurity statistics for 2024
6. COBALT-11 (2022) Biggest cybersecurity attacks in history
7. AFRICOM (2022) Defending against hybrid threats. Vol 15(2)
8. ASD (2017) ASD cyber threat report 2022–2023
9. Harutyunyan H, Martirosyan S, Kostopoulos J (2017) Challenges in cyber space. Noravank GKH. Scientific and Educational Foundation, Yerevan, Yerevan
10. Proletarsky AV, Surovov AM, Smirnova EM, Rudenkov NA (2016) Information security technologies in computer networks. National Open University, Russian
11. FBI (2024) Internet crime report, 2021. https://www.ic3.gov/Media/PDF/AnnualReport/2021_IC3Report.pdf. Accessed 9 Aug
12. Cohen F (2010) What makes critical infrastructures critical? Int J Crit Infrastruct Prot 3:53–54. https://doi.org/10.1016/j.ijcip.2010.06.002
13. Birkett DM (2017) Water critical infrastructure security and its dependencies. J Terror Res 8:1. https://doi.org/10.15664/jtr.1289
14. Pimentel D, Houser J, Preiss E, White O, Fang H, Mesnick L, Barsky T, Tariche S, Schreck J, Alpert S (1997) Water resources: agriculture, the environment, and society. Bioscience 47:97–106. https://doi.org/10.2307/1313020
15. Wolf AT (2007) Shared waters: conflict and cooperation. Annu Rev Environ Resour 32:241–269. https://doi.org/10.1146/annurev.energy.32.041006.101434
16. Gleick PH (1993) Water and conflict: fresh water resources and international security. Int Secur 18:79. https://doi.org/10.2307/2539033
17. CISA (2024) Water and wastewater systems. https://www.cisa.gov/topics/critical-infrastructure-security-and-resilience/critical-infrastructure-sectors/water-and-wastewater-sector. Accessed May 2
18. Owen T (2000) Fundamentals of modern UV-visible spectroscopy: primer. Agilent Technologies
19. Gleick P, Heberger M (2014) Water conflict chronology, vol 8. Island Press, Washington DC
20. Kroll D (2006) Securing our water supply: protecting a vulnerable resource. PennWell Corporation, Tulsa, OK
21. Clark RM, Hakim S, Panguluri S (2018) Protecting water and wastewater utilities from cyber-physical threats. Water Environ J 32:384–391. https://doi.org/10.1111/wej.12340
22. Tuptuk N, Hazell P, Watson J, Hailes S (2021) A systematic review of the state of cyber-security in water systems. Water 13:81. https://doi.org/10.3390/w13010081
23. Futehally I (2019) Blue peace bulletin—water and violence:- attacks on water infrastructure. Blue Peace Bulletin, p 14
24. IPDF (2024) Wars for resources. Indo-Pacific Defence Forum. https://ipdefenseforum.com/ru/2022/12/%D0%B2%D0%BE%D0%B9%D0%BD%D1%8B-%D0%B7%D0%B0-%D1%80%D0%B5%D1%81%D1%83%D1%80%D1%81%D1%8B. Accessed 11 Aug
25. Statista (2023) Number of cyber crimes reported across India from 2012 to 2022

26. MSSP Alert. Top 10 Cyberattacks of 2023 (2024). Retrieved from https://www.msspalert.com/news/top-10-cyberattacks-of-2023. Last accessed May 5, 2024.
27. USEPA (1996) Safe drinking water act, 1996. US EPA, Washington, DC
28. ALEC (2017) Critical infrastructure protection act. American Legislative Exchange Council
29. Vaseashta A, Carmen M (eds) (2021) Water safety, security and sustainability: threat detection and mitigation, Advanced sciences and technologies for security applications. Springer International Publishing, Cham. https://doi.org/10.1007/978-3-030-76008-3
30. Vaseashta A, Gevorgyan G, Kavaz D, Ivanov O, Jawaid M, Vasović D (2021) Exposome, biomonitoring, assessment, and data analytics to quantify universal water quality. In: Vaseashta A, Maftei C (eds) Water safety, security and sustainability. Springer International Publishing, Cham, pp 67–114. Advanced Sciences and Technologies for Security Applications. https://doi.org/10.1007/978-3-030-76008-3_4
31. Vaseashta A (2006) Carbon nanotubes based devices and sensors. Mater Manuf Process 21:710–716. https://doi.org/10.1080/10426910600613595
32. Baldwin J, Noorali S, Vaseashta A. (2019). Wide spectrum bio-threats identification and classification. In Modernizing the future fight: accelerate & adapt. Wilmington, DE
33. Africa Defence Forum (ADF). Africa Defence Forum, Volume 15, Number 2 (2022). Retrieved from https://adfmagazine.com/ADF_V15N2_ENG.pdf, last accessed 2024/08/10.
34. Russian newspaper (RG). The number of cyberattacks on Russian organizations has increased significantly in 2023 (2023) (in Russian). Retrieved from https://rg.ru/2023/07/27/kolichestvo-kiberatak-na-rossijskie-organizacii-v-2023-godu-zametno-vyroslo.html. Last accessed May 5, 2024.

Chapter 12
A Review of the Models for the Series of Water Flow of the Buzău River in Romania

Alina Bărbulescu and Cristian Ștefan Dumitriu

Abstract Dams are critical infrastructures built for multiple purposes, such as electricity production, water supply, and irrigation. The appearance of big dams also significantly impacts population life, activity, and the natural ecosystems. River flow modifications are among their most obvious impacts on the environment. This article presents a short review of the results of the methodology used in the analyses and modeling of the water flow of one of the most important rivers in Romania and the changes in the water flow due to building the Siriu Dam. Various techniques, including IHA indicators, multifractal analysis, decomposition models, and Intrinsic Mode Functions (IMFs). Whereas the visualizations display evident oscillations in the river's monthly average discharge, the long-range dependence analysis indicates the various patterns of the river discharge. The study reveals that most techniques provide a good estimation of the series trend but fail to capture the maxima. All indicate the existence of two different patterns of the analyzed series, indicating 1984 as a change point.

Keywords Water flow · Alteration · Artificial intelligence methods

12.1 Introduction

Given the planet's population's continuous growth, the demand for water is increasing. Dam reservoirs represent a solution for managing water resources. While the development of dam projects may face opposition, it is important to focus on finding solutions that are socially acceptable and sustainable [1].

A. Bărbulescu
Transilvania University of Brașov, Brașov, Romania

C. Ș. Dumitriu (✉)
Technical University of Civil Engineering of Bucharest, Bucharest, Romania
e-mail: cristian.dumitriu@utcb.ro

D. Radu et al. (eds.), *Countering Hybrid Threats Against Critical Infrastructures*, NATO Science for Peace and Security Series B: Physics and Biophysics, https://doi.org/10.1007/978-94-024-2304-4_12

Great dams, primarily designed to produce energy and provide adjustable electricity generation and storage, serve a multitude of functions. They also contribute to food production through irrigation, supply drinking water, reduce flood flows, and support fish farming, making them a component of water resource management.

Despite their advantages, large dams have a significant detrimental impact on biological diversity. Large land areas are flooded when building these structures, leading to the fragmentation of habitats and the isolation of different species. This disruption affects the exchange of nutrients between ecosystems and hinders the natural migration routes of various organisms. In addition, the reduced water and sediment flows downstream profoundly influence the delicate balance of river estuaries, which serve as crucial spawning grounds for many fish species. Furthermore, the long-term effects of dam construction make entire ecosystems more vulnerable to a wide range of threats, including the pervasive implications of climate change. Moreover, on average, large dam construction projects exceed their initial cost estimates by 96% and their projected completion times by 44% [2–4]. Researchers have also challenged the dams' environmental neutrality, emphasizing the greenhouse gases emitted by dams and the potential long-term environmental costs [5–7].

One of the most evident impacts of a dam's apparition is the modification of the river discharge on which it is built. This article presents various approaches to this aspect in a case study from Romania.

12.2 Methodology

12.2.1 Buzău River Catchment and Analyzed Series

The study river's catchment, Buzău, belongs to Buzău County, in the Romanian Curvature Carpathians, a zone with a temperate-continental climate (Fig. 12.1). Its catchment spans 5264 square kilometers and boasts an average elevation of 1043 meters. Historical data from 1955 to 2010 indicate an average temperature and precipitation of 6 °C, and 500–1000 mm, respectively [8, 9].

Built on the Buzău River Valley in 1972, between the Podu Calului and Siriu mountains, the Siriu dam is the second largest rockfill with a clay core dam in Romania. It was intended for electricity production in the Nehoiaşu Hydropower Plant, water supply, and agricultural use [10]. It is 122 meters high and has a crest length of 570 m, and it delimitates an accumulation lake with a depth of 120 m, 11.5 km length, 420 ha surface, and a storage capacity of up to 155 million m^3 of water.

This dam plays a crucial role in water management by draining approximately 56% of the river's water and its tributaries [11]. It constitutes a water resource used to supply water to the population downstream, for agricultural and industrial uses [8].

Fig. 12.1 Maps of Romania and Buzău County to which the study river belongs

The Buzău River, with its unique geographical traits, presents a complex challenge. Any changes in its water flow can have far-reaching effects on its surrounding area, impacting both biodiversity and human communities. The agriculture-based economy of the local communities along the Buzău River makes the river's well-being extremely important. Even small changes in its flow patterns can lead to various consequences, disrupting traditional farming practices and affecting economic productivity [12].

After 1948, the Buzău catchment was affected by catastrophic floods, the most noticeable in 1975, with a 2100 m^3/s discharge. Other periods of floods were in 1971, 1980, 1981, 2004, 2005 [13]. However, after the construction of the dam, a significant decrease in the frequency of floods was noticed. The events recorded after 1994 were mainly due to the Buzău's tributaries downstream of Nehoiu (which represent 44% of its tributaries). The dam has indeed made a significant impact on reducing flood frequency in the region, ensuring the safety of the local communities [13].

The series we are working with is formed by the Buzău River discharge recorded from January 1, 1955, to December 31, 2010. Romania's map and that of the Buzău County are presented in Fig. 12.1. The overview of the methodology used in the previous studies refers to the daily and monthly data series, while the new results are obtained with the last one as input.

12.2.2 Methodology

The previous studies involved using the methodology shortly enumerated in what follows.

1. Performing statistical analysis [14–16], including:

    (a) Computation of the basic statistics, building histograms and boxplots.
    (b) Assessing the existence of a monotonic trend of the time series vs. its randomness by the Mann-Kendall test and its seasonal version [17]. In the affirmative case, the linear trend slope was determined by Sen's nonparametric procedure [18].
    (c) Test the series stationarity against its nonstationarity utilizing the KPSS test [19].

2. Assessing the water flow alteration utilizing the IHA indicators [20], implemented in IHA software [21]. Based on the average daily water flow, the program provides 33 indicators, grouped in:

    (i) monthly average values,
    (ii) minimum and maximum for one - to ninety-day means,
    (iii) day of year corresponding to the values computed in (ii),
    (iv) pulses in the water conditions' pulses,
    (v) water conditions' increases, decreases, and reversals.

3. Assessing the series multifractality [12] using the partition function, Renyi dimension and $f(\alpha)$—spectrum [22–25].
4. Decomposing the data series [12] by:

    (a) Classical Decomposition Method - additive and multiplicative [26],
    (b) Empirical Mode Decomposition (EMD) [27].

5. Modeling the data series by:

    (a) Regression [16, 20],
    (b) Box-Jenkins methodology: ARIMA models [28],
    (c) Artificial Intelligence methods [29–31]:

        1. Single: Multilayer Perceptron (MLP), Long-short Term Memory (LSTM), Backpropagation Neural Network (BPNN), Extreme Learning Machine (ELM), Generalized Regression Neural Networks (GRNN),
        2. Hybrid: Sparrow-Search Algorithm (SSA-BP), Convolution Neural Network—LSTM (CNN-LSTM), Particle Swarm Optimization (PSO-LSTM).

Here we propose the decomposition of the monthly data series using the Holt-Winters multiplicative damped method (HWMDM), and a seasonal step of 12 months. It is based on the following equations:

$$\hat{y}_{t+h} = \left[l_t + \left(\varphi + \varphi^2 + \ldots + \varphi^h \right) b_t \right] s_{t+h-12(k+1)}, \tag{12.1}$$

with

$$l_t = \frac{\alpha y_t}{s_{t-12}} + (1-\alpha)(l_{t-1} + \varphi b_{t-1})(s_{t-12} \neq 0) \tag{12.2}$$

$$b_t = \beta (l_t - l_{t-1}) + (1-\beta)\varphi b_{t-1}, \tag{12.3}$$

$$s_t = \gamma y_t / (l_{t-1} + \varphi b_{t-1}) + (1-\gamma) s_{t-12}, \tag{12.4}$$

where $\hat{y}_{t+h}$ is the estimated value of the recorded value y_{t+h} and k is the integer part of $(h-1)/m$, if $h > m$. The parameters φ, α, β, γ must be computed. The last three are used for smoothing l_t (the level), b_t, (trend), and s_t (seasonality index), while the last is for controlling the increase (so, for damping) [31, 32].

12.3 Results and Discussion

For assessing the change in the river discharge after the operation of the Siriu dam, the entire series, S, and the subseries, S2, which contains only values after January 1984, were modeled. R software was utilized to perform the analysis. The parameters in the models, together with the goodness of fit indicators—mean absolute error (MAE), mean square error (MSE), and mean absolute percentage error (MAPE)—are contained in Table 12.1 for both scenarios.

With respect to all goodness of fit indicators, the results in the second scenario are better. It confirms the hypothesis that there are different behaviors of the entire time series and the subseries S2.

Figure 12.2 contains the chart of the series forecast for 24 months in both scenarios. The black curve represents the series record, the blue one represents the forecast, and the grey part—light and dark—are the confidence intervals at 80 and 95% confidence levels. The Forecast in Fig. 12.2(b) is more reliable, following the series shape and level, compared to Fig. 12.2(a), where the level is higher and the amplitude smaller. Since using the entire series results in a worse model than when utilizing only S2 shows differences between the seasonal patterns of S1 and S2.

In the classical decomposition, the residual's MAE, MSE, and MAPE in the forecast using S2 were 8.3273, 144.7643, and 50.5265, respectively, confirming that

Table 12.1 The smoothing parameters and goodness-of-fit indices when used for forecasting the HWMDM method with S and S2 series

Series	alpha	beta	gamma	phi	MAE	MSE	MAPE
S	0.2837	0.0014	0.0001	0.974	9.413	182.2178	49.597
S2	0.3383	0.0001	0.0001	0.964	7.869	143.475	43.250

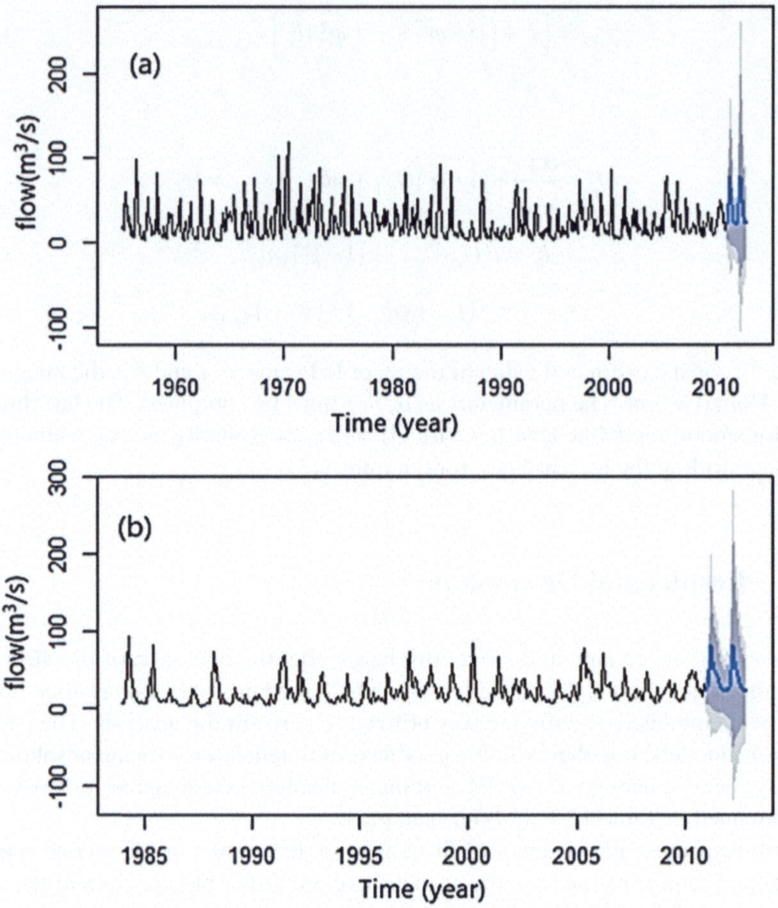

Fig. 12.2 The forecast of the Buzau River's flow using (**a**) S series and (**b**) S2 series. The black curve is S in (**a**) and S2 in (**b**). The blue curve is the forecast and the grey part—light and dark—are the confidence intervals at the confidence levels of 80 and 95%

the actual approach is better. This method is recommended because of its simplicity. It also has the advantage that the seasonality index is updated at each step, together with the series trend and level, considering the last 12 records. Therefore, the forecast accuracy is higher than when the seasonality indices are determined based on all the recorded values, and no update is done. Moreover, by visual examination of the output, it provides a quick understanding of the forecast accuracy.

12.4 Conclusion

Based on the understanding that critical infrastructure, such as dams, can significantly alter the environment, our study is twofold in its aim. First, we provide a brief overview of the methods successfully used to evaluate the alteration of the water flow of the studied river. Second, we highlight a new instrument that can be used for this purpose. The article emphasizes the use of various methods, from classical decomposition to the more sophisticated artificial intelligence algorithms that can provide the same kind of information, but the depth depends on the study's goal.

References

1. Boyé H, de Vivo M (2016) The environmental and social acceptability of dams. Field Act Sci Rep 14:32–37
2. Kotzé P (2022) The world's dams: doing major harm but a manageable problem? https://news.mongabay.com/2022/04/the-worlds-dams-doing-major-harm-but-a-manageable-problem/. Accessed 4 Sept 2024
3. Diehn SA (2020) The environmental impact of mega-dams. https://www.dw.com/en/five-ways-mega-dams-harm-the-environment/a-53916579. Accessed 4 Sept 2024
4. Arora A Dams: economic assets or ecological liabilities? https://earth.org/dams-economic-assets-or-ecological-liabilities/. Accessed 4 Sept 2024
5. Deemer BR, Harrison JA, Li S, Beaulieu JJ, DelSontro T, Barros N, Bezerra-Neto JF, Powers SM, dos Santos MA, Vonk JA (2016) Greenhouse gas emissions from reservoir water surfaces: a new global synthesis. BioScience 66(11):949–964
6. St. Louis VL, Kelly CA, Duchemin É, Rudd JWM, Rosenberg DM (2000) Reservoir surfaces as sources of greenhouse gases to the atmosphere: a global estimate: reservoirs are sources of greenhouse gases to the atmosphere, and their surface areas have increased to the point where they should be included in global inventories of anthropogenic emissions of greenhouse gases. BioScience 50(9):766–775
7. Song C, Gardner KH, Klein SJW, Pereira Souza S, Mo W (2018) Cradle-to-grave greenhouse gas emissions from dams in The United States of America. Renew Sustain Energy Rev 90:945–956
8. Chendeş V (2011) Water resources in curvature Subcarpathians. Geospatial assessments. Editura Academiei Române, Bucureşti. (In Romanian with English abstract)
9. Minea SI (2011) The rivers from Buzău River catchment. Hydrographical and hydrological constructions. Editura Alpha MDN, Buzău. (In Romanian)
10. Top 5 dams in Romania. https://fermoar.ro/barajul-siriu/. Accessed 3 Sept 2024
11. Diaconu D (2008) Water resources from of the Buzău River catchment. Editura Universitară, Bucureşti. (In Romanian)
12. Bărbulescu A, Mohammed N (2024) Study of the river discharge alteration. Water 16:808
13. The management plan of flood risk. Buzău—Ialomiţa Water Basin Administration. http://buzau-ialomita.rowater.ro/wp-content/uploads/2021/02/PMB_ABABI_Text_ actualizat.pdf. Accessed 3 Sept 2024
14. Mocanu-Vargancsik CA, Bărbulescu A (2018) Analysis on the variability of Buzău River discharges. Ovidius Univ Ann Ser Civil Eng 20:51–56
15. Mocanu-Vargancsik CA, Bărbulescu A (2019) On the variability of a river water flow, under seasonal conditions. 2019 IOP Conf Ser: Earth Environ Sci 344:012028. https://doi.org/10.1088/1755-1315/344/1/012028

16. Mocanu-Vargancsik CA, Tudor G (2020) On the linear trends of a water discharge data under temporal variation. Case study: the upper sector of the Buzău River (Romania). Forum Geogr XIX(1):37–44
17. Hippel K, McLeod AI (1994) Time series modelling of water resources and environmental systems. Elsevier Science B.V, Amsterdam
18. Sen PK (1969) Estimates of the regression coefficient based on Kendall's tau. J Am Stat Assoc 63(324):1379–1389
19. Kwiatkowski D, Phillips PCB, Schmidt P, Shin Y (1992) Testing the null hypothesis of stationarity against the alternative of a unit root. J Econ 54(1–3):159–178
20. Minea G, Bărbulescu A (2014) Statistical assessment of hydrological alteration of Buzău River induced by Siriu dam (Romania). Forum geogr 13(1):50–58
21. Indicators of hydrological alteration. https://www.conservationgateway.org/ConservationPractices/Freshwater/EnvironmentalFlows/MethodsandTools/IndicatorsofHydrologicAlteration/Pages/indicators-hydrologic-alt.aspx. Accessed 3 Sept 2024
22. Ferraris L, Gabellani S, Parodi U, Rebora N (2003) Revisiting multifractality in rainfall fields. J Hydrometeorol 4(2):544–551
23. Renyi A (1955) On a new axiomatic theory of probability. Acta Math Hungar 6:285–335
24. Hentschel HGE, Procaccia I (1983) The infinite number of generalized dimensions of fractals and strange attractors. Phys D 8:435–444
25. Halsey T, Jensen M, Kadanoff L, Procaccia I, Shraiman BI (1986) Fractal measures and their singularities: the characterization of strange sets. Phys Rev A 33(2):1141–1151
26. 6.3. Classical decomposition. https://otexts.com/fpp2/classical-decomposition.html. Accessed 3 Sept 2024
27. Huang NE, Shen Z, Long SR, Wu MC, Shih HH, Zheng Q, Yen N-C, Tung CC, Liu HH (1998) The empirical mode decomposition and the Hilbert spectrum for nonlinear and non-stationary time series analysis. Proc R Soc Lond A 454:903–995
28. Zhen L, Bărbulescu A (2024) Comparative analysis of convolutional neural network-long short-term memory, sparrow search algorithm-backpropagation neural network, and particle swarm optimization-extreme learning machine for the water discharge of the Buzău River, Romania. Water 16:289
29. Bărbulescu A, Zhen L (2024) Forecasting the river water discharge by artificial intelligence methods. Water 16(9):1248
30. Bărbulescu A (2017) Statistical assessment and model for a river flow under variable conditions. In: Lekkas TD (ed) Proceedings of the 15th international conference on environmental science and technology, Rhodes. https://cest2017.gnest.org/sites/default/files/presentation_file_list/cest2017_00715_poster_paper.pdf. Accessed 3 Sept 2024
31. Holt CC (2004) Forecasting seasonals and trends by exponentially weighted. Int J Forecast 20(1):5–10
32. Winters PR (1960) Forecasting sales by exponentially weighted moving averages. Manag Sci 6(3):324–342

Chapter 13
$ZnFe_2O_4$/ZnO:Ga/Glass Heterostructure as Peroxidase Enzyme Mimetics for Hydrogen Peroxide Sensing

Anatolie Sidorenko, Tatiana Gutsul, Elena Monaico, Maria Lupu, Serghey Zavrajnyi, and Ashok Vaseashta

Abstract We report here a method for synthesizing ZnO:Ga and $ZnFe_2O_4$ films and $ZnFe_2O_4$/ZnO:Ga heterostructure and studying them as peroxidase enzyme mimetics for hydrogen peroxide determination and simulation-based detection has been proposed. A method for forming ZnO:Ga films by the DC magnetron sputtering method and $ZnFe_2O_4$ films and $ZnFe_2O_4$/ZnO:Ga heterostructures by the RF-sputtering method using a target consisting of $ZnFe_2O_4$ nanoparticles has been described. Thus, the procedure for synthesizing a heterostructure, which is based on the principle of self-assembly in a glow discharge plasma generated by a magnetron sputtering source, the cathode of which is a target consisting of zinc ferrite nanoparticles, has been optimized. Using the SEM and AFM methods, it has been revealed that with an increase in the argon pressure to 8.1 Pa, intergranular voids are formed in the resulting films due to the occurrence of the shading effect. These conditions provide the formation of a structure with a developed surface. This in turn typically leads to an increase in the valence electron concentration due to a large number of broken interatomic bonds and the formation of a large number of free oxygen sites on the film surface, which are donor centers involved in the redox process. Experimental results showed that the $ZnFe_2O_4$/ZnO:Ga film heterostructure exhibits peroxidase-like activity in a linear range of 0.017–0.240 μM. In addition, the color

A. Sidorenko (✉) · T. Gutsul · M. Lupu · S. Zavrajnyi
Technical University of Moldova, IEEN, "D. Ghitu", Chisinau, Moldova

E. Monaico
Technical University of Moldova, Chisinau, Moldova

A. Vaseashta (✉)
Applied Research Division, International Clean Water Institute, Manassas, VA, USA

IEEN, Academy of Sciences of Moldova, Chisinau, Moldova

Institutul de Cercetare, Universitatea din Bucureşti, Bucureşti, Romania
e-mail: prof.vaseashta@iee.org

D. Radu et al. (eds.), *Countering Hybrid Threats Against Critical Infrastructures*, NATO Science for Peace and Security Series B: Physics and Biophysics, https://doi.org/10.1007/978-94-024-2304-4_13

change dependent on hydrogen peroxide concentration can offer a convenient approach to detecting H_2O_2 both with the naked eye and spectrophotometrically. Therefore, the $ZnFe_2O_4$/ZnO:Ga film heterostructure is a promising candidate for the colorimetric detection of H_2O_2, ascorbic acid, and similar oxidizers.

Keywords Mimetic of Peroxidase · Film heterostructure $ZnFe_2O_4$/ZnO:Ga · Spinel Ferrite · H_2O_2 Sensing

13.1 Introduction

Nanozymes are nanomaterials with outstanding enzyme-like activity. These nanocomposites can directly activate and accelerate the oxidation of peroxidase substrates in the presence of hydrogen peroxide (H_2O_2) [1] and can be used for the design of H_2O_2 detectors. Compared with natural enzymes, nanozymes have substantial advantages, viz., low cost, high stability, durability, controllable activity, and reusability, thus they have been widely applied in industrial, medical, and biological fields. Various physicochemical properties and prospective applications extend to fields such as advanced electronics, microwave devices, and biotechnology. As an example, spinel nanoferrites and their synthesis methods are highlighted due to their versatility, as nanomaterials [2]. Due to the widespread use of hydrogen peroxide in industry and due to its high chemical activity and toxicity, a simple, inexpensive, fast, and sensitive method of H_2O_2 analysis is needed, because it could be used for bioanalysis in the area of food safety, security, and environmental protection [3]. The literature describes various methods for producing ZnO/$ZnFe_2O_4$ powder nanocomposites, widely known as successful nanozymes. The resulting nanocomposites act like peroxidase in the catalysis of 3,3′, 3,5′-tetramethylbenzidine (TMB) and can directly activate and accelerate the oxidation of peroxidase substrates in the presence of hydrogen peroxide (H_2O_2) [4, 5].

The first works on the production of a thin nanostructured ZnO/$ZnFe_2O_4$ film relate to the preparation of a mixture of zinc oxide and zinc ferrite nanopowders by immersion in suspension, with subsequent drying and annealing [6, 7]. Another advanced method was presented in [8], where spinel nanofibers decorated with zinc oxide nanoparticles and possessing their peroxidase activity were obtained by electrospinning. As is known, there is practically no information in the literature about the peroxidase activity of ZnO/$ZnFe_2O_4$ film structures and the detection of hydrogen peroxide, as well as about their possession of the properties of the peroxidase enzyme mimetic. In the present work, the possibility of the $ZnFe_2O_4$/ZnO:Ga/glass layered heterostructure to exhibit the properties of the peroxidase enzyme mimetic was studied and advanced technology for its formation was developed.

13.2 Materials and Methods

13.2.1 Chemical Synthesis

Iron(II) sulfate (≥99.7%), a saturated iron(III) chloride solution (≥99.0%), sodium acetate, and polyethylene glycol (PEG, MW: 6000) as a macromolecular nonionic surfactant were purchased from Sigma-Aldrich. Ethanol (≥96.5%) and acetone (≥99.9%) were purchased from EuroLab. Deionized water was used in the experiment. Hydrogen peroxide H_2O_2 (30%) and 3,3′,5,5′-tetramethylbenzidine (TMB, ≥ 98%) were purchased from Sigma-Aldrich. A 0.2M acetate buffer with a pH of 4.0 was used. All chemicals were used as received without further purification. The ultrasonic treatment was conducted on a unit with the following parameters: an ultrasonic bath (ISOLAB Laborgerate GmbH), an ultrasonic power of 120 W, a heating power of 180 W, and a frequency of 40 kHz.

13.2.2 Synthesis of $ZnFe_2O_4$ Nanoparticles (NPs)

Zinc ferrite ($ZnFe_2O_4$) NPs were synthesized using iron(III) chloride and zinc(II) chloride by the solvothermal method according to a modified synthesis procedure described in [9]. The starting salts were dissolved in an ethylene glycol solution with the addition of water-soluble polyethylene glycol PEG 6000 as a macromolecular nonionic surfactant acting as a stabilizer by ultrasonic treatment. Further, the solution was alkalized with a sodium acetate solution to a pH of 10 and transferred to a fluoroplastic reactor; the synthesis was run in an autoclave at 185 °C for 16 h. The resulting black powder was separated from the mother liquor, washed with deionized water, acetone, and ethanol, dried in a vacuum at 100 °C, and then annealed at 250 °C.

13.2.3 Molding of a $ZnFe_2O_4$ Ferrite Target

The resulting $ZnFe_2O_4$ NP powder was placed in a special mold for molding targets and subjected to mechanical compression at a pressure of 250 kg/cm^2; the resulting target had the form of a tablet with a diameter of 25 mm and a thickness of 2 mm.

13.2.4 ZnO:Ga Thin Film by DC Magnetron Sputtering of the Target Material

The synthesis of ZnO:Ga thin films deposited on glass was conducted on a VUP-4 vacuum unit by direct current (DC) magnetron sputtering, i.e., by the DC discrete method. The films were deposited on a glass substrate subjected to special cleaning in a Piranha solution, then washed with deionized water, acetone, and ethanol, and dried at a temperature of 100 °C. After that, the deposition of layers from a polycrystalline target of zinc oxide doped with 2% gallium (Alfa Aesar) was conducted. The DC power of the magnetron was 1.2 W; the pressure in the chamber was 0.27 Pa; sputtering was performed in an atmosphere of argon of special purity grade (99.998%) at a rate of 0.5–4.0 Å/s. The substrate temperature was 200 °C.

13.2.5 $ZnFe_2O_4$/Glass Films and $ZnFe_2O_4$/ZnO:Ga/Glass Film Heterostructures by RF Magnetron Sputtering of the Target Material

The second layer of the heterostructure consists of a thin layer of $ZnFe_2O_4$ ferrite NPs; it was formed by high-frequency magnetron sputtering at a frequency of 13.56 MHz and a magnetron power in a range of 40–100 W. The $ZnFe_2O_4$/ZnO:Ga/glass film heterostructures were grown at a temperature of 200 °C; the temperature of the target in the erosion (bombardment) zone reached 700 °C. The gas used in the working chamber was argon with a purity of 99.998% and a pressure of 0.81 Pa. The deposition rate of the thin layer was varied in a range of 0.5–4.0 Å/s.

13.3 Characterization

13.3.1 Morphology

The morphology and chemical composition of ZnO:Ga thin layers were studied using a Zeiss Sigma and Tescan Vega TS 5130MM scanning electron microscope equipped with an Oxford Instruments INCA energy dispersive X-ray system operating at 20 kV. The morphology of the $ZnFe_2O_4$ NP powder was characterized by scanning electron microscopy (SEM-Philips XL30 SFEG and TESCAN VEGA 5124). Scanning electron microscope (SEM) images were recorded using a SUPRA 55VP-32-49 electronic microscope operating at 30 kV in the secondary and backscattered electron imaging modes in a high vacuum mode. Upscale. Media, Img2Go, Fotor, ImageJ Software were used for the graphical processing of SEM.

To analyze the structural parameters of the ZnO:Ga, $ZnFe_2O_4$/glass, and $ZnFe_2O_4$/ZnO:Ga layers, the X-ray diffraction spectra were recorded on a Rigaku Miniflex

600 Desktop X-ray diffractometer using CuK_α radiation ($\lambda = 1.54059$ Å) at an accelerating voltage of 40 kV and a current of 15 mA in a range of $2\theta = 2°–140°$ at room temperature. Scanning was performed in the θ–2θ Bragg–Brentano focusing geometry. Processing of the recorded X-ray diffraction patterns (background subtraction, $K\alpha_1$ –$K\alpha_2$ doublet separation, etc.) and calculation of the diffraction line profile parameters were conducted using the SmartLab Studio II software. The presence of crystalline phases was detected by comparing the experimental X-ray diffraction patterns with the reference data database using the ICSD Database software, Version 1.2.0, 50,567 XRD, page 113, 2003. The *a* and *c* lattice parameters of the hexagonal zinc oxide phase and the cubic zinc ferrite phase were calculated from the position of the last two indexed lines of the X-ray diffraction patterns. The surface morphology of the ZnO:Ga, and $ZnFe_2O_4$/ZnO:Ga layers was studied by atomic force microscopy (AFM) on a NanoStation II^+ microscope operating in the noncontact mode with Si(111) type on a scanning area of 5 × μm², 10 × 10 μm².

13.3.1.1 Peroxidase-Like Activity of the $ZnFe_2O_4$/ZnO:Ga/Glass Film Heterostructure

Colorimetric analysis was performed using a T-80 UV-VIS spectrophotometer in a wavelength range of 652 nm. The probes were analyzed in the form of films; they were assembled in an array on a special holder, which makes it possible to change the area of interaction of the test sample, and placed in chemical beakers to implement the reaction of the test components. After that, the solution was transferred to a quartz cell and placed in the holder of the device where optical measurements were conducted. The testing procedure is standard and complies with the protocol [10].

The colorimetric determination of hydrogen peroxide was conducted using 3,3′,5,5′-tetramethylbenzidine (TMB) as a substrate and the synthesized film samples as a catalyst. Catalytic oxidation was run using an acetate buffer solution as a reaction medium. First, 25 μL of a methanol solution of 0.01 M TMB was added to 2 mL of an acetate buffer; after that, 100 mL of the catalyst in the form of a film structure was added into an aqueous solution of hydrogen peroxide in a concentration range of 0.017–0.240 μM using a 200 mM H_2O_2 solution. The above system was allowed to react for 10 min; after that, the absorption spectrum of the solution as a function of hydrogen peroxide concentration was recorded.

13.3.2 Optical

The optical characteristics of the grown ZnO:Ga/glass, $ZnFe_2O_4$/glass, and $ZnFe_2O_4$/ZnO:Ga/glass films were measured on a T-80 UV-VIS spectrophotometer in a wavelength of 350–850 nm and the results are described in the section below.

13.4 Results and Discussion

The morphology of the $ZnFe_2O_4$ nanopowder was studied by SEM, as shown in Fig. 13.1.

The morphology and sizes of the ZnFe2O4–polymer NPs synthesized in the presence of a polymer have a spherical form and a size of 5–10 nm. Since the NPs have a high surface energy, they undergo rapid interparticle interaction and coarsening and aggregate into spherical structures with a diameter of 80–120 nm (Fig. 13.1). EDX determined composition with a Zn:Fe ratio is 1:2.5. Figure 13.2 shows a powder diffraction pattern of the $ZnFe_2O_4$ nanoparticles synthesized by the solvothermal method in accordance with the procedure described in [9].

It is evident from the figure that the observed diffraction peaks at $2\theta = 18.3°$, 29.9°, 35.2°, 37.2°, 42.7°, 53.5°, 56.7°, and 62.1° correspond to the (111), (220), (311), (222), (400), (422), (511), and (440) crystallographic planes of the spinel structure with a cubic phase structure of zinc ferrite (Franklinit), space group Fd3m, and a lattice constant of $a = b = c = 8.4362$ Å; no other impurity peaks are observed, $\alpha = 90°$. The results are in good agreement with the spectra from the database of standard powder diffraction pattern samples of zinc ferrite (JCSD no. 87989).

The size of the synthesized $ZnFe_2O_4$ crystallites (d) was determined from the full width at half maximum of the diffraction peaks using the Scherrer equation: $d = K\lambda/(\beta\cdot\text{Cos}\theta)$, where K is dimensionless particle's form coefficient, $\lambda = 1.54059$ Å is the X-ray wavelength, β is the width of the reflex at half height, and θ is the diffraction angle. The zinc ferrite crystallite size was 85 ± 2 nm. These values are consistent with the SEM data shown in Fig. 13.3 and correspond to the sizes of the nanoparticles synthesized by the solvothermal method.

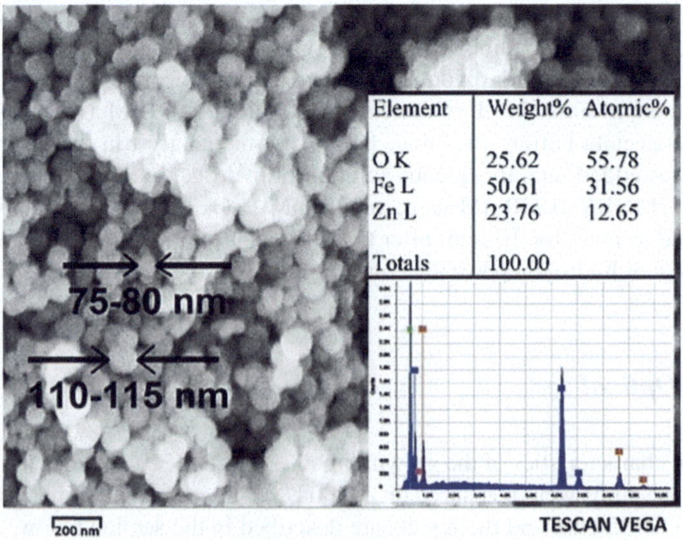

Fig. 13.1 Scanning electron microscopy image of $ZnFe_2O_4$ NPs

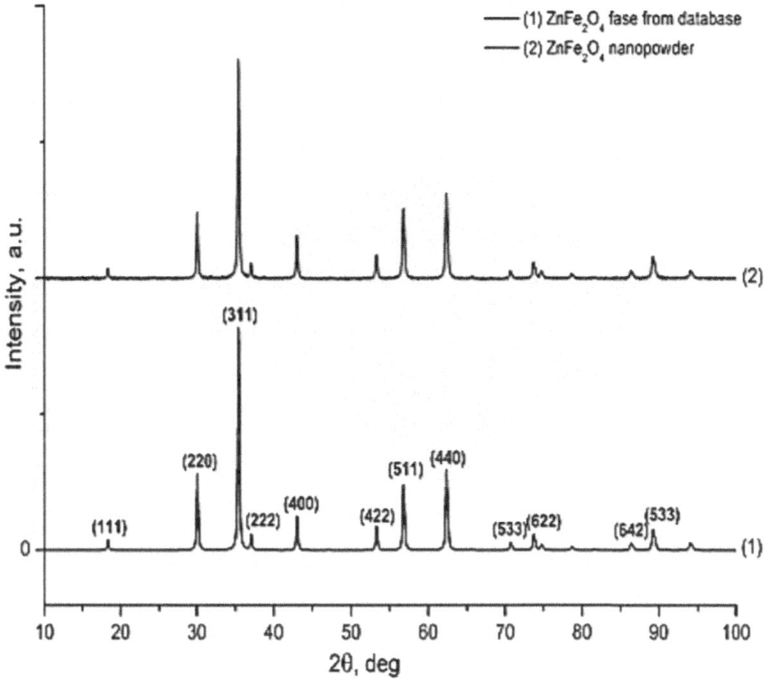

Fig. 13.2 X-ray diffraction pattern of the $ZnFe_2O_4$ nanopowder

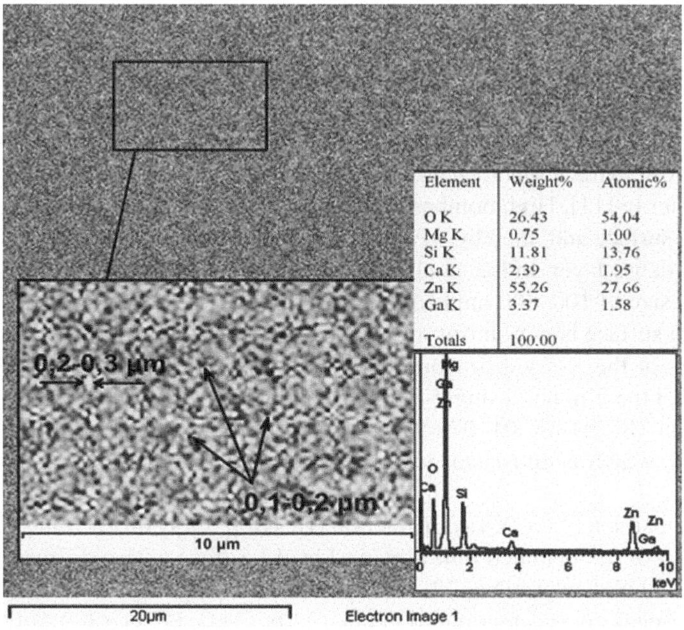

Fig. 13.3 Scanning electron microscopy and EDX pattern of ZnO:Ga films

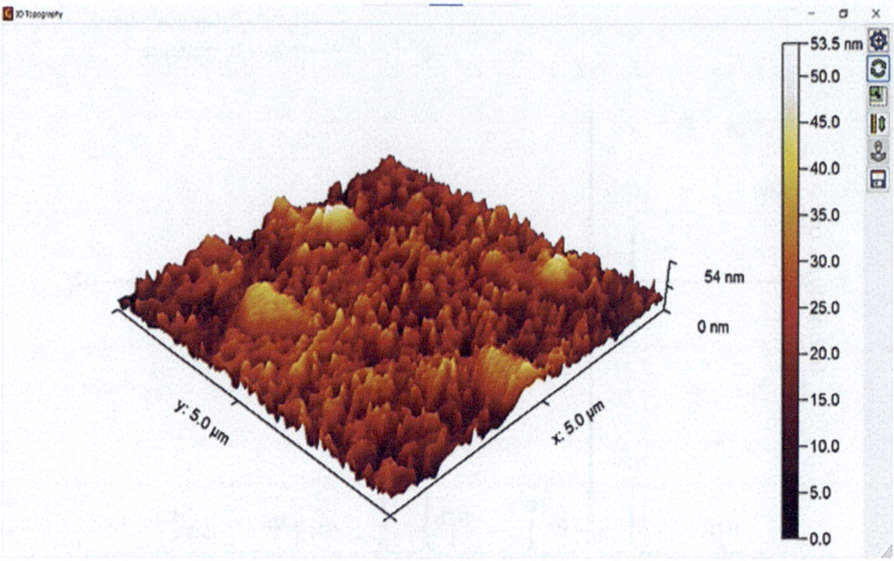

Fig. 13.4 Atomic force microscopy image of the ZnO:Ga thin film obtained using an Ar pressure of 0.27 Pa

Results of studying the morphology and composition of the ZnO:Ga film structure by SEM and EDX methods are shown in Fig. 13.3. It is evident that the film is represented by well-formed crystallites with a small amount of intergranular voids with a size of 0.1–0.3 µm. The composition of the film corresponds to zinc oxide: zinc, 27.66 at %; gallium, 1.58 at %; and oxygen, 54.04 at %; it is attributed to both zinc oxide and Mg, Si, and Ca oxides that are part of the glass substrate.

During the formation of a ZnO:Ga film by RF sputtering at a low argon pressure of 0.27 Pa, the number of collisions is small; therefore, the incidence angles are typically close to the normal, and the bombardment energy is generally high, as was shown earlier in [11]. High bombardment energy leads to high mobility of adatoms on the film surface and, thereby, to a high film density; as a consequence, according to AFM, a dense layer with a small number of depressions has well-formed structures and a size of 100–300 nm is observed, as shown in Fig. 13.4.

The film surface is non-uniform with an arithmetic roughness of Ra = 20 nm.

Analysis of the X-ray diffraction patterns of the gallium-doped zinc oxide film suggests that the film has a single-phase hexagonal structure of ZnO:Ga of wurtzite modification (ICDD no. 01–078-2585) with lattice parameters a = 3.271 Å and c = 5.23 Å, which is consistent with the data of other authors [12], as shown in Fig. 13.5.

The comparison of the X-ray diffraction (Desktop X-ray diffractometer Miniflex 600 Rigaku) of the obtained thin films in Fig. 13.5 (top) with a diffraction pattern from the JCPDS database No. 22–1012 presented in Fig. 13.5 (middle) shows the presence of peaks corresponding to (111), (220), (311), (222), (400), (422), (511),

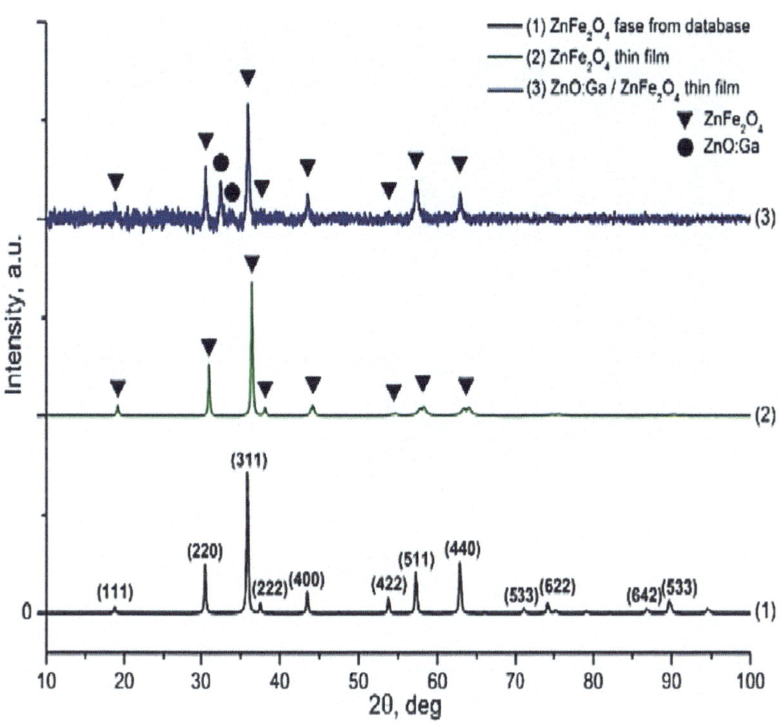

Fig. 13.5 XRD of $ZnFe_2O_4$/glass thin film (2), $ZnFe_2O_4$/ZnO:Ga/glass heterostructure [3]

(440) phases of cubic spinel (Franklinite). In Fig. 13.5 (bottom), specific peaks related to ZnO:Ga in the $ZnFe_2O_4$/ZnO:Ga/glass heterostructure are presented. The ZnO:Ga sample shows the specific peaks at $2\theta = 31.7°$ (100), 34.4° (002) and 56.5° (110) corresponding to the hexagonal structure of wurtzite (ICDD No.01–078-2585).

Thus, the ZnO:Ga gallium-doped zinc oxide film formed by DC magnetron sputtering has a thickness of 550 nm, a transparency of 85%, and a resistance of 2.8×10^{-3} Ω·cm. The resulting ZnO:Ga film can be used as a base for the formation of a $ZnFe_2O_4$ layer to synthesize a $ZnFe_2O_4$/ZnO:Ga/glass film heterostructure.

Figure 13.6 shows the results of studying the morphology of the $ZnFe_2O_4$/ZnO:Ga/glass film heterostructure by SEM. The presence of flexible chain aggregates and ring aggregates that form a toroid, along with the presence of V-formed depressions with void size of 0.2–0.9 μm, is observed.

It is known that during the formation of a film by the RF-magnetron sputtering method at a high argon pressure, the number of collisions that occur in the space between the target and the substrate with the growing film increases; this factor causes a decrease in the average angle of incidence; that is, a shading effect takes place; it leads to the formation of intergranular voids in the growing layer. A similar effect was described by the authors of [13, 14], who synthesized $Zn_{0.87}Fe_{2.13}O_4$ films by the RF-magnetron sputtering method; in that case, the argon pressure was

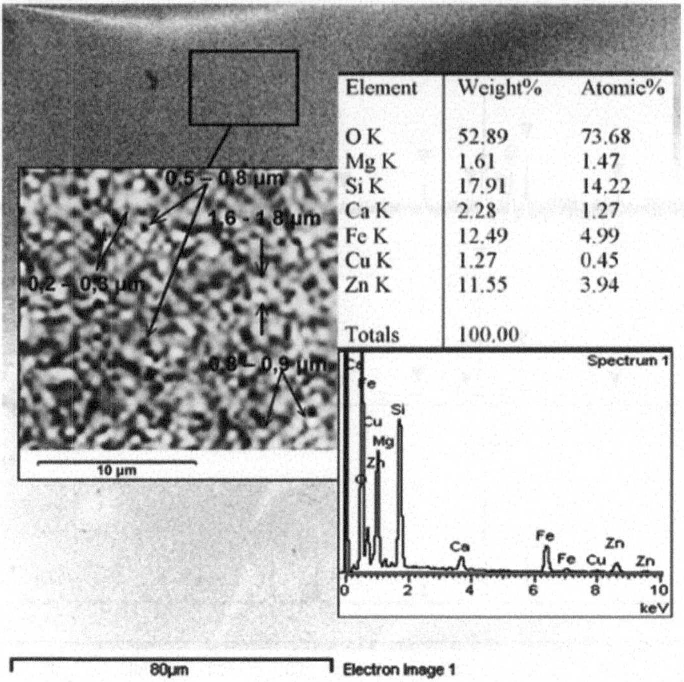

Fig. 13.6 Scanning electron microscopy image of the $ZnFe_2O_4$/ZnO:Ga/glass heterostructure

1.5–2 Pa, as in the case of the formation of $Co_1Mn_{0.65}Fe_{1.35}O_4$ and $ZnFe_2O_4$ films. In this study, the shading effect is observed even at an argon pressure of 0.81 Pa; this fact is attributed to the design features of the magnetron, as shown in Figs. 13.7 and 13.8.

The results of the compositional analysis conducted by EDX (Fig. 13.6 (inset)) show the presence of elements O, Fe, and Zn in the film. The Zn: Fe ratio is 1:1.27, which is close to the theoretical ratio in the $ZnFe_2O_4$/ZnO:Ga heterostructure. Thus, the $ZnFe_2O_4$/ZnO:Ga heterostructure formed by RF magnetron sputtering has a thickness of 850 nm and a resistance of $2 \times 10^7\ \Omega$.

Figure 13.9 shows the transmission spectra in a wavelength range of 250–800 nm for ZnO:Ga films grown by DC magnetron sputtering and for $ZnFe_2O_4$ and the ZnO:Ga/$ZnFe_2O_4$ film heterostructure synthesized by RF-magnetron sputtering. Based on the results obtained, the band gap was calculated using Tauc's formula [15], which shows the relationship between absorption coefficient (α) and photon energy (hν), as in the following equation: $(\alpha h\nu)^2 = A(h\nu—Eg)$, where A is an energy-dependent constant and E_g is the optical band gap of the material.

The optical band gap energy values (E_g) were calculated by extrapolating the linear portion of $(\alpha h\nu)^{½}$ vs. hν plots for all the synthesized samples, as shown in Fig. 13.10 (a, b, c).

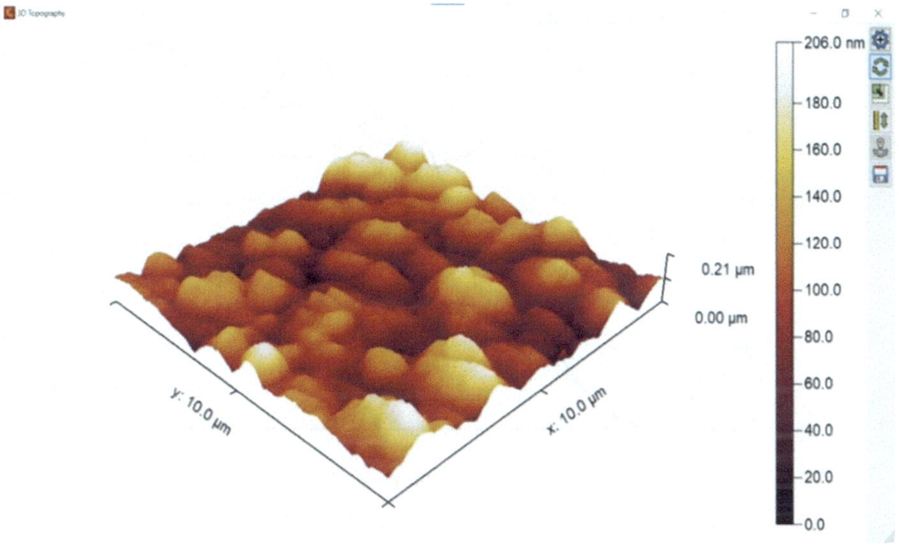

Fig. 13.7 Atomic force microscopy image of the $ZnFe_2O_4$/glass structure obtained using an Ar pressure of 0.81 Pa

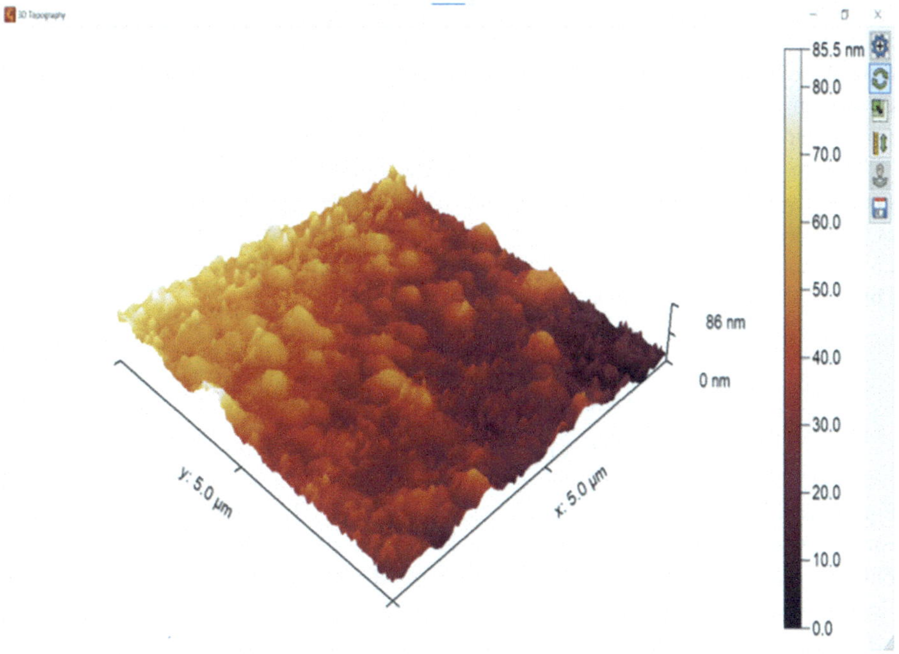

Fig. 13.8 Atomic force microscopy image of the $ZnFe_2O_4$/ZnO:Ga/glass heterostructure obtained using an Ar pressure of 0.81 Pa

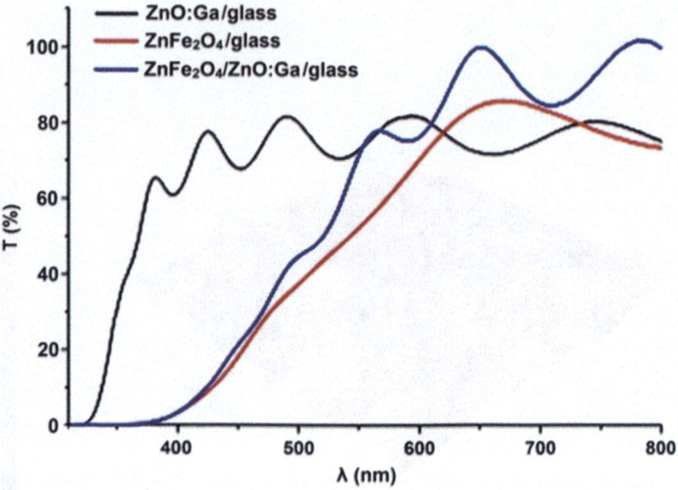

Fig. 13.9 UV–Vis transmission absorption spectra of the ZnO:Ga, $ZnFe_2O_4$, and $ZnFe_2O_4$/ZnO:Ga films

According to the results of the studies, the band gap energy value of the $ZnFe_2O_4$/ZnO:Ga film heterostructure (1.89 eV) is lower than that of pure ZnO (3.2 eV) and pure $ZnFe_2O_4$ (1.92 eV). Thus, this will significantly facilitate photocatalytic processes for the $ZnFe_2O_4$/ZnO:Ga heterostructure as a peroxidase enzyme mimetic. The shift of the band gap to the visible region of the spectrum, together with the developed surface, gives rise to a synergistic effect and increases the catalytic activity of the heterostructure, as shown by the Tauc's plots for ZnO:Ga, $ZnFe_2O_4$, and $ZnFe_2O_4$/ZnO:Ga.

The intrinsic peroxidase-like activity of the $ZnFe_2O_4$/ZnO:Ga/glass film heterostructure was studied by the method described earlier. The film structure was used as a catalyst; in the presence of hydrogen peroxide, the chromogenic substrate - the TMB dye, undergoes oxidation, which leads to a change in the color of the solution. Three controlled tests were conducted:

$$\mathrm{TMB} + \mathrm{H_2O_2} \tag{13.1}$$

$$\mathrm{TMB} + \mathrm{ZnFe_2O_4} / \mathrm{ZnO} : \mathrm{Ga} / \mathrm{glass} \tag{13.2}$$

$$\mathrm{TMB} + \mathrm{H_2O_2} + \mathrm{ZnFe_2O_4} / \mathrm{ZnO} : \mathrm{Ga} / \mathrm{glass} \tag{13.3}$$

Figure 13.11 shows that, in the presence of TMB + H_2O_2, the colorless solution containing the test sample became blue, while the TMB + H_2O_2 and TMB + $ZnFe_2O_4$/ZnO:Ga/glass solutions did not change their color significantly. The obtained results clearly show that the $ZnFe_2O_4$/ZnO:Ga/glass film heterostructure exhibits a peroxidase-like catalytic activity, as was shown for the first time in the study using magnetite nanoparticles [14].

The colorimetric determination of hydrogen peroxide was conducted using 3,3′,5,5′-tetramethylbenzidine (TMB) as a substrate and the synthesized film

Fig. 13.10 Tauc's plot [15] analysis of UV-Vis absorption spectra of $(\alpha h\nu)^{1/2}$ $(cm^{-1}eV)^{1/2}$ versus photon energy $h\nu$(eV) of ZnO:Ga, $ZnFe_2O_4$, and $ZnFe_2O_4$/ZnO:Ga films heterostructure

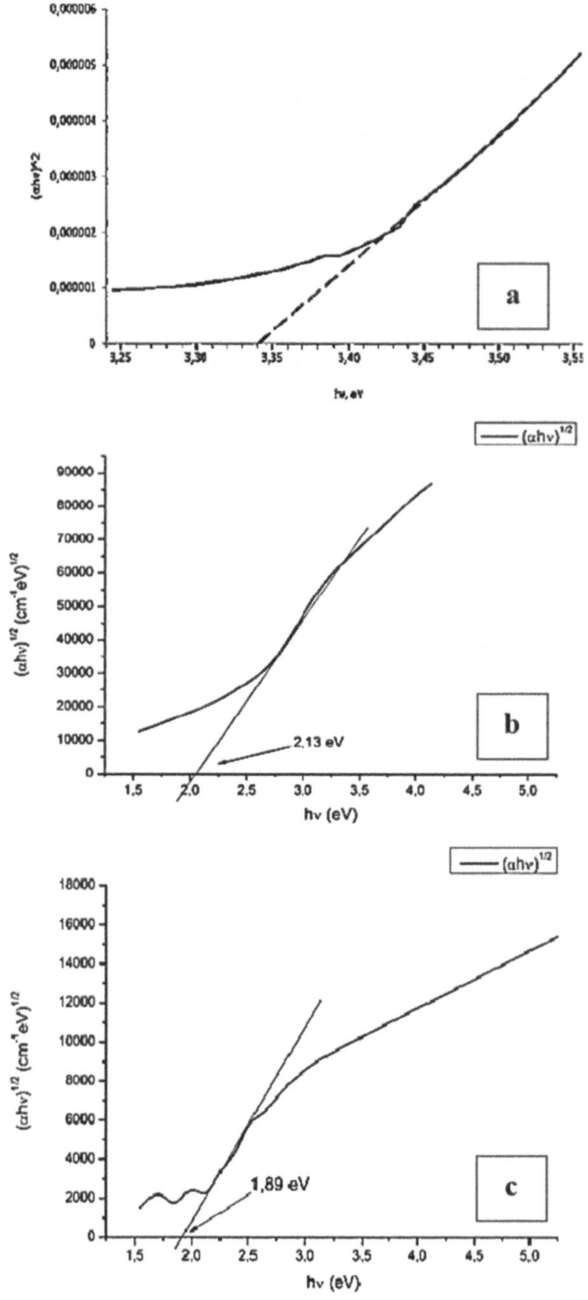

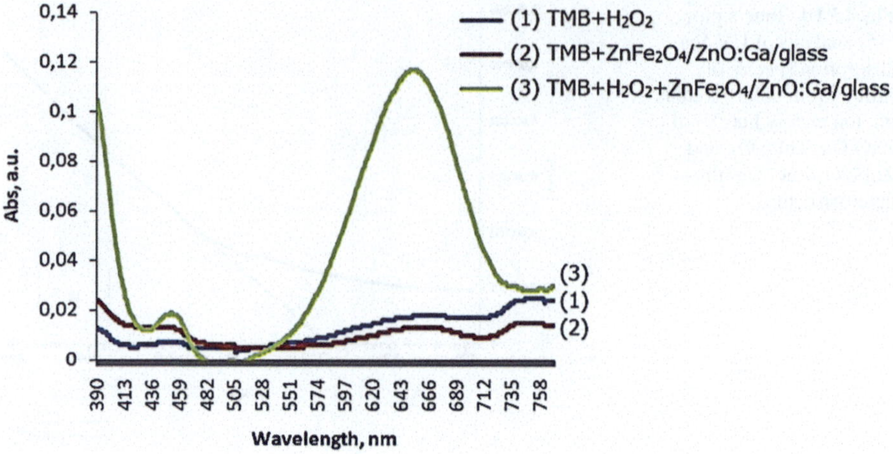

Fig. 13.11 Absorption spectra of colorimetric reactions under different conditions

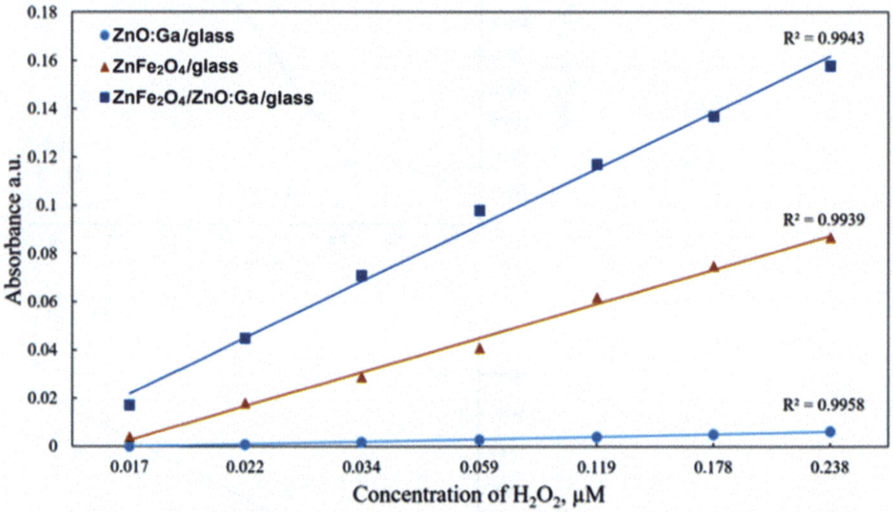

Fig. 13.12 Linear calibration plots of ZnO:Ga, $ZnFe_2O_4$ films and $ZnFe_2O_4$/ZnO:Ga heterostructures for H_2O_2 detection in the 0.17–0.2 μM range, $\lambda = 625$ nm

samples as a catalyst. Catalytic oxidation was run using an acetate buffer solution as a reaction medium. First, 25 μL of a methanol solution of 0.01 M TMB was added to 2 mL of an acetate buffer; after that, 100 mL of the catalyst in the form of a film structure was added into an aqueous solution of hydrogen peroxide in a concentration range of 0.017–0.240 μM using a 200 mM H_2O_2 solution.

The above system was allowed to react for 10 min; after that, the absorption spectrum of the solution as a function of hydrogen peroxide concentration was recorded, as shown in Fig. 13.12.

Thus, it was shown that the synthesized film heterostructure exhibits peroxidase enzyme mimetic activity being involved in the catalytic oxidation of the dye; this property can be used in the colorimetric detection of hydrogen peroxide, ascorbic acid, glyphosate, and other objects [16].

13.5 Conclusions

To summarize, the porous film heterostructure of $ZnFe_2O_4/ZnO$:Ga was successfully obtained by gradual sputtering using DC magnetron sputtering of ZnO:Ga film and RF-sputtering of $ZnFe_2O_4$ film, using a target consisting of $ZnFe_2O_4$ nanoparticles. The hierarchical structure of the film was shown by SEM and AFM methods and the internal peroxidase-like activity was identified. From a safety and security standpoint, the determination of H_2O_2 at high concentrations, a strong oxidizer, is critical if not properly controlled. Its accurate determination is essential for safe handling and storage and provides an alternative to other nanomaterials [17, 18] and flexible sensors [19]. The study demonstrates an accurate measurement design that ensures effective use, minimizes risks, and contributes to regulatory compliance.

References

1. Wang Z, Zhang R, Yan X, Fan K (2020) Structure and activity of nanozymes: inspirations for *de novo* design of nanozymes. Mater Today 41:81–119. https://doi.org/10.1016/j.mattod.2020.08.020
2. Sharma SK (2021) Spinel nanoferrites: synthesis, properties and applications. Springer International Publishing, Cham
3. Trujillo RM, Barraza DE, Zamora ML et al (2021) Nanostructures in hydrogen peroxide sensing. Sensors 21:2204. https://doi.org/10.3390/s21062204
4. Huang W, Lin T, Cao Y et al (2017) Hierarchical $NiCo_2O_4$ hollow sphere as a peroxidase mimetic for colorimetric detection of H_2O_2 and glucose. Sensors 17:217. https://doi.org/10.3390/s17010217
5. Chen S, Gu Z, Hu J, et al A method of synthesis of ultrathin palladium nanosheet with peroxidase mimetic activity for the colorimetric detection of H_2O_2
6. Mei H, Zhou S, Lu M, Cheng L (2020) Tetrapod-like $ZnO/ZnFe_2O_4$ based heterostructure for enhanced ethanol detection. J Alloys Compd 840:155583. https://doi.org/10.1016/j.jallcom.2020.155583
7. Falak P, Hassanzadeh-Tabrizi SA, Saffar-Teluri A (2017) Synthesis, characterization, and magnetic properties of ZnO-Zn Fe_2O_4 nanoparticles with high photocatalytic activity. J Magn Magn Mater 441:98–104. https://doi.org/10.1016/j.jmmm.2017.05.044
8. Zhao M, Huang J, Zhou Y et al (2013) Controlled synthesis of spinel $ZnFe_2O_4$ decorated ZnO heterostructures as peroxidase mimetics for enhanced colorimetric biosensing. Chem Commun 49:7656. https://doi.org/10.1039/c3cc43154a
9. Manohar A, Krishnamoorthi C, Naidu KCB, Pavithra C (2019) Dielectric, magnetic hyperthermia, and photocatalytic properties of Zn Fe_2O_4 nanoparticles synthesized by solvothermal reflux method. Appl Phys A Mater Sci Process 125:477. https://doi.org/10.1007/s00339-019-2760-0

10. Gao L, Zhuang J, Nie L et al (2007) Intrinsic peroxidase-like activity of ferromagnetic nanoparticles. Nature Nanotech 2:577–583. https://doi.org/10.1038/nnano.2007.260
11. Navaseepthy D, Rebekkah A, Viswanahan C, Ponapandia N (2017) N-doped graphene/ZnFe2O4: a novel nanocomposite for intrinsic peroxidase-based sensing of H_2O_2. Mater Res Bull 95:1–8. https://doi.org/10.1016/j.materresbull.2017.06.033
12. Lalanne M, Soon JM, Barnabé A et al (2010) Preparation and characterization of the defect–conductivity relationship of Ga-doped ZnO thin films deposited by nonreactive radio-frequency–magnetron sputtering. J Mater Res 25:2407–2414. https://doi.org/10.1557/jmr.2010.0300
13. Capdeville S, Alphonse P, Bonningue C et al (2004) Microstructure and electrical properties of sputter-deposited Zn0.87 Fe_2O_4 thin layers. J Appl Phys 96:6142–6146. https://doi.org/10.1063/1.1808904
14. Sandu I, Presmanes L, Alphonse P, Tailhades P (2006) Nanostructured cobalt manganese ferrite thin films for gas sensor application. Thin Solid Films 495:130–133. https://doi.org/10.1016/j.tsf.2005.08.318
15. Tauc J, Grigorovici R, Vancu A (1966) Optical properties and electronic structure of amorphous germanium. Phys Status Solidi B 15:627–637. https://doi.org/10.1002/pssb.19660150224
16. Jiang B, Duan D, Gao L et al (2018) Standardized assays for determining the catalytic activity and kinetics of peroxidase-like nanozymes. Nat Protoc 13:1506–1520. https://doi.org/10.1038/s41596-018-0001-1
17. Vaseashta A (2006) Carbon nanotubes based devices and sensors. Mater Manuf Process 21:710–716. https://doi.org/10.1080/10426910600613595
18. Vaseashta AK (2013) Advanced sensors for safety and security. Springer, Dordrecht
19. Giaretta JE, Duan H, Oveissi F et al (2022) Flexible sensors for hydrogen peroxide detection: a critical review. ACS Appl Mater Interfaces 14:20491–20505. https://doi.org/10.1021/acsami.1c24727

Chapter 14
Advancing Critical Infrastructure Resilience to Hybrid Threats through a Systems Approach, Standardization, and MCDA

Dejan Vasović, Goran Janacković, Marija Milenković, and Ashok Vaseashta

Abstract Critical infrastructure, including energy grids, public utilities, transportation networks, telecommunications systems, financial services, etc., is highly susceptible to hybrid threats due to its interconnected nature and reliance on digital technology. Disruptions to critical infrastructure can have cascading effects on society, economy, environment, and national security overall. The fight against hybrid threats and the protection of critical infrastructure are of paramount importance in the European Union (EU), because these threats, through the aforementioned cascading effects, can destabilize economies, and erode public trust. Ensuring the resilience of critical infrastructure represents a vital element for safeguarding public safety and social well-being. In this sense, enhancing resilience and diversifying supply chains are key strategies for reducing reliance on external sources. With no less importance is the fact that addressing hybrid threats and protecting critical infrastructure require cross-border collaboration and information sharing both among EU member states and accessing countries like the Republic of Serbia, prominent international partners (like ISO), and relevant stakeholders (like NATO). In the Republic of Serbia, a formal legislative framework for critical infrastructure protection was established by the adoption of the Law on critical infrastructure in 2018. With a significant reference to international experiences, in the Republic of Serbia, there are still many challenges in the field of protection of critical infrastructure, especially from hybrid threats. In this sense, the goal of this paper is to present the possibility of improving the resilience of critical infrastructure through the standardization of

D. Vasović (✉) · G. Janacković
University of Niš, Faculty of Occupational Safety, Niš, Republic of Serbia
e-mail: dejan.vasovic@znrfak.ni.ac.rs

M. Milenković
University of Niš, Niš, Republic of Serbia

A. Vaseashta
International Clean Water Institute, Manassas, VA, USA

D. Radu et al. (eds.), *Countering Hybrid Threats Against Critical Infrastructures*, NATO Science for Peace and Security Series B: Physics and Biophysics, https://doi.org/10.1007/978-94-024-2304-4_14

management activities, the operationalization and objectification of the sources of priority activities using the multi-criteria decision-making method, and the integration of the requirements of various interested parties through a systems approach. The main conclusion and contribution of the paper refer to the need for more intensive harmonization of practice in the Republic of Serbia with that in the EU.

Keywords Critical infrastructure · Resilience · Systems approach · MCDA

14.1 Introduction

Technical systems play an essential role in shaping modern and future society, influencing nearly every aspect of human life, from communication and transportation to healthcare and industry. Their importance is marked by several key factors that collectively drive societal progress, economic growth, and the sustainability of our global civilization. For instance, automation and robotics significantly enhance productivity by performing tasks more quickly and accurately than humans. This leads to economic growth by increasing the output of goods and services while reducing costs [1]. Also, information and communication technologies (ICTs) enable global trade and collaboration, breaking down geographical barriers and creating new markets. The Internet of Things (IoT) as well as the Internet of Behavior (IOB) in particular have become the vital infrastructure for the global economy, facilitating e-commerce, remote work, and digital entrepreneurship [2]. From the environmental perspective, advanced technical systems, including satellite monitoring, sensors, and data analytics, are vital for tracking environmental changes, predicting natural disasters, and implementing effective nature conservation strategies. With further reference to monitoring, advanced technical systems play a vital role in monitoring and controlling public health threats. During pandemics, data-driven models help predict the spread of disease, enabling timely interventions and the efficient allocation of resources [3]. With no less importance is the fact that as societies become more dependent on digital systems, cybersecurity becomes of paramount importance. Security-technical systems become essential for protecting critical infrastructure (CI), sensitive data, and national security primarily from cyber threats.

In the context of the mentioned critical infrastructure, it is necessary to know that modern society-supporting technical systems are integral components of critical infrastructure, serving as the backbone of essential services that societies rely on for survival, safety, and economic stability [4]. These systems encompass a wide range of technologies, including energy supply networks, water treatment facilities, healthcare systems, communication networks, transportation systems, etc. Their connection as critical infrastructure elements can be scientifically and quantitatively analyzed through several dimensions or sectors, either from a technical or educational perspective [5]. The CI transportation system facilitates the movement of goods and people, supporting trade, commerce, and the labor market. Disruptions in these infrastructures can lead to significant economic losses, as seen in events like power outages or transportation strikes, which can halt production, disrupt supply chains, and cause widespread financial damage. The health and safety of the population are directly tied to the reliability of critical infrastructure. Healthcare systems, including hospitals, emergency services, and public health networks, are dependent on continuous access to electricity, water, and communication. In emergencies, such as natural disasters or pandemics, the resilience of these infrastructures becomes even more important. For example, during the COVID-19 pandemic, robust healthcare infrastructure, supported by reliable communication networks, was essential in managing the crisis, distributing vaccines, and disseminating public health information [3]. Water supply and sanitation systems are also critical for public health. Clean drinking water and proper waste management prevent the spread of diseases and ensure a healthy population [6]. The failure of these systems can lead to public health crises, as evidenced by the Flint water crisis in Michigan, where lead contamination in the water supply had severe health implications for residents [7].

Nowadays, cybersecurity is increasingly a concern as critical infrastructure becomes more interconnected and reliant on digital systems. Cyberattacks on power grids, water treatment facilities, or communication networks can have devastating effects, disrupting services, causing economic damage, and potentially leading to loss of life. The importance of securing critical infrastructure against such threats has led to the development of specialized cybersecurity frameworks and the integration of resilience measures into infrastructure design and operation [8].

When talking about critical infrastructure protection, it is necessary to know its nature of a comprehensive and multidisciplinary process aimed at safeguarding the essential systems and assets that support a nation's economy, security, public health, and overall well-being. Given the interdependence of modern infrastructure systems and the increasing complexity of threats, including natural disasters, cyberattacks, and terrorism, critical infrastructure protection—CIP requires a systematic and coordinated approach. The process involves risk assessment, prevention, detection, response, and recovery measures, which are implemented across various stages to ensure the resilience and continuity of critical infrastructure [9].

As the complexity of threats evolves (often ending with hybrid threats—HT), CIP must also adapt to address new and emerging challenges. One such challenge is the rise of the mentioned hybrid threats, which combine traditional military tactics with unconventional methods like cyberattacks, disinformation, and economic manipulation. The intersection of CIP and hybrid threats requires a balanced understanding of how these threats can target critical infrastructure and what measures can be taken to mitigate their impact.

14.2 Systems Approach and Critical Infrastructure Resilience

The systems approach is based on a comprehensive analysis of problems, goals, and alternatives, with an emphasis on the analysis of relationships in the system and interactions with the environment, based on which alternative strategies for realizing the goals of the system are identified and evaluated [10–12]. The importance of considering the whole, interdisciplinarity, and multidisciplinarity are the basic characteristics of this approach.

The resilience of critical infrastructure is vital to the stability and security of society, as this infrastructure fundamentally affects the functioning of society and national security. Disruptions in critical infrastructure can have effects on different sectors, so resilience as a property of systems to withstand, recover, or adapt to disturbances is the basis of disaster management [13, 14].

There is a strong link between the systems approach and the resilience of critical infrastructure. The systems approach involves understanding and managing complex systems based on the analysis of their interrelated elements, behavior, and interactions in the system, as well as the system with its environment. A systems approach enables an understanding of complexity and interdependence [11, 15]. Disturbances in one part of the system can have effects on other parts of the system, so by analyzing interdependencies, potential deficiencies of the system are identified. System dynamics modeling allows us to identify the critical components of the system and to analyze the behavior of the system in the presence of disturbances.

Thus, the systems approach enables a holistic understanding of risk by observing not only individual elements of the infrastructure but the system as a whole. The analysis of dynamic risks is particularly important because the disruption of one part of the system often leads to consequences in other parts, which is especially important when multiple hazards occur simultaneously. Most often, resilience in systems is achieved through redundancy and flexibility. The systems approach allows for reserve resources or alternative modes of operation to be foreseen. Holism is a fundamental principle for understanding complex systems because a complex whole can exhibit properties that are not easily explained by understanding the parts.

System analysis is a general system research methodology whose goal is to define problems in the system and determine ways to solve them. It considers problems related to system goals, available resources, constraints, interactions in the system, as well as interactions with the environment. Fig. 14.1 shows a simplified algorithm of system analysis with representative stages.

System analysis is carried out through several basic stages and functions. The basic stages of system analysis (problem formulation; definition of goals, limitations, and criteria; system development; system evaluation) represent purposeful units [11]. These stages consist of activities and system analysis functions in which the collected information about the system of interest is used. The basic functions are information gathering, model formation, and information synthesis. Returning branches in the algorithm implies rejecting the initial assumption (stage I), as well as improving the description of the problem by setting additional goals and hypotheses (from stage V back to stage II) as a basis for improving the existing models and evaluation criteria.

The need for data collection, processing, and use is very important in complex systems that make up critical infrastructure. Collecting information in critical infrastructure systems is considered particularly important. A sufficient level of detail enables a better understanding of problems, a more realistic assessment of functioning, and potential threats. Each stage of the system analysis is carried out in

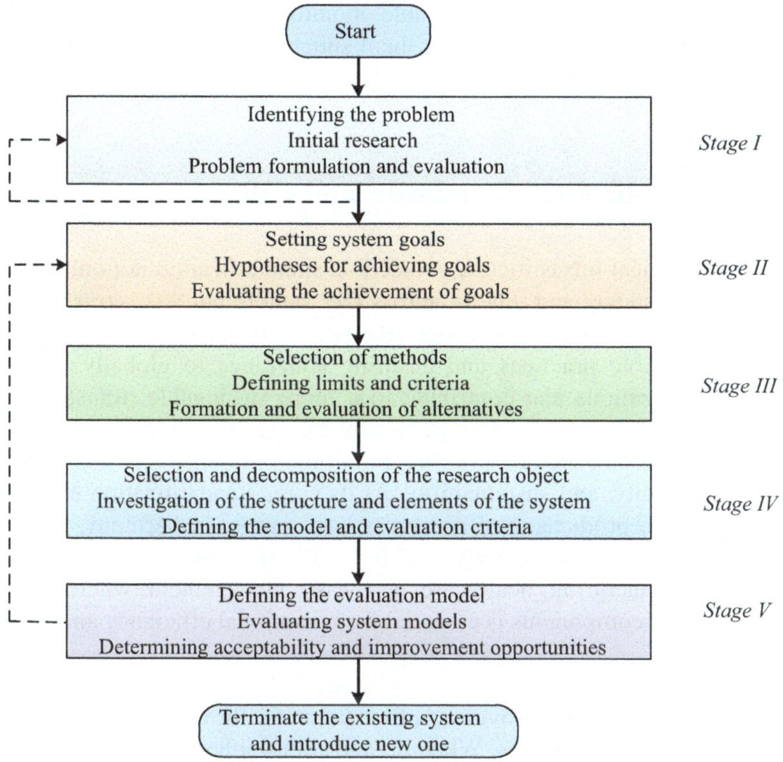

Fig. 14.1. A simplified algorithm of system analysis

accordance with a pre-defined approach (research model) on the basis of which appropriate information is collected, and a conclusion is drawn through the synthesis of the collected information, within and using the formed model.

The functions of system analysis are interrelated. The model of any aspect of system research does not have its full meaning without concrete information that is incorporated into it and reflects the real state of the system under investigation. Likewise, the collected information does not have full significance if it is not interpreted through the adopted model.

The functioning and development of critical infrastructure systems, primarily due to their complexity, impact on the functioning of the community, and the involvement of a large number of interested parties, is accompanied by a large number of problems. That is why every systemic research begins with the phase of formulating the problem within which potential threats are recognized and clearly formulated. During the functioning of the system, additional information on potential threats is collected, which is the basis for improving the initially defined models and selected criteria [11, 16, 17]. A particularly important property of critical infrastructure is resilience. Resilience indicators can be classified into different categories for measuring numerous dimensions (technical, economic, environmental, and social) [18–20]. These indicators enable monitoring and quantifying different aspects, and their classification makes them suitable for the application of multi-criteria decision-making methods.

14.3 Standardization, Standards, and CIP

In the field of critical infrastructure protection, standardization not only enhances operational consistency and quality across CIP sectors but also serves as a vital tool for promoting environmental/societal sustainability and managing risks. By fostering responsible practices and ensuring adherence to globally recognized standards, organizations can contribute to a more sustainable, transparent, and accountable global CI security. Standardization plays a critical role across a diverse range of industries and sectors, offering a multitude of benefits that enhance efficiency, compatibility, and sustainability. At its core, standardization ensures that disparate systems, products, and processes can operate in harmony, facilitating interoperability. This is particularly vital in highly interconnected sectors such as technology, manufacturing, healthcare, and waste management, where the seamless integration of components is essential for operational efficiency and effectiveness [12].

From a technical perspective, standards establish universally recognized best practices and specifications, providing a foundation for the consistent production of high-quality goods and services. When organizations adhere to these standardized processes and procedures, they not only ensure uniformity in their outputs but also enhance overall customer satisfaction and foster greater trust. Standardization thus serves as a mechanism for maintaining excellence and reliability, which are essential for the competitive edge in today's global market. For instance, in terms of environmental sustainability, standardization plays a pivotal role. By promoting eco-friendly practices, standards contribute to long-term environmental stewardship. Standardized frameworks in environmental and risk management provide organizations with structured approaches to identifying, assessing, and mitigating risks more effectively. These frameworks enable businesses to anticipate potential threats proactively, thus safeguarding their operations and protecting their reputation. Standards like ISO 14001 for environmental management systems empower organizations to continuously assess and improve their environmental performance, enabling them to communicate their sustainability efforts transparently to stakeholders [21, 22].

Standardization is fundamental to the protection of critical infrastructure (CI), which includes essential systems and assets suitable for standardization efforts. These infrastructures form the backbone of modern society, and their disruption can lead to severe societal, economic, and environmental consequences. Standardization offers a systematic approach to safeguarding these critical systems through the development of uniform protocols, technologies, and practices that enhance resilience, interoperability, and security. Standardization plays a pivotal role in ensuring that the technologies, protocols, and procedures used by various sectors and organizations are compatible.

Resilience is a key aspect of critical infrastructure protection, referring to the ability of systems to anticipate, withstand, and recover from disruptions, whether due to natural disasters, cyberattacks, or technical failures. Standardized procedures for risk assessment and management, such as those outlined in the ISO 31000 risk management framework, provide organizations with a structured approach to identify vulnerabilities, evaluate risks, and implement mitigation strategies. Standardization ensures that resilience-building measures are consistently applied across different sectors and regions, enhancing the ability of infrastructures to withstand and recover from both anticipated and unforeseen events [23].

Standards such as ISO 22301 for business continuity management and ISO 27001 for information security management contribute directly to the resilience of CI by ensuring that organizations have established protocols for responding to incidents, protecting sensitive information, and maintaining operational continuity. Standardization in this context provides a common language and methodology for all stakeholders involved, from local governments to private sector organizations, fostering a coordinated and efficient response to crises [24].

Internationally recognized standards such as the Common Criteria for Information Technology Security Evaluation (ISO/IEC 15408) ensure that security measures across different systems can integrate seamlessly, reducing gaps in protection and minimizing the risk of cascading failures across infrastructures. This cross-sector compatibility, enabled by standardization, is essential for coordinating defense mechanisms against threats such as cyberattacks, which can target multiple infrastructures simultaneously [25].

Moreover, standardization facilitates cross-sector collaboration by aligning the efforts of diverse stakeholders, including public authorities, private entities, and international organizations. The alignment of technical specifications, communication protocols, and safety practices allows for more effective information sharing and joint action in the face of shared risks. For example, standards for cyber incident reporting, like those under NIST's Cybersecurity Framework, enable rapid dissemination of information about potential threats, allowing for timely and coordinated responses across sectors [26].

Standardization is also essential in establishing comprehensive security protocols to protect CI from threats such as terrorism, cyberattacks, and physical sabotage. Security standards provide a baseline for assessing the security posture of infrastructure systems, ensuring that critical assets are safeguarded against both intentional and unintentional disruptions. For instance, ISO 27001 offers a framework for implementing and managing information security within organizations, which is particularly vital for protecting critical infrastructures from cyber threats [27].

Standardized security measures contribute to the identification and assessment of risks, facilitating a more holistic understanding of vulnerabilities across interconnected systems. This ensures that CI operators not only focus on individual infrastructure assets but also consider the broader implications of a failure within the interconnected network. For instance, standards for physical security, such as ISO 28000 for supply chain security, address the protection of infrastructure from physical threats, while cyber-specific standards like the ISO/IEC 27032 provide guidelines for securing critical online systems against cyber risks [28].

In addition to security, standardization also supports a risk-based approach to CI protection. The integration of risk management principles into CI planning and operations, as outlined in ISO 31000, ensures that operators take a proactive approach to identifying, analyzing, and mitigating potential threats. This comprehensive view of risk enhances the resilience of critical systems, promoting the adoption of preventative measures that minimize disruptions and facilitate rapid recovery [23].

As critical infrastructure systems become more digitized and interconnected, the role of technology in CI protection has grown exponentially. Standardization is vital for guiding the development and implementation of new technologies that can enhance infrastructure protection. For instance, emerging technologies like the Internet of Things (IoT) are increasingly being integrated into CI systems for real-time monitoring, predictive maintenance, and automated responses to threats (mentioned and elaborated in more detail in Sect. 14.5). However, without standardized protocols for data security, interoperability, and reliability, the deployment of such technologies can introduce new vulnerabilities.

Standards such as ISO/IEC 27001 for information security and ISO/IEC 62443 for industrial control systems security help ensure that new technologies are secure and compatible with existing infrastructure, enabling CI operators to harness innovation while minimizing risks [29]. Furthermore, standardization provides a pathway for the global dissemination of technological advancements, allowing countries and industries to adopt best practices and cutting-edge solutions in CI protection more rapidly [27].

Critical infrastructure protection is a global concern, with threats often transcending national borders, particularly in sectors like energy, telecommunications, and transportation. International standards, such as those developed by the International Organization for Standardization (ISO) and the International Electrotechnical Commission (IEC), play a key role in fostering global cooperation and policy alignment in CI protection. These standards create a common framework for international collaboration, allowing countries to align their strategies, share intelligence, and develop joint defence mechanisms. For example, international cooperation on cybersecurity for critical infrastructure is guided by standards like ISO/IEC 27032, which addresses cybersecurity management across interconnected systems. Such standardization efforts enable countries to work together to defend against global threats like cyberterrorism, providing a unified approach to safeguarding critical assets that span multiple jurisdictions [27].

14.4 Role of MCDA in CIP Management Activities

The modern way of life and the development of techniques and technology have accelerated the transfer of information and facilitated the processes of production, traffic, and financial transactions. At the same time, protecting critical infrastructure and preventing the obstruction or hacking of information has become more challenging due to the use of technologies like artificial intelligence in disrupting legal information flows and enabling various types of diversions. In this regard, one of the tools that is very useful for addressing hybrid threats (physical, cyber, and other forms of destabilization) is the application of multi-criteria analysis. This approach is effective in identifying and ranking key parameters (criteria that affect the security of sensitive infrastructure, the selection of resource factors needed for the implementation of protection measures, potential damage, and protection costs), data collection of factors related to a given sensitive system (threat history, expert assessments, and analyses), assigning weights to each criterion based on its importance (according to expert judgments, expertise, study analyses, and historical threat data), as well as evaluating alternatives (selecting strategies based on essential criteria and weighted factors). Combining results from different multi-criteria analyses, with a final control analysis, makes it easier for decision-makers to choose the most favorable optimal solution, strategy, and system of measures to protect critical infrastructure [19].

The use of multi-criteria analysis enables a systems approach to decision-making, particularly in situations where it is necessary to combine the common interests of infrastructure users and consider different qualitative and quantitative criteria and factors affecting the final decision. Every user wants to protect their infrastructure system in the best, fastest, and cheapest way. Multi-criteria analysis combines different types of system information, resources, perspectives, and threats. The method simplifies complex decisions by consolidating all relevant weights of criteria, reducing the subjectivity of decisions, and emphasizing transparency in understanding, comparing alternatives, and explaining decisions. It includes scaling (converting different values into a common scale), normalization (assigning values to eliminate the influence of different units of measurement), and assigning appropriate weights to criteria in the decision-making process [13].

The resilience of critical infrastructure can be described by indicators classified into technical, social, economic, organizational, and environmental dimensions. Different methods are used to assess resilience, from quantitative to quantitative approaches and index methods. These assessments aim to identify potential problems and improve resilience and decision-making in critical situations. Multi-criteria analysis (MCDA) methods help decision-makers identify key influencing factors as well as prioritize specific activities or investments based on multiple decision-making elements, such as social impact, costs, and risks [30].

There are several methods of multi-criteria analysis. Each of them has its own specifics, advantages, and disadvantages in application. For example, ELECTRE and TOPSIS are used in fields such as engineering, economics, management, and ecology, as well as the environmental impact of various projects and policies. PROMETHEE is applied in the evaluation of health technologies, services, and policies, industry optimization of production processes, supplier selection, transport, economics and finance, evaluation and selection of investment decisions, as well as project and portfolio evaluation. These methods can be combined with others to assess the dynamic properties of systems. For example, the Monte Carlo method is used to simulate and analyze complex systems and can be combined with TOPSIS for decision-making in case of uncertainty and variability in the data. Monte Carlo performs risk modeling, and multi-criteria methods like VIKOR or TOPSIS can be used to rank different risk scenarios. By combining several methods, we get more reliable results that would help stakeholders in the decision-making process [31].

In the context of critical infrastructure examples, multi-criteria methods could help combine comprehensive information regarding selected parameters, with a focus on essential sectors that need special attention. The methods take into consideration the history of hybrid threats like cyber-attacks, disinformation, physical sabotage, economic coercion, and the impact of various policies and regulations considering the vulnerability of infrastructure and the complex interdependences within the system and its supply chain, including the influence of the human factor. By managing maximum security and protection in specific sectors—cyber and digital infrastructure, public safety, water, the environment, energy, transportation (water and air), and space research—multi-criteria analysis enables the combination of diverse information and compromise solutions. This allows users to simplify complex problems in the best possible way, and make timely and effective decisions. Protection systems can be implemented with the help of various tools, including systems for detecting intrusions, monitoring incoming and outgoing network traffic, reducing suspicious activities, innovating data encryption solutions and backup solutions, regularly educating people, and setting adequate regulations for data protection.

14.5 Emerging Technologies and AI Role in CIP

Emerging technologies play an increasingly pivotal role in the protection of critical infrastructure. Emerging technologies, driven by advances in artificial intelligence (AI), machine learning, blockchain, and the Internet of Things (IoT), offer new ways to safeguard these vital infrastructures. Miniaturization of IoT devices plays a significant role in enhancing connectivity at the system level and boosting resiliency. Enhanced connectivity is through distributed network architecture, accessibility, scalability, and processing of the data at the edge, also termed edge computing, thus reducing the need for real-time data exchange across different nodes. The improvement in resiliency is achieved through redundancy and fault tolerance, decentralization, and improved energy efficiency.

AI and machine learning are at the forefront of efforts to protect critical infrastructure. These technologies enable real-time monitoring of large-scale systems, identifying anomalies and predicting potential threats before they escalate into crises. For example, in energy grids, machine learning algorithms can analyze vast amounts of data from sensors placed across the network, detecting early signs of malfunction or sabotage. This predictive capability is particularly useful in thwarting cyberattacks, as AI can continuously learn from new threats and adapt its defense mechanisms accordingly [32]. A particularly effective application of AI is in automated incident response systems. By quickly analyzing incoming data, AI can trigger pre-defined responses to isolate or neutralize threats. This reduces response time, which is critical in preventing cascading failures in interconnected systems, such as a power outage affecting communication networks [33].

Another emerging technology that holds promise for critical infrastructure protection is digital twin technology. Digital twins are virtual replicas of physical systems that allow operators to simulate and analyze the behavior of infrastructure in a safe, virtual environment. These simulations can predict how systems will respond to different stressors, such as increased demand, cyberattacks, or natural disasters. For example, a digital twin of a power plant could model the effects of extreme weather conditions or a sudden increase in energy demand, enabling operators to take pre-emptive actions to avoid failures. By offering a detailed, real-time view of infrastructure systems, digital twins help in planning predictive maintenance, reducing the likelihood of breakdowns, and improving overall system resilience [34, 35].

IoT devices, which include sensors and smart devices, allow for real-time monitoring and control of critical infrastructure. In smart cities, IoT devices are already being used to manage traffic flow, water distribution, and waste management. By collecting vast amounts of data from physical systems, IoT technologies provide operators with real-time insights into infrastructure performance, helping to quickly detect and respond to issues. For example, IoT in water infrastructure enables continuous monitoring of water quality, pressure, and usage, helping to prevent leaks, contamination, and other disruptions. Similarly, in the energy sector, IoT sensors monitor the performance of power plants and grids, ensuring efficiency and detecting faults or potential threats. However, the proliferation of IoT devices also introduces new vulnerabilities, as each connected device represents a potential entry point for cyberattacks. To mitigate this risk, cybersecurity frameworks must evolve in parallel, ensuring that IoT devices are secure, regularly updated, and monitored for potential breaches [36, 37].

14.6 Concluding Remarks

As society moves forward, the importance of critical infrastructure will only grow, making it central to the development of a resilient, inclusive, and sustainable future. The connection of modern society-supporting technical systems as critical infrastructure elements is essential for ensuring the stability and continuity of societal functions. By quantitatively assessing the resilience, interdependence, risk, and societal impact of these systems, we can better understand their importance and identify areas for improvement. In a future increasingly reliant on complex technical systems, the ability to manage and protect critical infrastructure will be a key determinant of societal well-being and economic prosperity. The importance of critical infrastructure lies in its role as the backbone of modern society. As we face new challenges and opportunities, from climate change to digital transformation, investing in and safeguarding our critical infrastructure will be predominant in building a resilient and prosperous future. Protecting critical infrastructure from hybrid threats requires a multi-layered approach that addresses both the physical and digital vulnerabilities of key systems. By enhancing cybersecurity, countering disinformation, building resilience, and fostering collaboration, governments and infrastructure operators can mitigate the risks posed by these complex and evolving threats. As hybrid threats continue to develop in sophistication and scale, proactive and adaptive but also standardized CIP strategies will be essential to safeguarding the infrastructure that supports modern society.

Acknowledgements A section of this article presents the results of research supported by the Ministry of Science, Technological Development and Innovation of the Republic of Serbia, supported by contract #: 451–03-66/2024–03/200148.

References

1. Bartusiak A, Kühne M, Nitschke O, Lässig J, Nicolai S, Bretschneider P (2023) First step into automation of security assessment of critical infrastructures. Sustain Energy Grids Netw 36:101139. https://doi.org/10.1016/j.segan.2023.101139
2. Caldwell T (2018) Plugging IT/OT vulnerabilities—part 1. Netw Secur 2018(8):9–14. https://doi.org/10.1016/S1353-4858(18)30078-3
3. Vaseashta A, Maftei C, Radu D (2023) Exploring disruption trajectories from COVID-19 on education and the impact of policies: lessons learned and path forward. In: Gjorchev J, Malcheski S, Rađenović T, Vasović D, Živković S (eds) Transformation and efficiency enhancement of public utilities systems: multidimensional aspects and perspectives. Business Science Reference, Hershey, pp 314–340. https://doi.org/10.4018/978-1-6684-7730-4.ch012
4. Vasovic D, Janackovic G, Musicki S (2020) Model of effective civil-military collaboration in natural disaster risk management. In: Gocić M, Aronica G, Stavroulakis G, Trajković S (eds) Natural risk management and engineering. Springer tracts in civil engineering. Springer, Cham. https://doi.org/10.1007/978-3-030-39391-5_2
5. Nikolić V, Vasović D (2015) Tailor made education: environmental vs. energy security and sustainable development paradigm. In: Čaleta D, Radović V (eds) Comprehensive approach as "Sine Qua Non" for critical infrastructure protection. NATO advanced research workshop. NATO science for peace and security series D: information and communication security, vol 39. IOS Press, Amsterdam, pp 201–214. https://doi.org/10.3233/978-1-61499-478-7-201
6. Vaseashta A, Gevorgyan G, Kavaz D, Ivanov O, Jawaid M, Vasović D (2021) Exposome, biomonitoring, assessment and data analytics to quantify universal water quality. In: Vaseashta A, Maftei C (eds) Water safety, security and sustainability. Advanced sciences and technologies for security applications. Springer, Cham. https://doi.org/10.1007/978-3-030-76008-3_4

7. Paine M, Kushma JA (2017) The flint water crisis and the role of professional emergency managers in risk mitigation. J Homel Secur Emerg Manage 14(3):1–9. https://doi.org/10.1515/jhsem-2017-0009
8. Janaćković G, Mušicki S, Vasović D (2019) Information technology risks: management, standards, and methods. In: IX international symposium engineering management and competitiveness, technical faculty "Mihajlo Pupin", Zrenjanin, Serbia, pp 179–184
9. Trump BD et al (2017) Social resilience and critical infrastructure systems. In: Linkov I, Palma-Oliveira J (eds) Resilience and risk. NATO science for peace and security series C: environmental security. Springer, Dordrecht. https://doi.org/10.1007/978-94-024-1123-2_9
10. Chouinard P, Giddings J (2023) A systems approach to critical infrastructure resilience. In: Masys AJ (ed) Safety and security science and technology. Advanced sciences and technologies for security applications. Springer, Cham. https://doi.org/10.1007/978-3-031-21530-8_3
11. Savić S, Stanković M, Janaćković G (2021) Teorija sistema i rizika (Theory of systems and risk). Akademska misao (Academic Mind), Belgrade
12. Vasović D, Janaćković GL, Vaseashta A (2021) Application of the systems approach and system standards in water safety plan development and implementation. In: Vaseashta A, Maftei C (eds) Water safety, security and sustainability. Advanced sciences and technologies for security applications. Springer Nature Switzerland AG, Cham, pp 243–262. https://doi.org/10.1007/978-3-030-76008-3_10
13. Janaćković G, Savić S, Stanković M (2017) Framework for indicator-based optimization of disaster risk management in local communities. Facta Univ, Work Living Environ Prot 14(1):11–22
14. Rathnayaka B, Siriwardana C, Robert D, Amaratunga D, Setunge S (2022) Improving the resilience of critical infrastructures: evidence-based insights from a systematic literature review. Int J Disaster Risk Reduct 78:103123
15. Hayes RB (2022) A systems approach to a resilience assessment for agility. Syst Sci Control Eng 10(1):955–964
16. Amadei B (2020) A systems approach to building community capacity and resilience. Challenges 11(2):28
17. Lichte D, Torres FS, Engler E (2022) Framework for operational resilience management of critical infrastructures and organizations. Inf Dent 7(5):70
18. Ewa WO, Ugwu OO, Okafor FO (2024) Resilient–vulnerability analysis of critical infrastructure, key resources, assets, and facilities. Innov Infrastruct Solut 9(4):1–27. https://doi.org/10.1007/s41062-024-01405-9
19. Janaćković G, Vasović D, Malenović-Nikolić J, Mušicki S, Vranjanac Ž (2018) A vulnerability assessment of municipality areas to natural disasters based on group fuzzy analytic hierarchy process. J Environ Prot Ecol 19(4):1526–1535
20. Mottahedi A, Sereshki F, Ataei M, Nouri Qarahasanlou A, Barabadi A (2021) The resilience of critical infrastructure systems: a systematic literature review. Energies 14(6):1571
21. Vasović D, Janacković GL, Mušicki S (2017a) Integrative education model for resources and critical infrastructure protection based on risk assessment, resources valorization and threat ranking. In: Linkov I, Palma-Oliveira J (eds) Resilience and risk. NATO science for peace and security series C: environmental security. Springer, Dordrecht. https://doi.org/10.1007/978-94-024-1123-2_19
22. Vasović D, Malenović Nikolić J, Janaćković G, Radosavljević J, Vukadinović A (2017b) Environmental management systems: contemporary trends and practices. Acta Tech Corviniensis—Bull Eng 10(1):94
23. ISO—International Organization for Standardization homepage (2018). https://www.iso.org/iso-31000-risk-management.html. Accessed 9 Sept 2024
24. ISO—International Organization for Standardization homepage (2019). https://www.iso.org/standard/75106.html. Accessed 10 Sept 2024
25. ISO—International Organization for Standardization homepage (2022a). https://www.iso.org/standard/72891.html. Accessed 10 Sept 2024

26. NIST—National Institute of Standards and Technology homepage (2024). https://www.nist.gov/cyberframework. Accessed 11 Sept 2024
27. ISO—International Organization for Standardization homepage (2022b). https://www.iso.org/standard/27001. Accessed 11 Sept 2024
28. ISO—International Organization for Standardization homepage (2022c). https://www.iso.org/standard/79612.html. Accessed 11 Sept 2024
29. IEC—International Electrotechnical Commission homepage (2023). https://www.iec.ch/blog/understanding-iec-62443. Accessed 11 Sept 2024
30. Linkov I, Moberg E (2014) Multi-criteria decision analysis: environmental applications and case studies. CRC Press/Tailor & Francis Group, Boca Raton
31. Baker D, Bridges D, Hunter R, Johnson G, Krupa J, Murphy J, Sorenson K (2002) Guidebook to decision-making methods. WSRC-IM-2002-00002, Department of Energy, USA
32. Vaseashta A (2022) Nexus of advanced technology platforms for strengthening cyber-defense capabilities. In: Djugumanov I, Hadji-Janev M (eds) Vol. 155, practical applications of advanced technologies for enhancing security and defense capabilities: perspectives and challenges for the western balkans. IOS Press, The Netherlands, pp 14–31. https://doi.org/10.3233/NHSDP220003
33. Raval KJ, Jadav NK, Rathod T, Tanwar S, Vimal V, Yamsani N (2024) A survey on safeguarding critical infrastructures: attacks, AI security, and future directions. Int J Crit Infrastruct Protect 44:100647. https://doi.org/10.1016/j.ijcip.2023.100647
34. Taofeeq DM, Rotimi JOB, Shahzad W, Bamgbade JA (2024) Infrastructure digital twin technology: a new paradigm for the future construction industry. Technol Soc 77:102519. https://doi.org/10.1016/j.techsoc.2024.102519
35. Vaseashta A, Sussman P, Braman E (2014) Cyber security—threat scenarios, policy framework and cyber wargames. In: Vaseashta A, Susmann P, Braman E (eds) Vol. 38: cyber security and resiliency policy framework. IOS Press, Amsterdam. https://doi.org/10.3233/978-1-61499-446-6-1
36. Simon T (2017) Critical infrastructure and the Internet of things. Centre for International Governance Innovation and Chatham House. Paper series No. 46
37. Mušicki S, Janaćković G, Vasović D (2020) Standardization and system standards usage in the field of occupational and environmental safety. In: X international symposium engineering management and competitive-ness, technical faculty "Mihajlo Pupin", Zrenjanin, Serbia, pp 89–92

Chapter 15
High Energy Electron Irradiated Diamond Nanoparticles Influence on Yeast Cells' Viability

Galina Boka, Yuri Dekhtyar, Mark Gorohov, Galina Khrustalyova, Aleksandr Rapoport, and Mirko Rocca

Abstract The nanoscale version of diamond results to be very useful as functionalized material for medical devices such as coating layers for prosthesis, drug delivery, and implanted sensors. For this reason, the diamond nanoparticles (DNPs) should not be against the cells' viability, when the patients are treated with ionizing radiation. To study the cells' response after a radiation treatment in the presence of DNPs it was considered to use yeast cells, were considered because of their simplicity, and they are suitable as eukaryotic cells. The aim of this work is to study the influence of DNPs radiated with electron beam radiation treatment and on the viability of the yeast cells being the organism model. DNPs were radiated with a dose rate of 1 to 6 Gy/min, the dose from 15 to 30 Gy, and energy from 6 to 20 MeV, with irradiated DNPs mixed with yeast cells, increasing viability of the cells four times; on the other hand, considering this mixture of DNPs and yeast cells allow to understand if the yeast cells' viability could be influenced by the ionization treatment with the presence of DNPs in solution. The post-treatment of the DNPs and yeast cells solution ended up with a decrease of the yeast cells' viability of 20%. With this study it has been proved that some cells could survive and remain unaltered after a radiation treatment, with instead a viability enhancement, highlighting potential safety and health issues if considering more dangerous cells.

Keywords Diamond nanoparticles · Electron beam radiation · Yeast cells · Toxicity

G. Boka
Oncology Centre of Latvia, Riga, Latvia

Y. Dekhtyar · M. Gorohov · M. Rocca (✉)
Riga Technical University, Riga, Latvia
e-mail: Mirko.Rocca@edu.rtu.lv

G. Khrustalyova · A. Rapoport
Institute of Microbiology and Biotechnology, University of Latvia, Riga, Latvia

D. Radu et al. (eds.), *Countering Hybrid Threats Against Critical Infrastructures*, NATO Science for Peace and Security Series B: Physics and Biophysics, https://doi.org/10.1007/978-94-024-2304-4_15

15.1 Introduction

Nowadays, diamonds have started to increase their importance as material for different fields of biomedical applications like sensors, coating for medical devices, imaging techniques, therapy, etc. [1]. The possibility of artificially producing diamonds increased their application in the biomedical field, thanks to their size manipulation leading to the possibility of using them as diamond nanoparticles (DNPs) [2]. The DNPs maintain the same characteristics of the diamond: hardness, stability, and conductivity; on the other hand, due to these characteristics [2]. It is possible to consider DNPs as a coating thin layer on TiO_2 implant [3]. In the last decades DNPs have been strongly studied to be implemented as drug delivery systems. Thanks to the possibility of coating the DNPs, their specificity to cells and proteins' receptors increase and let them trespass the blood-brain barrier, one of the most selective membranes in the human body, and open the possibility to introduce drugs using DNPs as a shuttle to reach the brain or specific located tumor target [4, 5]; in addition, it was proved that DNPs can support artificial neural network development [6]. In the study [7] it is presented that ultraviolet radiation exposed to DNPs in solution with yeast cells affects their growth characteristic. However, it is unknown if DNPs exposed to ionizing radiation and mixed with the cells may influence their viability and hence the aim of this article is to explore these characteristics.

15.2 Materials

15.2.1 Diamond Nanoparticles

Synthetic diamond nanoparticles made by AdamasNano (140 nm raw DND -15 mV, Zeta 2.0% ash) were used as powder.

15.2.2 Yeast Cells

The *Saccharomyces cerevisiae 77* were used as an organism study model and were taken from the Microbial Strain Collection of Latvia. The strain was maintained on agar medium (gL^{-1}): yeast extract, peptone, glucose (YPG), and agar at 4 °C. The yeast was cultivated at 30 °C in a nutrient-rich liquid medium (gL^{-1}): yeast extract, peptone, and dextrose (YPD). Culture optical density (OD) was monitored at a wavelength of 595 nm, using a Tecan's Sunrise Absorbance Microplate Reader (Switzerland).

15.3 Method

15.3.1 Irradiation

The DNPs were irradiated with high energy electrons from 6 to 20 MeV exploited for radiation therapy. The doses were 15 and 30 Gy with the corresponding dose rate 6 and 1 Gy/min. The "Variant TrueBeam" linear accelerator was employed for irradiation. The distance between the sample and the electron beam source was 100 cm with a radiation field of 10 × 10 cm. The three boluses having a thickness of 1.3 cm to a maximum of 3.0 cm were collocated over the samples to guarantee the maximum dose at the layer of the DNPs powder samples, which had a thickness of 1 mm.

15.3.2 Viability

15.3.2.1 Samples Preparation

The following samples were prepared:

1. Yeast cells mixed with distillate water (ratio 1:1).
2. Distillate water solution of yeast cells with non-radiated DNPs (ratio 1:1).
3. Distillate water solution of yeast cells with irradiated DNPs (ratio 1:1).
4. The solution #2 was also consider for post radiation treatment with 6 Gy/min, dose of 15 Gy and energy of 20 MeV (ratio 1:1).

The cultivated yeast biomass was collected from 10 mL suspension of Saccharomyces cerevisiae 77 after 24 h of growth and centrifugation. The supernatant was discarded after centrifugation and the biomass was diluted with distilled water up to 80 mL in a 100 ml volumetric cylinder to get an optical density (OD) corresponding to the absorbance value OD595 ≈ 0.5. The latter was measured with the Tecan's Sunrise Absorbance Microplate Reader (Switzerland). The prepared yeast suspension was added to a suspension with DNPs (concentration ~ 3 mg/mL) in a ratio of 1:1. The samples were held on an orbital shaker, using the orbital shaker-incubator ES-20/80 BIOSAN Latvia, and rotated at 50 rpm for 1 hour at 25 °C.

15.3.2.2 Viability Measurement

The visual observation was done in phase-contrast mode, using a 1000x magnification using a fluorescence microscope computer-controlled Olympus BX51 equipped with a 100-W mercury lamp, attached to a Peltier-cooled CCD camera DP71 (Olympus, Tokyo, Japan). The image acquisition was performed using (Cell)^{F} imaging software (Olympus, Tokyo, Japan). For the determination of the cell growth curve was considered an amount of 10 μl of suspension of the immobilized system of diamond nanoparticles and yeast was taken as aliquots to make 11 parallel repetitions for each sample. The aliquots were inoculated in a YPG media (190 µl), considering a total of 96-well microplate for the colony growth measurements. Using a Tecan's Sunrise Absorbance Microplate Reader (Switzerland) the cells grown aerobically with shaking (Kinetic interval: 538 s; shake duration between cycles: 470 s; shake settle time between cycles: 60 s) under controlled temperature conditions (30 °C). Optical density at wavelength 595 nm was then monitored in a Tecan's Sunrise Absorbance Microplate Reader (Switzerland) at 30 °C for 24 h. Mean values of optical density were plotted against incubation time for each treated sample as well as for the control. The irradiated DNPs powder was mixed with the yeast cells solution in a ratio of 1:1.

15.4 Results and Discussion

15.4.1 Viability

Figure 15.1 demonstrates that the radiation treatment of DNPs tends to enhance the formation of DNPS agglomeration and their connection to yeast cells.

From Fig. 15.1b, c one may observe that irradiated DNPs tend to cluster formation, the latter attaching the cells to characterize cell viability the Colony Formation Units (CFU) method was employed (the amount of cell colonies grown in the Petri dish from 1 ml of solution was counted) The results are shown in Fig. 15.2.

The presence of non-radiated and radiated DNPs increases the viability of yeast cells even four times. A kinetic study of the yeast cells' growth curve was performed in 24 h. Figure 15.3 presents growth diagrams.

Figure 15.4 presents first derivates of absorbance (A) on time (t) the diagrams in Fig. 15.3.

Figure 15.4 Demonstrates that the location of the maximum of the derivates is influenced by the irradiation of DNPs. The asymmetry coefficient (α%) is calculated as:

$$\alpha\% = \left(\frac{\mathrm{Max} - \mathrm{median}}{\mathrm{Max}}\right) \times 100 \tag{15.1}$$

where Max and median are based, respectively, on the maximum and central values relate to of the data distribution of the first derivate signal from Fig. 15.4, this way is possible to calculate the shift as α%. Figure 15.5 demonstrates the influence of irradiation of DNPs on α%.

Following Figs. 15.2 and 15.4 both radiated and non-radiated DNPs increase the viability of yeast cells. However, the DNPs prolong the required time for the colony to reach the colony maximum growth in 24 h as shown in Fig. 15.5, with an α% value between 2.5% to 5%. Figure 15.6 demonstrates the influence of irradiated

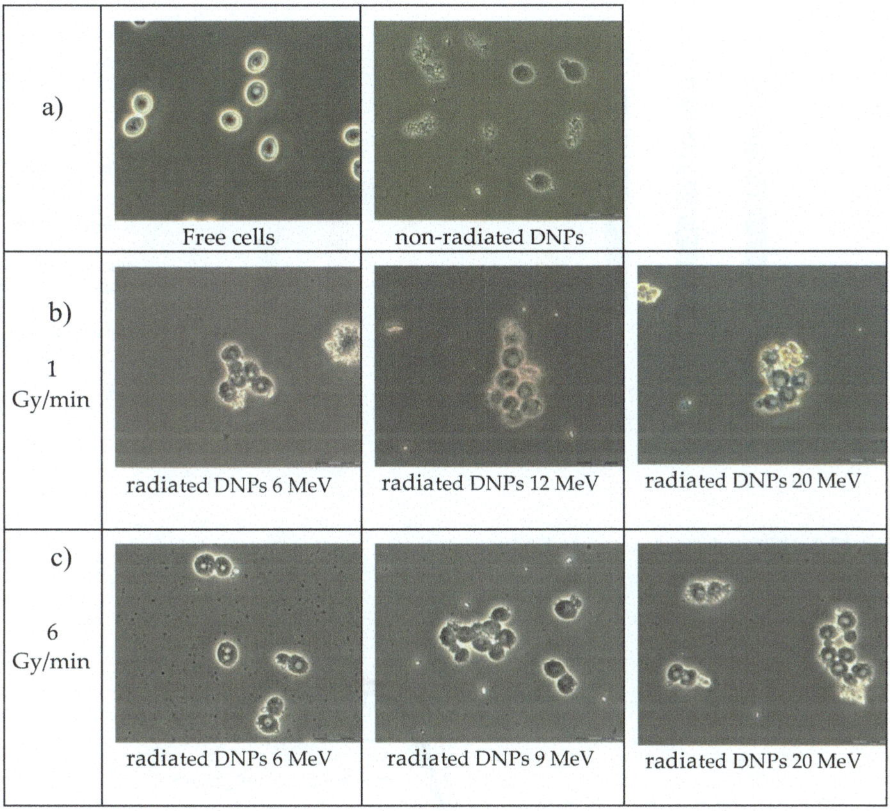

Fig. 15.1 Phase-contrast microscopy images of: **a**) free cells and untreated DNPs in yeast cell solution; **b**) radiated DNPs with dose of 30 Gy in yeast cells solution; **c**) radiated DNPs with dose of 15 Gy in yeast cells solution, immobilized on/in the Saccharomyces Cerevisiae 77 yeast cells. 1000× magnification, scale of 20 μm

DNPs mixed in water solution with the cells on their maximal place of growth (value of the maxima in Fig. 15.4).

Figure 15.6 gives a sense that the maximal growth pace of the cells decreases when the energy of irradiating electrons and delivered dose increases. Figure 15.7 shows the viability of the yeast cells non and mixed (1:1) with non-radiated DNPs. The solutions were irradiated with radiation treatment of 6 Gy/min, 15 Gy and 20 MeV.

Figure 15.7 highlights that with the radiation treatment to the solution made by only yeast cells and water the viability remains unchanged. However, the irradiation of yeast cells + DNPs solution decreased the viability by around 20%.

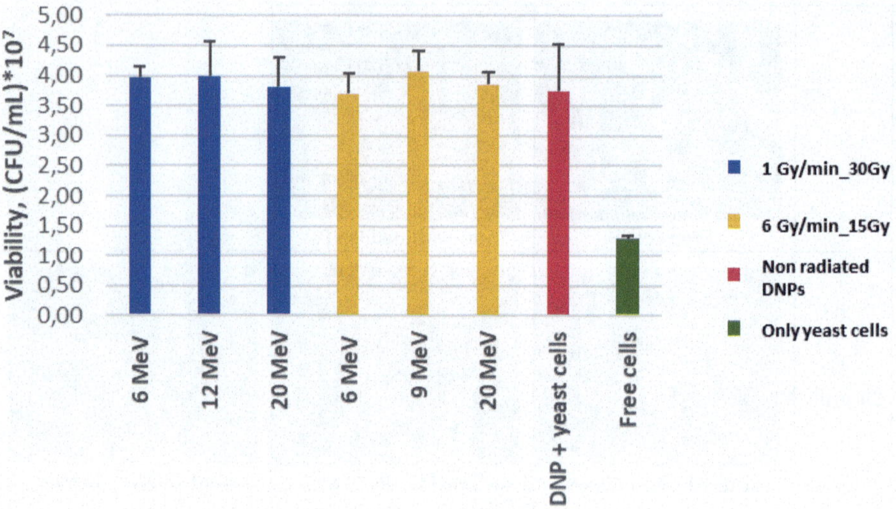

Fig. 15.2 CFU in dependence of the DNPs irradiation

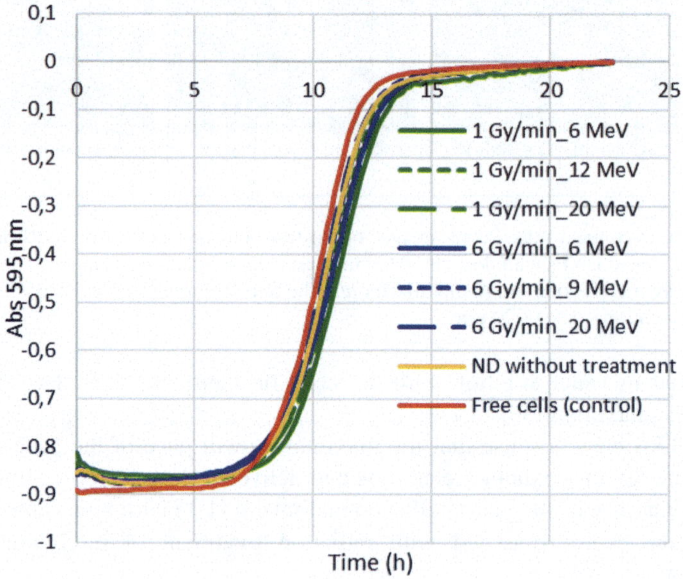

Fig. 15.3 Growth diagrams of yeast cells mixed with non-radiation and irradiated DNPS

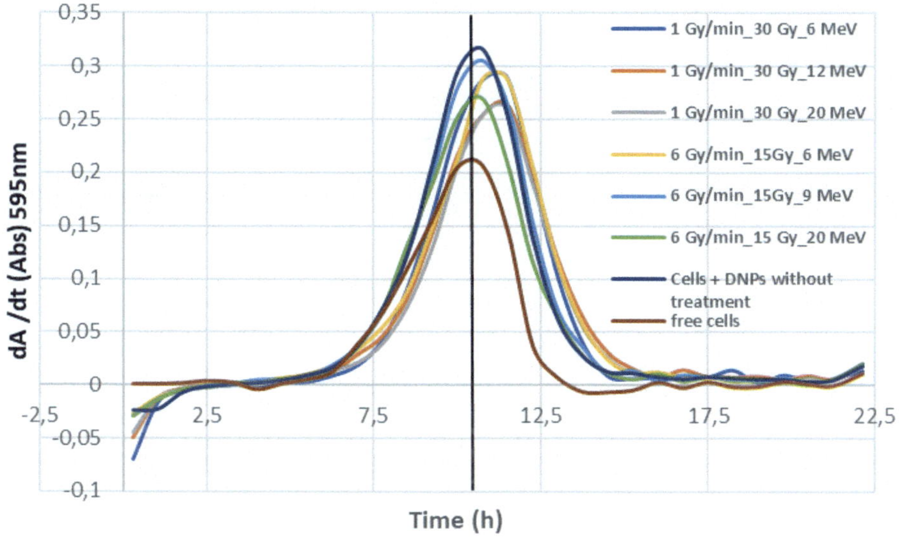

Fig. 15.4 Derivatives of absorbance

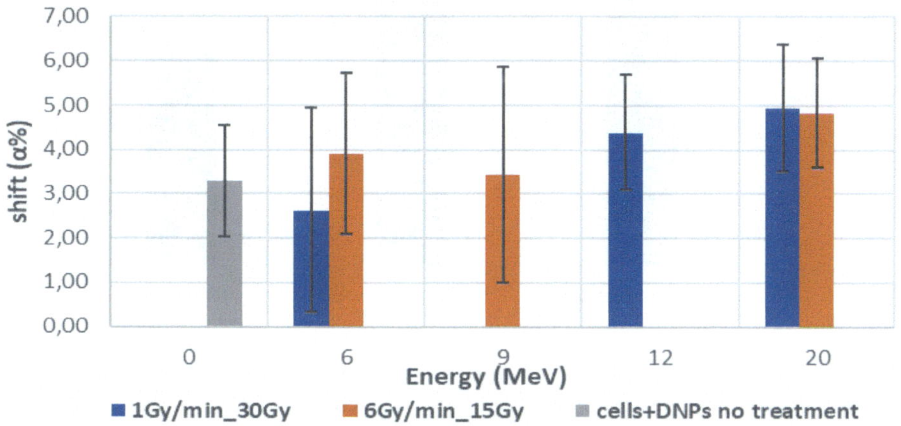

Fig 15.5 Asymmetry coefficient under irradiation of DNPs, mixed with yeast cells solution, comparing different dose exposition: non-radiated DNPs (grey); 30 Gy dose irradiated DNPs (blue) and 15 Gy irradiated DNPs (orange)

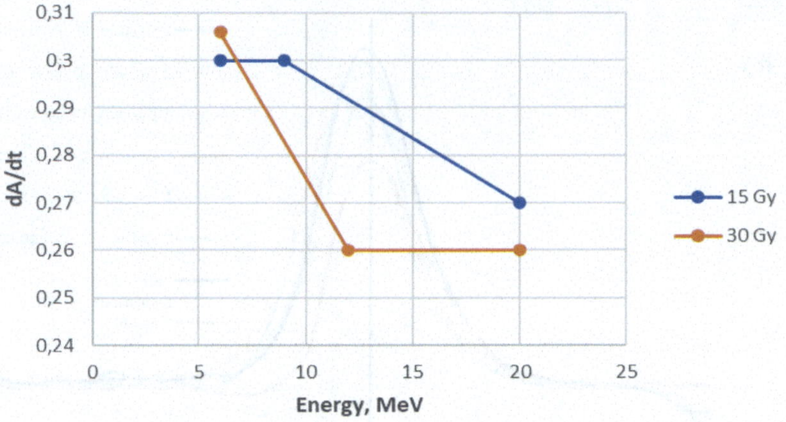

Fig. 15.6 Influence of irradiated DNPs mixed in water solution with the yeast cells on their maximal place of growth

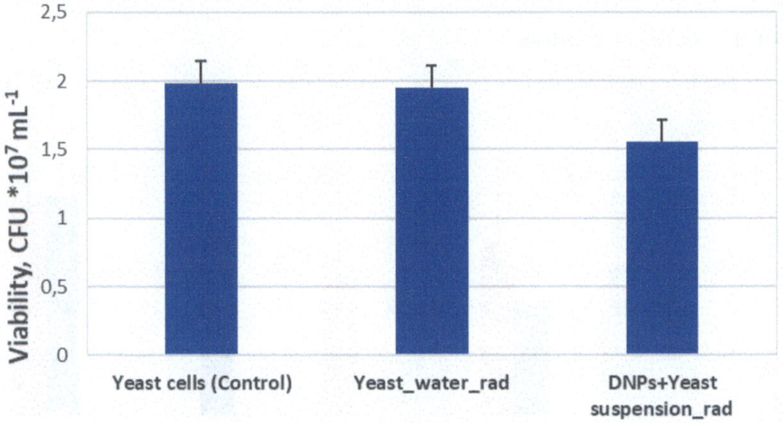

Fig. 15.7 Viability of the yeast cells in radiated solutions mixed with non-radiated DNPs

15.5 Conclusion

The irradiated DNPs mixed in water with yeast cells do not influence their viability, however, they influence their growth. The maximal growth pace of the cells decreases when the energy of radiation and delivered dose increases. The irradiation, considering a radiation treatment of 6 Gy/min 15 Gy and energy of 20 MeV, for the yeast cells mixed in water with non-radiated DNPs, decreases the viability of the cells by around 20%.

References

1. Turcheniuk K, Mochalin VN (2017) Biomedical application of nanodiamond. Nanotechnology 28:252001
2. Mani N, Rifai A, Houshyar S, Booth MA, Fox K (2020) Diamond in medical devices and sensors: an overview of diamond surfaces. Med. Devices Sens 3(6):e10127. https://doi.org/10.1002/mds3.10127
3. Tasat DR, Bruno ME, Domingo M, Gurman P, Auciello O, Paparella ML, Evelson P, Guglielmotti MB, Olmedo DG (2017) Biokinetics and tissue response to ultrananocrystalline diamond nanoparticles employed as coating for biomedical devices. J Biomed Mater Res B Appl Biomater 105(8):2408–2415. https://doi.org/10.1002/jbm.b.33777
4. Pandey PC, Shukla S, Pandey G, Narayan RJ (2021) Nanostructured diamond for biomedical applications. Nanotechnology. 32(13):132001. https://doi.org/10.1088/1361-6528/abd2e7
5. Gupta C, Prakash D, Gupta S (2017) Cancer treatment with nano-diamonds. Front Biosci (Schol Ed). 9(1):62–70. https://doi.org/10.2741/s473
6. Nistor PA, May PW (2024) Diamond thin films: giving biomedical applications a new shine. J R Soc Interface 14:20170382
7. Dekhtyar Y, Abols D, Avotina L, Stoppel A, Balakin S, Khroustalyova G, Opitz J, Sorokins H, Beshchasna N, Tamane P, Rapoport A (2023) Effects of diamond nanoparticles immobilization on the surface of yeast cells: a phenomenological study. Fermentation 9:162

Chapter 16
Planetary Precessional Transmission: Geometry and Contact Bearing Capacity, Kinematics and Profile Generation

Ion Bostan, Viorel Bostan, Maxim Vaculenco, Ion Bodnariuc, Sergiu Mazuru, Valeriu Dulgheru, Radu Ciobanu, Oleg Ciobanu, Iulian Malcoci, Nicolae Trifan, Dumitru Vengher, and Alina Bregnova

Abstract In this study, we propose a new gear with convex-concave multipair contact teeth contact for mechanical transmissions. The paper presents various kinematic structures of planetary precessional transmissions with multiple pair convex-concave gearings and the minimum difference of the flank curves at contact points of the teeth. We examined the influence of the gearings' geometric and kinematic parameters on the shape of the teeth profiles and the difference in the flank curvatures at the contact points of the pairs of simultaneously conjugated teeth and, respectively, on the bearing capacity of the multiple pair contact. It describes the mathematical modeling of teeth contact with relative sliding friction in the contacts of pairs of simultaneously conjugated teeth depending on the precession angle of the driving shaft. Technical solutions are proposed to reduce the energy losses in the gearings and, respectively, to increase the mechanical efficiency of the precessional transmission as a whole. Technological aspects of the methods of generating conical wheel teeth with a non-standard convex/concave profile developed and patented by the authors are presented, including generation in the Gleason system on numerically controlled multiaxial machine tools. Constructive-functional variants of precessional transmissions with various kinematic structures with multiple pair convex-concave gearings were described, including precessional moto reducers and

---

I. Bostan · M. Vaculenco (✉) · I. Bodnariuc · S. Mazuru · V. Dulgheru · R. Ciobanu · O. Ciobanu · I. Malcoci · N. Trifan · D. Vengher · A. Bregnova
Department "Fundamentals of Machine Design", Faculty "Mechanical Engineering and Transport", Technical University of Moldova, Chisinau, Moldova
e-mail: maxim.vaculenco@dip.utm.md

V. Bostan
Department "Software and Automation Engineering", Faculty "Computers, Informatics and Microelectronics", Technical University of Moldova, Chisinau, Moldova

D. Radu et al. (eds.), *Countering Hybrid Threats Against Critical Infrastructures*, NATO Science for Peace and Security Series B: Physics and Biophysics, https://doi.org/10.1007/978-94-024-2304-4_16

servo-motor reducers. Technical solutions for the development of precessional transmissions are protected by international patents USA and Germany.

Keywords Precessional transmissions · Gearings with conical wheels · Convex-concave contact · Multiple pair gearing · Precessional tool generation processes · Gleason system of conical wheel tooth generation

16.1 Transmissions with Multipair Gearings and Convex-Concave Contact of Teeth

Protection of critical infrastructure facilities is complex and requires many special devices of the know-how level. This chapter is devoted to one such devices—precession transmissions. The relevance of using precession transmissions in critical infrastructure facilities is a multi-faceted task due to the specifics of their design and functional features. Precession transmissions have no analogues in robotic submersible technology for extracting wealth from the bottom of the World Ocean, where unique ores are located for the needs of an important part of the critical infrastructure—energy facilities.

Another example of support and protection of critical infrastructure facilities is the implementation of precession gears in new aerospace technologies, in orbital and geostationary satellites—the use of precession gears with high and ultra-high gear ratios, as well as technologies with the transmission of motion and load in sealed spaces, implemented in gears with gearing multiplicity of teeth up to 100% pairs of teeth. This will facilitate the precise orientation of the satellite in space—a very important aerospace task [https://www.inmarsat.com/en/insights/corporate/2023/how-satellite-orbits-work.html].

In 1932, German engineer Braren L. K. developed the cycloidal transmission known as CYCLO (with a priority date of November 30, 1926). This transmission is notable for its high load-bearing capacity, which stems from the simultaneous engagement of many tooth pairs. It also boasts high mechanical efficiency, thanks to the use of higher pair tooth-roller kinematic couplings and can achieve transmission ratios of up to $i = 119$ in a single stage. Additionally, it features small dimensions and compact design, among other advantages [1–4].

In 1959, American engineer Musser C. W. patented a new method for transforming and transmitting motion and load, referred to as harmonic transmissions (with a priority date of March 21, 1955). This type of mechanical transmission features multiple internal gear pairs with a difference in the number of teeth of ±1, 2, 3, or 4, where one of the gears (typically the internal one) is flexible and undergoes continuous deformation. In 1963, Harmonic Drive LLC became the first company to produce harmonic transmissions on an industrial scale for the helicopter industry. Depending on the design of the wave generator (which can utilize 2, 3, or 4 deformation rollers), the gearing multiplicity can reach up to 25–30% of the total number of tooth pairs. Harmonic transmissions are known for their compactness, high load-bearing capacity, excellent kinematic precision, and ability to operate in confined spaces, making these features significant advantages [5–8].

A significant milestone in the evolution of mechanical transmissions with fixed axles and conjugate wheels is the invention of the helical gear, patented in the US in 1926 by Ernest Wildhaber [9, 10]. The technical innovations introduced by Wildhaber involve the design of gears with circular arc profiles that feature convex-concave point contact between the teeth and the origins of the displaced circular arcs, while maintaining minimal differences in the curvature of the flank profiles. Additionally, he proposed methods for generating the geometry of the rack, the profiled disc, and the notched cutter. In this way, Ernest Wildhaber established the foundational geometry and technology for generating helical gears with tooth flank profiles in circular arcs with punctiform contact, which later contributed to the development of a theoretical framework for these gears [11–13]. The loadbearing capacity of Wildhaber-Novicov transmissions, despite the convex-concave contact with a small difference in flank curvatures favorable from the point of view of resistance, according to solid body mechanics, is limited by the small number of simultaneously meshing teeth pairs ($\varepsilon_{\Sigma} < 2$).

In this context, it is crucial to advance mechanical transmissions that utilize multipair gearing while also featuring teeth that engage in convex-concave contacts with minimal differences in flank curvatures. So, this paper addresses these two technical solutions that can be achieved simultaneously in the transmissions with sphere-spatial interaction of teeth.

16.1.1 Transmissions with Identical Kinematic Structures Under Various Names

It is important to highlight that there is currently no comprehensive technological standardization of concepts and definitions in the area of gear transmissions. In some countries, these terms and definitions are grounded in structural aspects, in other cases, they rely on the functional principles of gearboxes, resulting in inconsistent terminology [14–21].

Mechanical transmissions that feature spherospatial interactions of teeth, although they share the same kinematic structure and generate the same relative movements of gears as described in [17, 22], are designated by various terms, including precessional transmissions [22] and nutational transmissions [17].

Some researchers refer to transmissions with gears that exhibit spherospatial interactions as nutational transmissions [14–21], which are characterized by the nutation angle of motion around a fixed point. However, the authors of this paper, along with others, argue that the appropriate term for these transmissions is precessional transmissions [22–32], characterized by the precession angle. This terminology is chosen because precession serves as a unifying parameter for all performance characteristics of the gear, covering factors such as tooth profile shape, contact geometry, the kinematics of the contact point, and the relative slip between engaging flanks. In comparison, the nutation angle is only one of five configuration parameters that influence the geometry of the transmission gear. The authors have consistently used the term "Precessional Transmissions" in around 220 inventions and patents, and in a habilitation thesis titled "Multipair Precessional Planetary Transmissions" presented at Bauman Moscow State Technical University in 1989 [31], the term is grounded in both structural aspects and functional principles.

Since 2000, the term "Precessional Transmissions" has become more widely adopted in the works of various authors [34–36] and others internationally.

16.1.2 Similarities Between "Precession" and "Nutation" Transmissions

Nutation and precession transmissions have a comparable kinematic structure. Both are single-stage systems that consist of a satellite gear with two crown gears: one that engages with a fixed gear and another that connects with a movable center gear. In precessional transmission, the difference in teeth count between the conjugate wheels is limited to ±1, while nutational transmission can vary by ±1 or ± 2 [18] (p. 5). The satellite wheels in both types are positioned on the inclined parts of the crankshafts, which, when rotated, cause the satellite wheels to move in a three-dimensional space, facilitating the relative movement of the tooth flank profiles and reducing rotation. The kinematic properties of both precessional (see Fig. 16.3) and nutational [18] (p. 4, Fig. 4) transmissions are influenced by the number of teeth and the variations between their conjugated wheels. Additionally, the load-bearing capacity of the tooth contact in these transmissions relies on the geometry of the tooth contact and the number of tooth pairs that are simultaneously engaged.

The authors [14] report that Robert Davison stumbled upon the idea of nutation motion while flipping a coin on a table. As the coin rolled and its rotation slowed, it exhibited an unusual motion. This movement, involving simultaneous rotation and rolling around three axes, exemplifies spherospatial motion with a fixed point that incorporates nutation motion. This concept has been utilized in creating speed reducers known as nutation bevel gears. In these nutational devices, the relative rotational movements are influenced by the number of teeth on the meshing gears and the differences in their counts.

The authors of the paper [14] assert that the nutation reducer is a groundbreaking mechanical transmission developed recently, founded on the principle of tooth interaction during nutation movement. They highlight that this nutation gear reducer features unique attributes, including single-stage motion reduction, tooth meshing at very steep angles, and enhanced overall bearing capacity. Additionally, they note that in nutational transmission, the theoretical interaction of the gear teeth occurs with pure rolling, suggesting lower heat generation between the teeth flanks and improved energy efficiency. The article also points out that the NuGear transmission has only been tested through virtual computer simulations [19]. Since the kinematics, structure, and relative movements of the teeth in both "nutating" and "precessing" gears are completely alike, the authors argue that claims about the novelty of "pure rolling" between the teeth flanks in nutation transmissions also apply to precession transmissions.

Two years prior to the publication of the paper [33], Ion Bostan's 1991 work, *"The wonderful World of reduction gears created by Ion Bostan"* noted that the concept of precession transmission originated from coincidental observations made by PhD Associate Professor Ion Bostan in August 1977 while attending a light music concert in Sopot, Poland. During the event, a laser beam projected a gymnastic ring that exhibited spherospatial motion with a fixed point. As the ring rolled with varying frequency and amplitude, the interference in its positions on the rolling surface resulted in the formation of a *variable profile* [33]. These profiles collectively inspired the idea of variable convex-concave shapes that depend on the revolutions of the crankshaft and the amplitude of the oscillations, thereby linking to the parameters of precession movement.

16.1.3 Aspects of the Development of Precessional Transmissions

In the development of precession transmissions, the most important was the invention of the spherospatial motion transmission with a fixed point with multipair convex-concave gears SU 1020667 A was invented in 1983 with the priority of 11.02.1981 and the invention of the method of generating tooth flanks with continuously variable curvature increasing on the height of the teeth SU 1663875 A1 was invented in 1988. To date, more than 220 inventions and patents have safeguarded the technical solutions for developing precessional transmissions. These patents cover aspects such as their design-functional structure, geometry, kinematics, tooth contact support, and the technological processes for creating flank surfaces. It should be noted that planetary precessional transmissions have kinematic structures K-H-V, 2K-H and their combinations can be with gears with bolts (TPB) and toothed gears (TPD) [22, 31]. Tooth flanks have non-standard variable convex-concave profiles with minimal difference in flank curvatures and multiple pair conjugated in gears with bolts A^{B}_{CX-CV}, gears with straight teeth A^{D}_{CX-CV}, and with inclined teeth $A^{D,\beta}_{CX-CV}$ with spherospatial interaction with a fixed point [27–29].

Thus, TPB and TPD are distinguished by gears with convex-concave geometry and by the kinematics of the multiple pair contact of the teeth, which ensure the increase of the bearing capacity and the reduction of energy losses in the tooth gears. Due to the possibility of varying the correlation of the number of teeth of the conjugated wheels with the difference ±1, the 2K-H type precessional transmissions ensure the achievement of transmission ratios in extended ranges without analogies among transmissions known worldwide.

A specific area of effective use refers to planetary precessional transmissions, developed for Robotized ECFMC from the bottom of the Planetary Ocean under conditions generated by high hydrostatic pressures of (40–60) MPa and low temperatures of (2–4) °C, at which the kinematic viscosity of the compensating oil in the cavities of the drive mechanisms increases up to (2500–3000) cst.

Within the framework of the interstate research program "OCEAN" (Poland, Russia, German Democratic Republic, Czechoslovakia, Bulgaria, including Finland) based in Warsaw, the Robotized ECFMC POB—Concept of CMEA0F Countries with 12 precessional actuation mechanisms, developed at the Technical University of Moldova, was developed, the Republic of Moldova, scientific coordinator Ion Bostan [22, 31].

Figure 16.1a shows the actuation mechanism of the upper casing of ECFMC POB developed based on submersible TPB of the 2 K-H type with the transmission ratio i = −575 and nominal power P = 74 kW (3D model). The TPB is located in the drum-shaped casing 1 mounted on two supports with sliding bearings 2. The Satellite 3 with two conical roller crowns is installed on the crankshaft 4 coupled to two axial submersible electric motors 5 with the power of P = 37 kW each and to a spare electric motor 6 coupled using a conical gear 7 [22, 31]1F.

To equalize the torque, one of the AMP-37 type electric motors is coupled to the crankshaft by means of a torsion shaft 8, and to ensure functional safety in accidental cases, the crankshaft is actuated with the spare electric motor 6. Electric submersible motors 5 and 6 and the precessional submersible reducer are equipped with compensators 9 to equalize the pressure inside them with the hydrostatic pressure outside.

Another field of TPD implementation refers to the cosmic flight technique, which requires gearing without backlash, a requirement that in planetary precessional transmission is achieved by multiple pair gears, assembled with guaranteed pre-tightening, including by ensuring the high torsional rigidity of the precessional mechanism, also by ensuring the possibility the transmission of the rotational motion through the wall or through deformable waffles.

Figure 16.1b shows the 2K-H type precessional electromotor-reducer with the transmission ratio i = −299, equipped with a solar electric motor and transducers for tracking, controlling, and directing the angular positioning of the input and output shaft of the scanning platform of a space flight device. The electromodule is characterized by a relatively small starting moment T = 320 gcm with pre-tightening in the gearing, high torsional stiffness $\varphi = 1.5 \times 10^5$ Nm/rd [22, 31].

The first transmission with precessional gearings (TPD) was registered in 1989 with priority from 13.05.1986 [32].

Next, we will address issues related to the development of toothed precessional transmissions (TPD).

One of the main reasons for studying the contact problems of mechanical transmissions with gears is the loss of their bearing capacity due to wear and deterioration of the conjugated surfaces of the teeth, especially with linear or point contact.

The wear being a complex superposition of mechanical, thermal, and chemical processes and not having yet a unitary theory of the expression of the phenomenon,

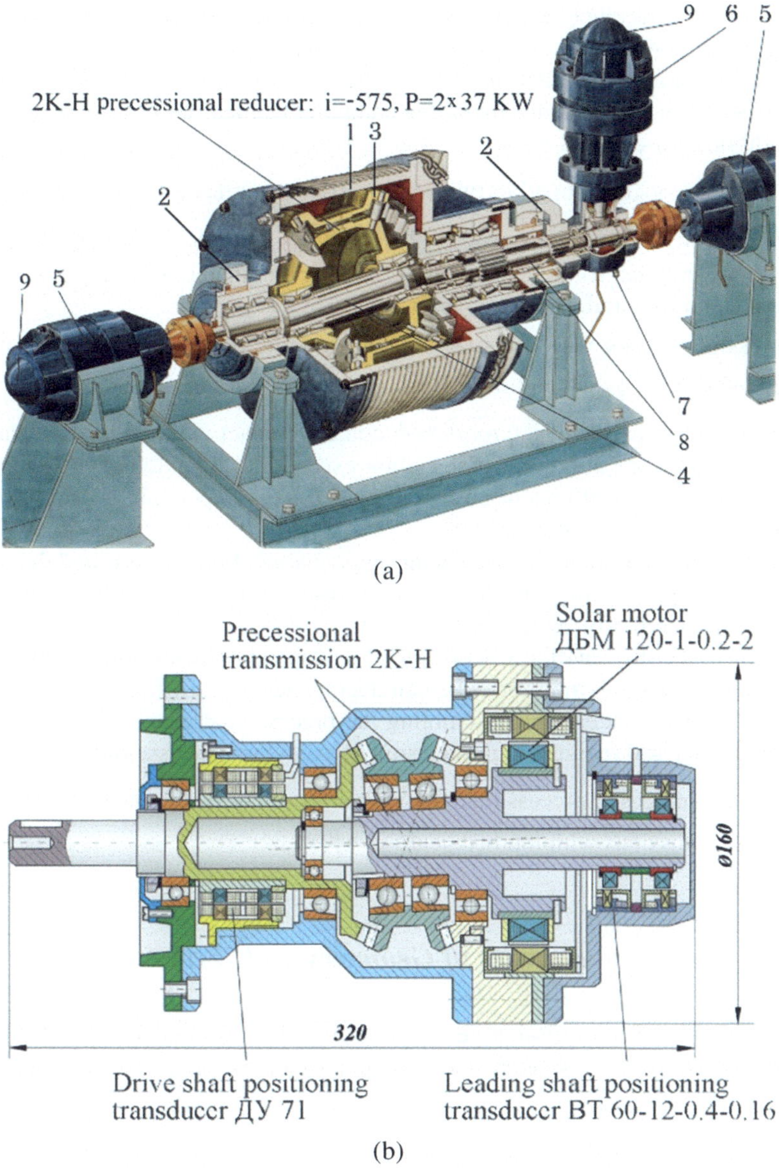

Fig. 16.1 Planetary precessional transmissions with multipair convex-concave contact in TPB gears (**a**) and in TPD gears (**b**)

which predicts the evolution of wear in all aspects of transmissions, it is proposed to study the PT regarding:

– the tribological processes in gearing with different geometries of the *tooth—tooth* contact with the presence of sliding for different materials of the teeth (metal-ceramic, plastics, composite materials, etc.);

- the surface wear rate of the *conjugated teeth* with spherospatial motion and various lubrication regimes specific to fluid friction;
- the deterioration of contact surfaces by *pitting*, performing comparative analyzes of experimental test results, including computer simulations of the contact under conditions close to real ones.

Next, the A^D gears developed in two ways will be examined: the first one with the straight teeth gearing A^{D}_{CX-CV} with frontal multiple pair coverage and the second one with the inclined teeth gearing $A^{D,\beta}_{CX-CV}$ with combined coverage, consisting of the frontal and longitudinal components.

Precessional gears with bolts A^{B}_{CX-CV} have some disadvantages compared to toothed ones A^{D}_{CX-CV}, which limit their range of applications, including:

- The bearing capacity of the *tooth-bolt* conical contact is limited by the median radius of the conical bolts, which cannot exceed half the pitch of the teeth, and their contact in most concurrently conjugated pairs depending on ψ is convex-convex or convex-rectilinear.
- Conical bolts require high manufacturing precision of dimensions and their individual axial positioning, on which the uniformity of the load distribution between pairs of concurrently geared teeth depends.
- The crowns of the satellite wheel formed by conical bolts make it irrational, difficult and sometimes impossible to manufacture gears with a diameter of $\leq$50 mm.
- The cost of producing and assembling the precessional gear with bolts is relatively higher and requires high precision of execution and assembly.

These disadvantages of precessional transmissions can be overcome by the development of transmissions with toothed gearing A^{D}_{CX-CV}.

16.1.4 The Particularities of Geometry and Teeth Contact

In the precessional transmission gearings discussed in this paper, we note the following features of the geometry and kinematics of the contact points of the tooth flank profiles:

- The profiles of teeth conjugate in precessional gears form convex-concave contacts with minimal difference in flank curvatures.
- The teeth profiles of the central gears in precessional transmissions are executed with continuously increasing variable curves from the root to the tip.
- Conjugate teeth in multipair gears with pure rolling greater than relative sliding with friction.
- Teeth conjugate in gears with multipair contacts of up to 100% pairs of teeth.

The scientific research featured in this article essentially aims to answer the question: How to combine in a single mechanical transmission extended kinematic means, high load-bearing capacity and the least possible energy losses?

The accomplishment if these three performances in a single mechanical transmission was tried by proposing and developing:

- a new kinematic principle of transformation and transmission of movement with sphero-spatial interaction of conjugated teeth;
- multiple par gearing with up to 100% pairs of teeth simultaneously in contact;
- geometry of the concave-concave contact of the conjugated teeth with the minimum difference in the radii of the profile curvature.

16.1.5 Geometry of the Linear and Punctiform Contact in the Precessional Gear

The creation of precessional gears comes down to ensuring the transformation of motion according to the principles of the fundamental law of gearing and load transmission through surfaces with punctiform or linear contact of the teeth of the toothed wheels with concurrent axes.

The classical theory of approximate contact between deformable bodies under contact conditions of tooth-tooth kinematic couples is based on the following assumptions:

- Contact surfaces are perfectly continuous and smooth;
- Bodies in contact are homogeneous, isotropic, and linearly elastic;
- Initially the contact is punctiform or linear, and the dimensions of the contact footprints after loading with a load are small in relation to the dimensions of the contracting bodies;
- Stress distribution in the contact zone follows from the Boussinesq elastic half-space theory and therefore the tangential stresses are zero.

The linear contact between the conjugated teeth, usually with an involute, cycloidal, epihipocloidal, octoidal profile, in a circle arc, archimedical spiral, is made between convex-convex, convex-concave, convex/concave plane surfaces, with external or internal gearing. The gears with linear contact between the teeth are designed with cylindrical and conical wheels (orthogonal or under angle) with straight, inclined and curvilinear teeth. The worm transmissions also refer to the gears with linear contact between the teeth, with cylindrical and globoidal worms with rectilinear and involute profiles, the precessional transmissions with conical bolts, the cycloid transmissions with cylindrical bolts, the harmonic, spiroidal, with rack ones, etc. [23].

The gears with punctiform contact between conjugated teeth, were designed, researched, and patented by the British engineers Boston and Bramley-Mur (1921), the American engineer Wildhaber (1923–1947), the Italian engineer Roana (1947) and the Russian engineer Novikov (1959), are considered of a great perspective from the point of view of the high load-bearing capacity, the small gauges and the low consumption of materials. So far, these transmissions have not been found widespread in the industry, due to insufficient research on the advantages of punctiform contact between the teeth and the lack of technologies to generate surfaces with punctiform contact [23].

In the precessional planetary transmissions of power, the contact of the surfaces of the teeth flanks in multiple-pair gearing is formed by two different profiles: the first convex/concave, usually of the teeth of the central wheels, and the second in the shape of a circle arc of the conical bolts in the satellite crowns. In the kinematic precessional planetary transmissions, the teeth of both conjugated wheels are usually made of teeth for the central wheels—with a convex/concave profile, and for the satellite crowns—with a circle arc profile.

The geometry of the teeth contact in the precessional gear is distinguished by the fact that the teeth of the conjugated wheels have different profiles, described by convex-concave curves, usually, for the central wheels and in a circle arc for the satellite crowns. Another unique aspect, specific only to precessional transmissions, resides in the variability of the convex/concave profile of the teeth dependent on five geometric parameters of the gear and on the working regime of the transmission, such as the one of reducer, multiplier or differential.

Thus, the contact in the multiple-pair precessional gear is made between convex-convex surfaces for the tip area of the teeth of the central wheels, between convex-concave surfaces for the bottom area of the teeth and convex/concave-plane for the transition area from the bottom profile to the tooth tip profile (see Fig. 16.2a).

In Fig. 16.2a, p. 1–3, examples of kinematic couplings with linear contact are presented, and in p. 4–7 with punctiform contact, characteristic for the teeth of the wheels of precessional gearings [23].

The linear contact between conjugated teeth in precessional gearing depending on the precession angle ψ transforms from contact featuring convex-concave geometry with a slight difference in curvatures to contact with convex-rectilinear geometry and later, with convex-convex geometry.

In the upper-class kinematic couplings, the initial punctiform or linear contact, after the application of the bearing load, turns into an imprint with the geometry of an ellipse with the semi-axes "a" and "b" (see Fig. 16.2).

The contact σ_z and deformable stresses depend on the load P, the elastic properties of the material, and the geometric shape of the flank surfaces of the teeth, and are determined by the classical Hertzian contact theory [23].

The linear-elastic properties of tooth contact are characterized by the Poison coefficients ν_1 and ν_2 and the Young's moduli E_1 and E_2 of the tooth materials. The generalizing shape parameters of the teeth are the equivalent radius of curvature ρ_e and the parameter of the difference of flank curvatures (ρ_{ki} –r) conjugated in the contacts k_i depending on the precession angle ψ.

Currently, the researchers are developing new approaches to the research and design problems of the mechanisms, especially regarding the contact problems in gears, based on the finite element method (FEM).

In upper-class kinematic couplings, before the application of the load, the geometrically conjugated surfaces contact at a point or on a common line. The mechanisms derived therefrom, performing distinct functions of transformation and

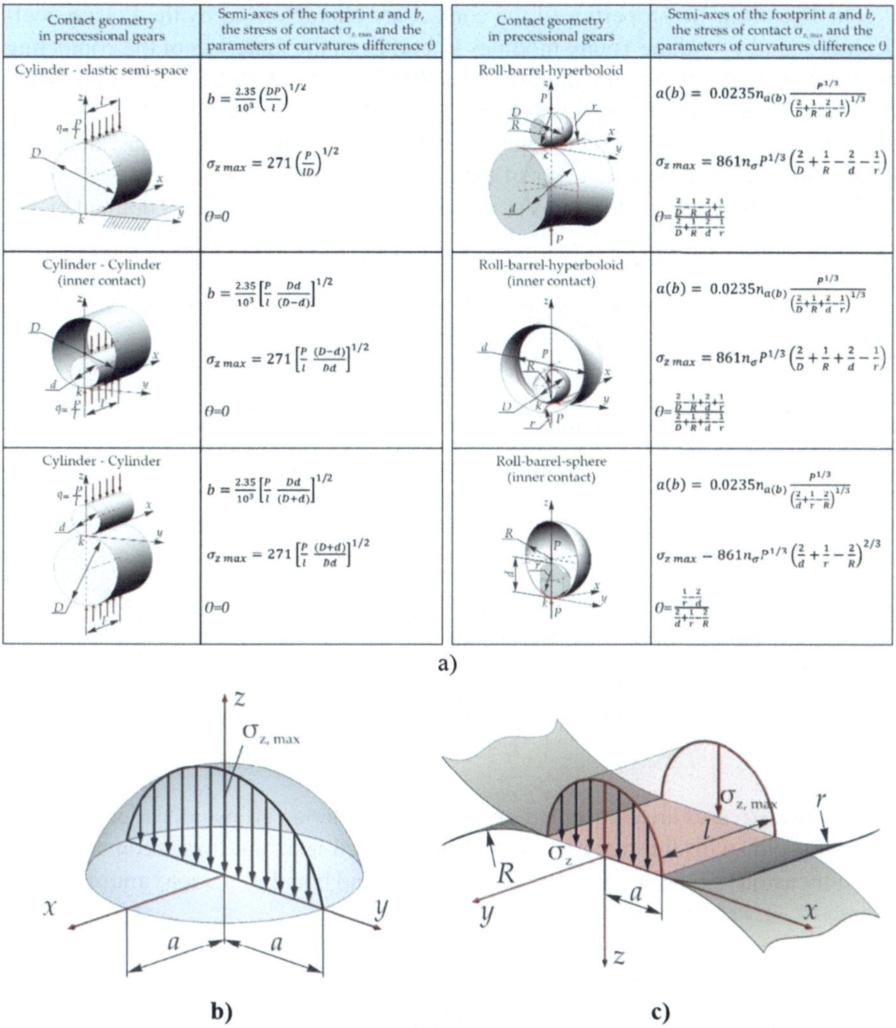

Fig. 16.2 Punctiform and linear contact geometry in precessional gearing (**a**), punctiform contact footprint (**b**) and stresses distribution in linear teeth contact (**c**) [2]

transmission of motion and energy, charge the constituent kinematic couplings with bearing load. After the application of the load, the punctiform contact between the conjugated parts becomes an imprint with the geometric shape of an ellipse with the semiaxes a and b, for example, in the particular case of the contact between two spheres, the imprint is a circle ($a = b$) (see Fig. 16.2b), and in the case of a sphere and a cylinder—in an ellipse ($a \neq b$) [23].

The contact stresses σ_z and the deformations depend on the load P, the elastic properties of the material, and the geometric shape of the contacting bodies and are determined according to the classical theory of the hertzian contact.

The linear-elastic properties of the contact are characterized by the Poison coefficients ν_1 and ν_2 and the Young modules E_1 and E_2 of the materials of the contacting bodies. For steels, at the temperature of t = 20 °C, $\nu = 0.29–0.3$, $E = 2.1 \times 10^5$ MPa, for brass, respectively $\nu = 0.3$, $E = 1.08 \times 10^5$ MPa.

The generalizing shape parameters of the contact bodies are the equivalent curvature radius ρ_e and the parameter of the difference of curves θ.

The equivalent curvature radius ρ_e in the case of external contact is defined by the relation

$$\rho_e = \left(\left(\frac{1}{r_1} + \frac{1}{R_1}\right) + \left(\frac{1}{r_2} + \frac{1}{R_2}\right)\right)^{-1} \tag{16.1}$$

and in the case of internal contact—by the relation

$$\rho_e = \left(\left(\frac{1}{r_1} + \frac{1}{R_1}\right) - \left(\frac{1}{r_2} + \frac{1}{R_2}\right)\right)^{-1} \tag{16.2}$$

where r_1 and R_1 are the radii of curvature of the first body, r_2 and R_2—of the second one, and the parameter of the difference of the curves θ is:

$$\theta = \rho_e \left(\left(\frac{1}{r_1} - \frac{1}{R_1}\right) + \left(\frac{1}{r_2} - \frac{1}{R_2}\right)\right) \tag{16.3}$$

The contact stresses are distributed on the imprint after an ellipsoid, called the pressure ellipsoid, with the maximum pressure reached near the initial contact point. The dimensions of the semiaxes of the ellipses a and b and of the maximum contact stresses $\sigma_{z,max}$ are determined by the formulas:

$$a = n_a \sqrt[3]{\frac{3\rho_e P}{2E_e}}, \tag{16.4}$$

$$b = n_b \sqrt[3]{\frac{3\rho_e P}{2E_e}}, \tag{16.5}$$

$$\sigma_{z,max} = \frac{n_\sigma}{\pi} \sqrt[3]{\frac{3PE_e^2}{2\rho_e^2}}, \tag{16.6}$$

where n_a, n_b and n_σ are coefficients that depend on the angle between the main planes of curvature, ρ_e is the equivalent curvature radius, and E_e is the module of equivalent elasticity of the materials of the contacting bodies defined by the relation

$$E_e = \left(\frac{1-\nu_1^2}{E_1} + \frac{1-\nu_2^2}{E_2} \right)^{-1} \tag{16.7}$$

The average value of the contact stresses σ_{zm} and the total elastic deformation δ are determined by the relations:

$$\sigma_{zm} = \frac{2}{3}\sigma_{z,max}, \tag{16.8}$$

$$\delta = n_\delta \sqrt[3]{\frac{3P^2}{4E_e^2\rho_e}}, \tag{16.9}$$

In the case of the linear contact (see Fig. 16.2c), after applying the load, the pressure ellipsoid degenerates into an elliptical cylinder (b = l), and the contact imprint is a narrow band of the width 2a, with the distribution of contact stresses σ_z on its width also according to the ellipsoid law. The semiaxis of the contact surface a, the maximum contact stress $\sigma_{z,max}$, the average contact stress σ_{zm} and the total elastic deformation δ are calculated by the formulas:

$$a = \sqrt{\frac{4\rho_e P}{\pi l \rho_e}}, \tag{16.10}$$

$$\sigma_{z,max} = \sqrt{\frac{PE_e}{\pi l \rho_e}}, \tag{16.11}$$

$$\sigma_{zm} = \frac{\pi}{4}\sigma_{z,max}, \tag{16.12}$$

$$\delta = \frac{2P}{\pi l}\left[\frac{1-\nu_1^2}{E_1}\left(\ln\frac{2R_1}{a}+0.407\right)+\frac{1-\nu_2^2}{E_2}\left(\ln\frac{2R_2}{a}+0.407\right)\right]. \tag{16.13}$$

It is obvious that in the precessional gear, the deterioration of the surfaces of the teeth flanks by pitting appears as a result of the action of the high stresses in the contact of the teeth, caused by the small values of the radii of curvature of the rollers, which, constructively, cannot be greater than half of the pitch of distributing them on the circle. Thus, due to the impossibility of increasing the radius of curvature of the rollers, the resistance of the gear to the contact pressure determines the value of the transmitted load. The phenomenon of pitting is described in the literature regarding toothed transmissions, bearings, variators, other mechanisms with superior kinematic couplings. In the calculation of stresses and deformations, the solution of the classical problem from the theory of the contact elasticity of two cylinders driven by external forces is used. At the same time, it is assumed that, for any phase of the gearing, the teeth can be replaced with two cylinders having the radii of curvature equal to the radii of curvature of the teeth at the contact point.

16.1.6 Peculiarities of Precessional Gear Resistance Calculation

Due to the specificity of the motion and load transformation with spherospatial interaction of the conjugated wheels in combination with the variable convex/concave profile of the teeth, in the precessional gears the multiplicity of simultaneous gearing is up to 100% pairs of teeth. The maximum multiplicity is conditioned by a concrete correlation of five defining parameters of the geometry of the precessional gear.

In order to determine the effective contact stresses, it is necessary to consider a few specific features of the precessional gears, namely:

- The degree of non-uniformity of the distribution of the load between the simultaneously geared teeth, expressed by the coefficient k_{HP};
- The degree of localization of the contact spot within the length limits of the teeth, expressed by the coefficient $k_{H\beta}$ of the non-uniformity of the distribution of the load along the length of the teeth.
- The dynamic character of the load, expressed by the coefficient of the load dynamics k_{HV};
- The influence of the radii of curvature ρ_1 and ρ_2 of the surfaces of the teeth flanks on the wheel diameters, expressed by the coefficient υ.

The values of the coefficients k_{HP} and $k_{H\beta}$ are chosen according to the recommendations approved by experimental tests on samples of precessional reducers of the types K-H-V and 2K-H based on the research method of the stress and deformation state of the teeth by tensometry and photometry.

The values of the coefficients k_{HP}, $k_{H\beta}$ and k_{HV} were further verified by analyzing the results of computer simulations on the CAD platform with the software products Autodesk Inventor, SolidWorks, CATIA, etc., using the programs Adams, Motion Inventor, ANSYS, Cosmos Works, MSG Visual Nastran, etc., and the proportionality coefficient υ is chosen according to the nomograms constructed based on the complex analysis of the influence on the wheel diameters of the number of teeth Z of the conjugated wheels and of the nutation angle θ, the conical axoid angle δ and the taper angle of the satellite crown bolts β.

The calculation of the resistance of the precessional gear is limited to the determination of the effective contact stresses σ_H and their comparison with the permissible limit σ'_{HP}, appreciated by considering the presence in the contact of the upper class kinematic coupling with rolling and sliding from the condition $\sigma_{HP} \leq \sigma'_{HP}$.

Through experimental tests, the influence of the sliding from the upper kinematic coupling on the resistance limit at the contact with rolling σ'_{HP} was appreciated, expressed by the relation

$$\sigma'_{HP} = \sigma_{HP}\left(0.28 + 0.72e^{-\frac{V_a}{K_3}}\right), \tag{16.14}$$

where σ_{HP} are the stresses permissible when pure rolling, V_a is the sliding speed, and K_3—dimensionless coefficient and which for steel is $K_3 = 30.5$.

The value of the effective hertzial stresses σ_H in the contact of the teeth of the multipair precessional gears with conical bolts is to meet the admissibility condition

$$\sigma_H = f\left(P, A, k_{HP}, k_{H\beta}, k_{HV}\right), \tag{16.15}$$

Considering the contact geometry and the specificity of the precessional gear, the effective contact stresses can be expressed by the relation

$$\sigma_H = Z_M\sqrt{\frac{2T_4(1-\nu)k_{HP}k_{H\beta}k_{HV}}{d_{mr}d_{m4}b_{\omega 4}Z_\varepsilon \cos\alpha_\omega}} \leq \sigma'_{HP}, \tag{16.16}$$

where $Z_M = \sqrt{E_e / \pi\left(1-\mu^2\right)}$ is the elastic constant of the materials of the conjugated toothed wheels.

Calculation of dimensioning of the precessional transmission, taking into account the admissibility condition $\sigma_{HP} \leq \sigma'_{HP}$ can be calculated in relation to the average diameter of the central wheel d_{m4}:

$$d_{m4} = 53\sqrt[3]{\frac{T_4(1-\nu)\cos(\delta+\theta)k_{HP}k_{H\beta}k_{HV}}{\left(\sigma'_{HP}\right)^2 \psi_{bd}Z_\varepsilon \operatorname{tg}\beta\cos\alpha_\omega}}, \tag{16.17}$$

where ψ_{bd} is the coefficient of the relative length of the tooth and Z_ε—number of pairs of teeth simultaneously geared.

The calculation of the precessional teeth gear at the contact pressure is described in 4.4.2 [22, 31].

It was found theoretically and experimentally that the absolute multiplicity (100% pairs of teeth) of the gearing under the mentioned conditions can only occur in the case of using a variable "convex/concave" profile of the teeth, depending on five geometrical parameters of the spatial gear. In the spatial precessional gear, the profile of the central wheel teeth is dependent on the angles of the conical axoid, nutation and conicity of the satellite bolts, the number of teeth of the central wheel and the ratio of the number of teeth of the conjugated wheels.

16.2 Kinematics of the 2K-H Precessional Transmissions

The 2K-H type precessional transmission features a satellite wheel 2 with two toothed crowns Z_2 and Z_3, which engage with both the stationary central wheel 1 and the mobile central wheel 4, the latter being firmly attached to the driven shaft 5 (see Fig. 16.3).

Figure 16.3 shows three structural schemes with different mechanisms for transforming the rotational motion of the crankshaft into the sphero-spatial motion with a fixed point of the satellite, namely: the satellite-wheel 2 is mounted between the

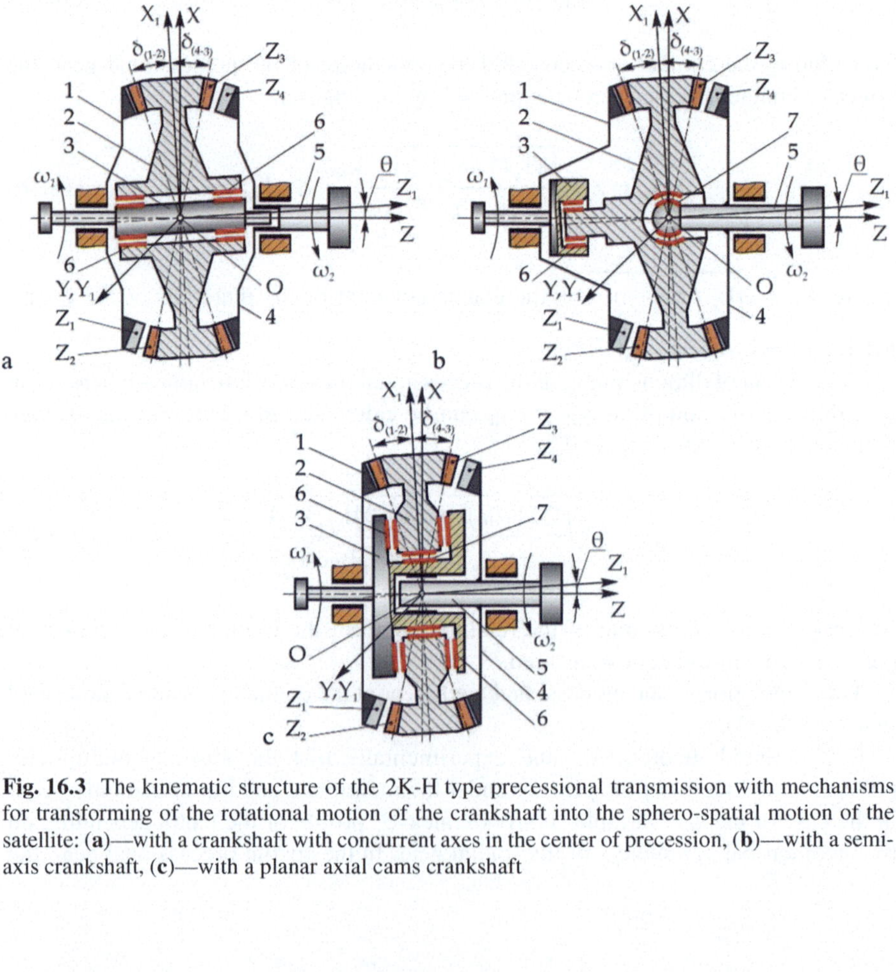

Fig. 16.3 The kinematic structure of the 2K-H type precessional transmission with mechanisms for transforming of the rotational motion of the crankshaft into the sphero-spatial motion of the satellite: (**a**)—with a crankshaft with concurrent axes in the center of precession, (**b**)—with a semi-axis crankshaft, (**c**)—with a planar axial cams crankshaft

supports 6 on the inclined portion of crankshaft 3 with concurrent axes in the center of precession between the central wheels 1 and 4 (see Fig. 16.3a) [18], the satellite wheel 2 equipped with a semi-axle shaft resting on the spherical support 7 located in the center of precession O and driven in rotation by the crankshaft 3 using the support 6 executed eccentrically or concurrently with the common axis of the central wheels 1 and 4 in the center of precession O (see Fig. 16.3b), the satellite wheel 2 rested on the support 7 and driven in rotation by the crankshaft 3 by means of the planar axial cams 6 inclined under the nutation angle θ (see Fig. 16.3c).

When the crankshaft 3 rotates at an angular velocity ω_1, the satellite wheel 2 undergoes spherospatial motion around a fixed point at the center of precession O. This occurs due to the interaction of the teeth in the gearings ($Z_1 - Z_2$) and ($Z_3 - Z_4$), causing the driven shaft 5 to rotate at an angular velocity ω_2 (see Fig. 16.3).

The transmission ratio is determined by the relation

$$i = -\frac{Z_2 \cdot Z_4}{Z_1 \cdot Z_3 - Z_2 \cdot Z_4} \tag{16.18}$$

where Z_2 and Z_3 are the numbers of teeth of the satellite-wheel 2 and Z_1 and Z_4 are the number of teeth of the central wheels 1 and 4.

The transmission ratio i in the 2K-H type transmission is determined by the relationship between the numbers of geared teeth $Z_{1(4)} = Z_{2(3)} \pm 1$.

Based on the configuration and correlation of the tooth numbers Z_1, Z_2, Z_3, and Z_4, and the gearings ($Z_1 - Z_2$) and ($Z_3 - Z_4$), if $Z_1 = Z_2 - 1$ and $Z_4 = Z_3 - 1$ with $15 \le Z_{1,2,3,4} \le 60$, transmission ratios ranging from ±20 to ±3600 are achieved (see Fig. 16.4).

When $Z_2 > Z_3$, the driving and driven shafts rotate in opposite directions (−). When $Z_2 < Z_3$, they rotate in the same direction (+).

Under these conditions, if $Z_1 = Z_2 - 1$ and $Z_4 = Z_3 + 1$, the 2K-H transmission provides transmission ratios from +8.3 to +30.3 with $15 \le Z_{1,2,3,4} \le 60$. Conversely, if $Z_1 = Z_2 + 1$ and $Z_4 = Z_3 - 1$, it provides transmission ratios from −7.3 to −29.3 (see Fig. 16.5).

Precessional transmissions featuring the gearing configuration A^{D}_{CX-CV}, are designed with a specific correlation between the tooth numbers of the conjugated wheels, specifically $Z_1 = Z_2 - 1$ and $Z_4 = Z_3 - 1$ (see Fig. 16.5a), can achieve transmission ratios from ±20 to ±3600, provided $16 \le Z_{1,2,3,4} \le 60$. The driving and driven shafts rotate in the same direction when $Z_2 < Z_3$, and in opposite directions when $Z_2 > Z_3$.

Transmissions with the tooth number correlation $Z_1 = Z_2 - 1$ and $Z_4 = Z_3 + 1$ (see Fig. 16.5b) achieve transmission ratios between +8.3 and +30.3 when $16 \le Z_{1,2,3,4} \le 60$, with both the driving and driven shafts rotating in the same direction for $Z_2 < Z_3$ and $Z_2 > Z_3$. In contrast, transmissions with the correlation $Z_1 = Z_2 + 1$ and $Z_4 = Z_3 - 1$ (see Fig. 16.5c) achieve ratios from −7.3 to −29.3, with the driving and driven shafts rotating in opposite directions for $Z_2 < Z_3$ and $Z_2 > Z_3$ [28, 29].

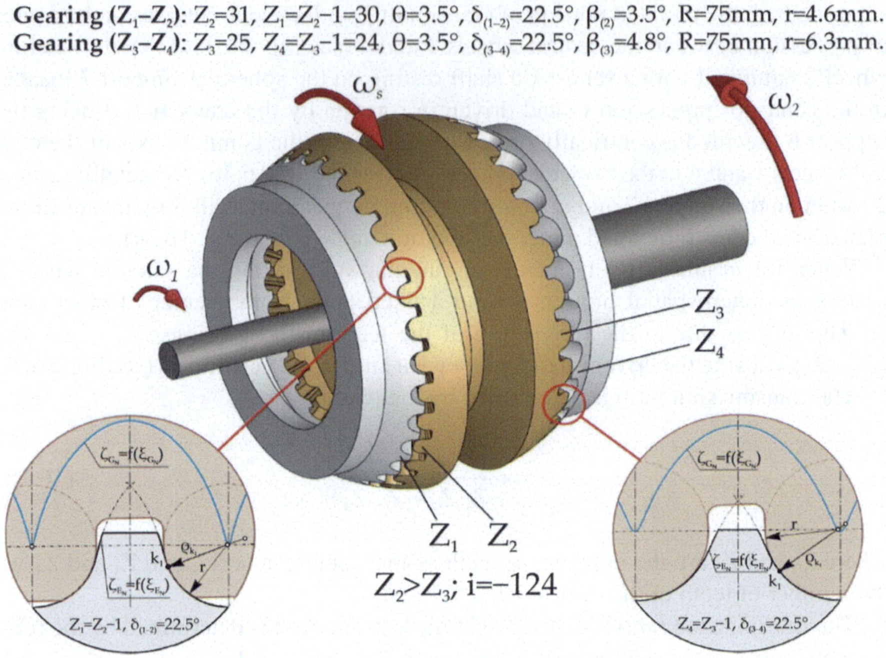

Fig. 16.4 The tooth profiles of the crowns in the gearings ($Z_1 - Z_2$) and ($Z_3 - Z_4$) within the precessional gear transmission

Thus, the 2K-H PT with gearings A^{D}_{CX-CV} shown in the Fig. 16.5 in their entirety can achieve transmission ratios in the range $\pm 8 \leq i \leq \pm 3600$ (if $16 \leq Z_{1,2,3,4} \leq 60$), with the rotation of the driving and driven shafts in one direction or in the opposite direction, with reducer or multiplier operation regime.

Based on the needs of mechanical transmission users [22, 30], the demand for reducers based on the transmission ratio i follows a typical logarithmic curve, peaking at i = 32.1. Additionally, ratios within the range of i = (8–50) account for approximately 75% of their total global usage (see Fig. 16.6).

The transmission ratio density diagram in Fig. 16.6 indicates that transmission ratios within the range of $\pm 8 \leq i \leq \pm 20$ cannot be attained using single-stage involute transmissions. Instead, these ratios require two to three-stage reducers, resulting in a significant increase in the reducer's size, the number of components, and ultimately, the cost compared to single-stage systems.

In contrast to traditional cycloidal or harmonic involute transmissions (see Fig. 16.6), 2K–H precessional transmissions allow for the achievement of small transmission ratios within a single stage. This is accomplished by selecting the number of teeth Z_1, Z_2, Z_3, and Z_4 and their relationships [30] (pp. 341–353) based on the desired transmission ratio ±i specified in the design requirements.

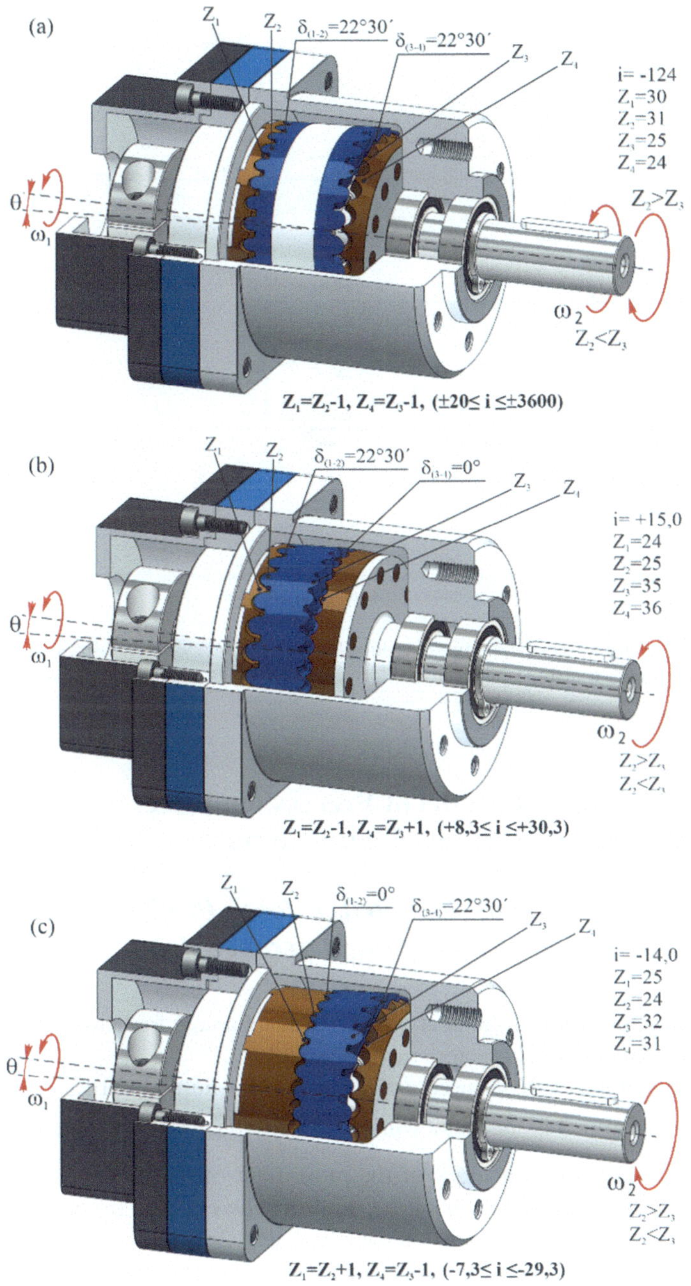

Fig. 16.5 Precessional transmissions with gearings A^{D}_{CX-CV} with the transmission ratio: (**a**) $i = -124$ ($\delta_{(1-2)} = \delta_{(3-4)} = 22^{\circ}30'$); (**b**) $i = +15.0$ ($\delta_{(1-2)} = 22^{\circ}30'$, $\delta_{(3-4)} = 0^{\circ}$); (**c**) $i = -14.0$ ($\delta_{(1-2)} = 0^{\circ}$, $\delta_{(3-4)} = 22^{\circ}30'$)

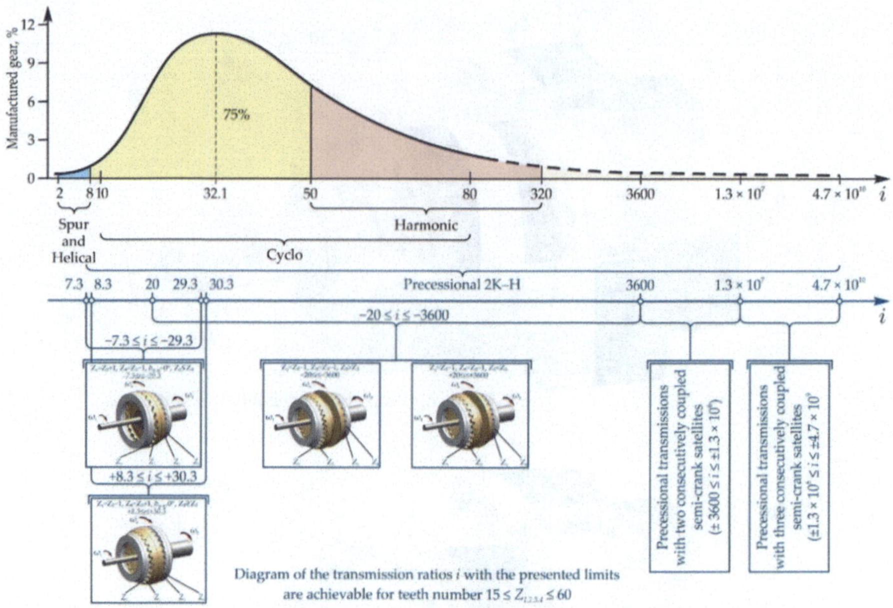

Fig. 16.6 The distribution of transmission ratios for mechanical systems utilized globally, as well as the potential transmission ratios in 2K–H precessional transmissions featuring toothed gears A^{D}_{CX-CV}

16.3 Mathematical Model of Precessional Gearing with Convex-Concave Tooth Contact

16.3.1 Specification of the Satellite Tooth Profile Shape and the Movement of the Reference Point

The profile of the teeth of the gear wheels of mechanical transmissions influences most of their functional characteristics.

The profile of the teeth of the central wheels in the multipair precessional gearing A^{D}_{CX-CV} is determined in the framework of the approach to the problems regarding:

- identification of the trajectory of the movement of the point of contact of the teeth with the velocity vector of the origin of the circular arc of the generating contour of the tool at a precession cycle $0 \le \psi \le 2\pi Z_2/Z_1$.
- identification of the winding of the generating contour of the arc-shaped tool and spherospatial movement with a fixed point at a precession cycle $0 \le \psi \le 2\pi Z_2/Z_1$;

These two approaches bring together the geometry of precessional gearing and technological aspects of the generation and aim to identify the tooth flank profile in multipart convex-concave contact gearing.

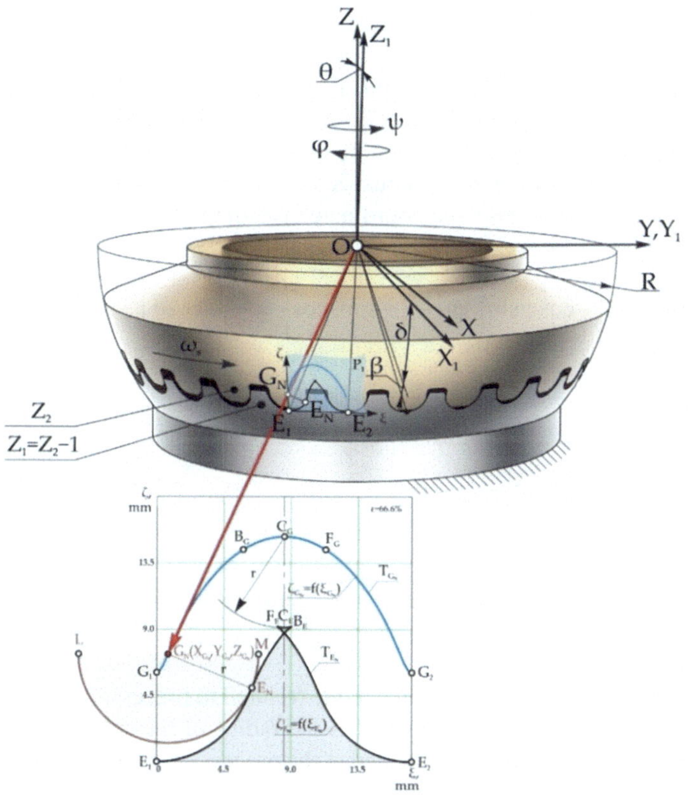

Fig. 16.7 The representation of the path traced by the origin G of the circle arc's radius in the $E_1\xi\zeta$, coordinates, as projected onto the plane P_1 for the gearing A^D_{CX-CV}

The Euler equations define the kinematic relations of the engagement between the central wheel of the precessional transmission and the satellite in spherospatial motion [30].

The theoretical advancements [30] related to the tooth–bolt gearing A^B, along with the models and methods used, are also thoroughly validated for the tooth–tooth gearing A^D that has been studied.

Precession gears featuring multipair convex-concave tooth contact [30] can be designed in two ways: the first involves conical bolts interacting with teeth that have convex-concave profiles A^B, while the second utilizes toothed gear systems A^D.

We suggest that the curve LE_NM, which represents a circular arc with radius r and originates from point G (see Fig. 16.7), which is designated as the reference point for the satellite teeth, represents the profile of the satellite teeth and establishes the convex-concave contact of the geared teeth with spherospatial motion..

In the mobile coordinate system $OX_1Y_1Z_1$ attached to the satellite, the location of the origin G of the circle arc's radius is defined by these specific coordinates:

$$X_{1G} = 0, Y_{1G} = -R\cos\delta, Z_{1G} = -R\sin\delta. \tag{16.19}$$

where δ represents the angle of the conical axoid.

In the Euler kinematic equations, considering (16.19) and the kinematic relationship between the angles φ and ψ given by $\varphi = -Z_1\psi/Z_2$, we derive the coordinates of the origin G of the circle arc's radius, X_G, Y_G, and Z_G, as functions of the crankshaft rotation angle ψ in the fixed coordinate system OXYZ:

$$\begin{aligned}
X_G &= R\cos\delta\left[-\cos\psi\sin\left(\psi\frac{Z_1}{Z_2}\right)+\sin\psi\cos\left(\psi\frac{Z_1}{Z_2}\right)\cos\theta\right]-R\sin\delta\sin\psi\sin\theta,\\
Y_G &= -R\cos\delta\left[\sin\psi\sin\left(\psi\frac{Z_1}{Z_2}\right)+\cos\psi\cos\left(\psi\frac{Z_1}{Z_2}\right)\cos\theta\right]\\
&\quad +R\sin\delta\cos\psi\sin\theta,\\
Z_G &= -R\cos\delta\cos\left(\psi\frac{Z_1}{Z_2}\right)\sin\theta-R\sin\delta\cos\theta.
\end{aligned} \tag{16.20}$$

where ψ is the crankshaft's positioning angle (related to precession), δ is the conical axoid's angle, θ is the nutation angle, R is the radius of the sphere centered at the origin of the coordinate system O located at the precession center, and φ is the satellite's rotation angle around its own axis.

The origin G of the circle arc's radius, which defines the teeth of the satellite, travels along the surface of a sphere with radius R centered at the precession point O, following a path represented by the coordinates X_G, Y_G, and Z_G.

The path of the origin G of the circle arc LE_NM from the sphere with radius R is projected onto the plane P_1 using spherical trigonometry principles [30] (p. 31). This allows us to derive the trajectory T_{G_N} o of the radius origin G on the plane P_1, represented by the relationship $\zeta_{G_N} = f\left(\xi_{G_N}\right)$ (see Fig. 16.7).

16.3.2 Identification of the Profile of the Central Wheel Teeth Based on the Motion Path of the Tooth Contact Point

By understanding the trajectory $\zeta_{G_N} = f\left(\xi_{G_N}\right)$ of the motion of the center of the circular arc G, represented by the coordinates X_G, Y_G, and Z_G (see Fig. 16.8), we can ascertain the location of the contact point E between the satellite wheel tooth profile and the circular arc corresponding to the central wheel teeth profile.

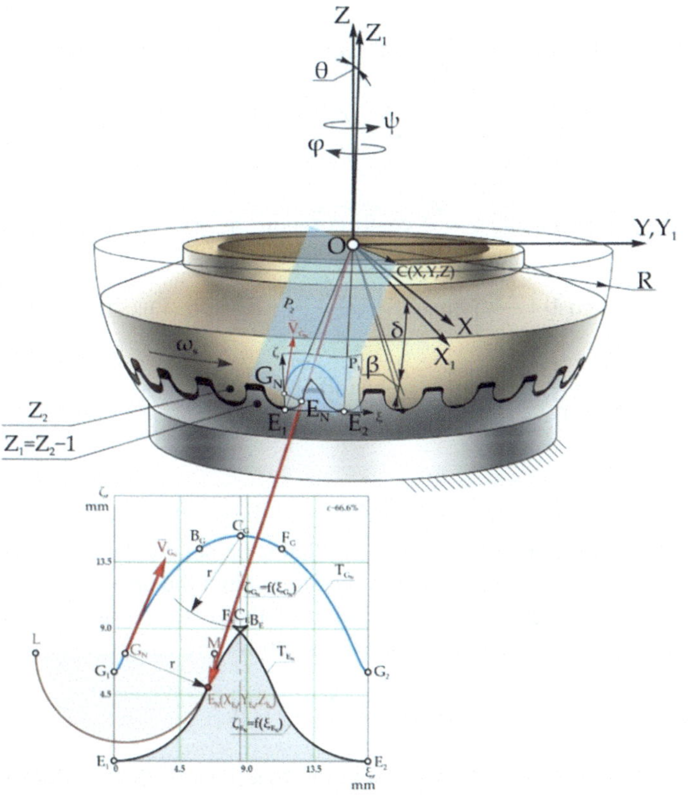

Fig. 16.8 Characterization of the central wheel teeth profile using the velocity vector V_G at the origin G of the circular arc in the gearing system A^{D}_{CX-CV}

The contact point E between the satellite wheel teeth and the central wheel, for any angular position ψ of the crankshaft, is situated at a distance equal to the radius r of the circular arc along the normal velocity vector V_G from its origin at point G (see Fig. 16.8). The collection of contact points E identified throughout a precession cycle, where $0 \leq \psi \leq 2\pi Z_2/Z_1$, forms the profile of the central wheel teeth.

Subsequently, to characterize the profile of the central wheel teeth, we calculated the projections of the velocity vector V_G onto the coordinate axes of the fixed system OXYZ.

To achieve this, we differentiated equation (16.20) with respect to time, resulting in:

$$V_{GX} = R\omega\cos\delta\left[\left(1-\frac{Z_1}{Z_2}\cos\theta\right)\sin\psi\sin\left(\psi\frac{Z_1}{Z_2}\right)\right.$$
$$\left.+\left(\cos\theta-\frac{Z_1}{Z_2}\right)\cos\psi\cos\left(\psi\frac{Z_1}{Z_2}\right)\right]-R\omega\sin\delta\cos\psi\sin\theta,$$
$$V_{GY} = -R\omega\cos\delta\left[\left(1-\frac{Z_1}{Z_2}\cos\theta\right)\cos\psi\sin\left(\psi\frac{Z_1}{Z_2}\right)\right.$$
$$\left.-\left(\cos\theta+\frac{Z_1}{Z_2}\right)\sin\psi\cos\left(\psi\frac{Z_1}{Z_2}\right)\right]-R\omega\sin\delta\sin\psi\sin\theta, \tag{16.21}$$
$$V_{GZ} = R\omega\frac{Z_1}{Z_2}\cos\delta\sin\left(\psi\frac{Z_1}{Z_2}\right)\sin\theta.$$

where $\omega = \dot{\psi}$ represents the angular velocity of the crankshaft.

We developed the equation of a plane P_2 that is perpendicular to the velocity vector $\mathbf{V_G}$ and passes through the precession center O and the origin of the circular arc radius G in order to determine the location of the teeth's contact point E on the spherical surface. The following is an expression for the equation for the plane P_2:

$$[\mathbf{OG}\times\mathbf{OC}]\times\mathbf{V_G} = 0, \tag{16.22}$$

where the vectors **OC** and **OG** represent the locations of the arbitrary point C on plane P_2 and the reference point of the satellite wheel tooth G on the circular arc, respectively, with respect to the fixed coordinate system's origin OXYZ (see Fig. 16.8).

A third-order determinant is used to express the vector product [**OG** × **OC**]. We obtain the following by enlarging it using the first row's elements:

$$[\mathbf{OG}\times\mathbf{OC}] = \begin{vmatrix} i & j & k \\ X_G & Y_G & Z_G \\ X & Y & Z \end{vmatrix} = i(Y_G Z - Z_G Y) + j(Z_G X - X_G Z)$$
$$+k(X_G Y - Y_G X). \tag{16.23}$$

where X, Y, and Z stand for the coordinates of an arbitrary point C on plane P_2, and X_G, Y_G, and Z_G for the coordinates of the reference point G of the satellite teeth on the circular arc.

The vector equation [**OG** × **OC**] × $\mathbf{V_G}$ as also developed; by applying (16.23), we get:

$$\begin{vmatrix} i & j & k \\ Y_G Z - Z_G Y & Z_G X - X_G Z & X_G Y - Y_G X \\ V_{GX} & V_{GY} & V_{GZ} \end{vmatrix} = 0; \tag{16.24}$$

and after expanding, we derive the following equations:

$$V_{GZ}\left(Z_G X - X_G Z\right) - V_{GY}\left(X_G Y - Y_G X\right) = 0, \tag{16.25}$$

$$V_{GX}\left(X_G Y - Y_G X\right) - V_{GZ}\left(Y_G Z - Z_G Y\right) = 0, \tag{16.26}$$

$$V_{GY}\left(Y_G Z - Z_G Y\right) - V_{GX}\left(Z_G X - X_G Z\right) = 0. \tag{16.27}$$

Equation (16.27) is derived from Eqs. (16.25) and (16.26) and can be considered the equation of plane P_2 shown in Fig. 16.8.

We assume that point E is located on both plane P_2 and the sphere with radius R in order to find the coordinates of the contact point E, where the profile of the satellite teeth crosses the circular arc with the profile of the central wheel teeth. The following equation results from the requirement that the coordinates of point E fulfill the equation of plane P_2:

$$V_{GX}\left(Z_G X_E - X_G Z_E\right) - V_{GY}\left(Y_G Z_E - Z_G Y_E\right) = 0. \tag{16.28}$$

The contact point E of the teeth also lies on the sphere with radius R, meaning its coordinates fulfill the equation of that sphere:

$$X_E^2 + Y_E^2 + Z_E^2 - R^2 = 0. \tag{16.29}$$

The taper angle β of the satellite's teeth along that arc is indicated in Fig. 16.7 by the angle formed by the position vectors of the tooth's reference point on the circular arc **OG** of the satellite wheel and the position vector of the teeth's contact point E **OE**. This leads to the following outcome:

$$\mathbf{OG}\cdot\mathbf{OE} = R^2\cos\beta. \tag{16.30}$$

or

$$X_G X_E + Y_G Y_E + Z_G Z_E - R^2\cos\beta = 0. \tag{16.31}$$

Utilizing Eq. (16.31), we established

$$X_E = \frac{R^2\cos\beta - Y_G Y_E - Z_G Z_E}{X_G}. \tag{16.32}$$

We entered Eq. (16.32) into Eq. (16.28) and calculated the coordinate Y_E of the contact point for teeth E

$$Y_E = k_1 Z_E - d_1. \tag{16.33}$$

Equation (16.33) was substituted into Eq. (16.32), and the expression for the coordinate XE of the next contact point was obtained:

$$X_E = k_2 Z_E + d_2. \tag{16.34}$$

in which

$$\begin{gathered} k_1 = \left[X_G \left(X_G V_{GX} + Y_G V_{GY}\right) + Z_G^2 V_{GX}\right] / \left(X_G V_{GY} - Y_G V_{GX}\right) Z_G, \\ d_1 = R^2 \cos\beta V_{GX} / \left(X_G V_{GY} - Y_G V_{GX}\right), \\ k_2 = -\left(k_1 Y_G + Z_G\right) / X_G, \\ d_2 = \left(R^2 \cos\beta + d_1 Y_G\right) / X_G. \end{gathered} \tag{16.35}$$

We obtain the following by solving the resulting equation for the coordinate Z_E of contact point E after substituting Eqs. (16.33) and (16.34) into Eq. (16.29) [30]:

$$Z_E = \frac{\left(k_1 d_1 - k_2 d_2\right) \pm \left[\left(k_1 d_1 - k_2 d_2\right)^2 + \left(k_1^2 + k_2^2 + 1\right)\left(R^2 - d_1^2 - d_2^2\right)\right]^{1/2}}{\left(k_1^2 + k_2^2 + 1\right)}. \tag{16.36}$$

The center wheel's tooth profile contour is equally spaced from the motion path of the circle arc's radius's origin G. It is also crucial to note that the criterion $Z_E < Z_G$ must be met for any crankshaft rotation angle ψ.

The coordinate Z_E can be written using the following equation after certain modifications are made to expression (16.36):

$$Z_E = \frac{\left(k_1 d_1 - k_2 d_2\right) - \left[\left(k_1 d_1 - k_2 d_2\right)^2 + \left(k_1^2 + k_2^2 + 1\right)\left(R^2 - d_1^2 - d_2^2\right)\right]^{1/2}}{\left(k_1^2 + k_2^2 + 1\right)}. \tag{16.37}$$

The Eqs. (16.33), (16.34), and (16.37) define the coordinates X_E, Y_E, and Z_E of the contact point E of the teeth, collectively forming the profile of the central wheel teeth positioned on the sphere during a precession cycle.

Because the precessional gear is cone-shaped and the generator lines intersect at the precession center, it's best to analyze the tooth profile using a normal section. For instance, we can examine plane P_1, which passes through points E_1 and E_2 and is perpendicular to plane OE_1E_2 (see Fig. 16.8).

The positions of points E_1 and E_2 on the spherical tooth profile are defined by their coordinates X_E, Y_E, and Z_E. These coordinates are calculated using Eqs. (16.33), (16.34), and (16.37), which take into account two specific precession angles: $\psi = 0°$ and $\psi = 2\pi Z_2/Z_1$. These angles represent the start and end of a complete precession cycle. Additionally, the equation of the plane is determined based on a particular condition.:

$$\left[E_1E_2 \times E_1E_N\right]\left[OE_1 \times OE_2\right] = 0. \tag{16.38}$$

The point labeled E_N is located on plane P_1.

The vector multiplications presented in Eq. (16.38) are computed using principles from matrix theory:

$$\begin{aligned}\left[\mathbf{E_1E_2 \times E_1E_N}\right] &= \begin{vmatrix} i & j & k \\ X_{E_2} - X_{E_1} & Y_{E_2} - Y_{E_1} & Z_{E_2} - Z_{E_1} \\ X_{E_N} - X_{E_1} & Y_{E_N} - Y_{E_1} & Z_{E_N} - Z_{E_1} \end{vmatrix} \\ &= i\left[\left(Y_{E_2} - Y_{E_1}\right)\left(Z_{E_N} - Z_{E_1}\right) - \left(Y_{E_N} - Y_{E_1}\right)\left(Z_{E_2} - Z_{E_1}\right)\right] \\ &+ j\left[\left(X_{E_N} - X_{E_1}\right)\left(Z_{E_2} - Z_{E_1}\right) - \left(Z_{E_N} - Z_{E_1}\right)\left(X_{E_2} - X_{E_1}\right)\right] \\ &+ k\left[\left(Y_{E_N} - Y_{E_1}\right)\left(X_{E_2} - X_{E_1}\right) - \left(X_{E_N} - X_{E_1}\right)\left(Y_{E_2} - Y_{E_1}\right)\right].\end{aligned} \tag{16.39}$$

and

$$\left[\mathbf{OE_1 \times OE_2}\right] = \begin{vmatrix} i & j & k \\ X_{E_1} & Y_{E_1} & Z_{E_1} \\ X_{E_2} & Y_{E_2} & Z_{E_2} \end{vmatrix} = n_X i + n_Y i + n_Z i. \tag{16.40}$$

in which

$$n_X = Y_{E_1}Z_{E_2} - Y_{E_2}Z_{E_1}, n_Y = Z_{E_1}X_{E_2} - Z_{E_2}X_{E_1}, n_Z = X_{E_1}Y_{E_2} - X_{E_2}Y_{E_1}$$

Substituting (16.39) and (16.40) into (16.38), we obtain the equation of plane P_1:

$$\begin{aligned}&\left[\left(Y_{E_2} - Y_{E_1}\right)\left(Z_{E_N} - Z_{E_1}\right) - \left(Z_{E_2} - Z_{E_1}\right)\left(Y_{E_N} - Y_{E_1}\right)\right]n_X \\ &+ \left[\left(Z_{E_2} - Z_{E_1}\right)\left(X_{E_N} - X_{E_1}\right) - \left(X_{E_2} - X_{E_1}\right)\left(Z_{E_N} - Z_{E_1}\right)\right]n_Y \\ &+ \left[\left(X_{E_2} - X_{E_1}\right)\left(Y_{E_N} - Y_{E_1}\right) - \left(Y_{E_2} - Y_{E_1}\right)\left(X_{E_N} - X_{E_1}\right)\right]n_Z = 0.\end{aligned} \tag{16.41}$$

We applied spherical trigonometry principles to map the tooth profile from a sphere of radius R onto plane P_1. This projection is defined by the coordinates of point E_N, which is where the plane intersects with a set of contour lines. These lines originate from the precession center O and include various points N from the spherical tooth profile. The equation describing the projector-director, which is a straight line connecting the precession center O and points E_N, is formulated as follows:

$$Y_{E_N} = \frac{X_{E_N} Y_E}{X_E}, Z_{E_N} = \frac{X_{E_N} Z_E}{X_E}. \tag{16.42}$$

Combining Eqs. (16.41) for plane P_1 and (16.42) for the contour line yields:

$$\begin{aligned}
&\left[\left(Y_{E_2} - Y_{E_1}\right)\left(\frac{X_{E_N} Z_E}{X_E} - Z_{E_1}\right) - \left(Z_{E_2} - Z_{E_1}\right)\left(\frac{X_{E_N} Y_E}{X_E} - Y_{E_1}\right)\right] n_X \\
&+ \left[\left(Z_{E_2} - Z_{E_1}\right)\left(X_{E_N} - X_{E_1}\right) - \left(X_{E_2} - X_{E_1}\right)\left(\frac{X_{E_N} Z_E}{X_E} - Z_{E_1}\right)\right] n_Y \\
&+ \left[\left(X_{E_2} - X_{E_1}\right)\left(\frac{X_{E_N} Y_E}{X_E} - Y_{E_1}\right) - \left(Y_{E_2} - Y_{E_1}\right)\left(X_{E_N} - X_{E_1}\right)\right] n_Z = 0.
\end{aligned} \tag{16.43}$$

To find the solution of Eq. (16.43), we can solve it with respect to X. This gives us:

$$X_{E_N} = \frac{X_{E_1} \Delta_1}{\Delta_2}. \tag{16.44}$$

in which

$$\begin{aligned}
\Delta_1 &= \left(X_{E_2} - X_{E_1}\right)\left(Y_{E_1} n_Z - Z_{E_1} n_Y\right) + \left(Y_{E_2} - Y_{E_1}\right)\left(Z_{E_1} n_X - X_{E_1} n_Z\right) \\
&\quad + \left(Z_{E_2} - Z_{E_1}\right)\left(X_{E_1} n_Y - Y_{E_1} n_X\right), \\
\Delta_2 &= \left(X_{E_2} - X_{E_1}\right)\left(Y_E n_Z - Z_E n_Y\right) + \left(Y_{E_2} - Y_{E_1}\right)\left(Z_E n_X - X_E n_Z\right) \\
&\quad + \left(Z_{E_2} - Z_{E_1}\right)\left(X_E n_Y - Y_E n_X\right).
\end{aligned} \tag{16.45}$$

The values for coordinates Y_{EN} and Z_{EN} were calculated using the formula provided in relation (16.42).

We created the coordinate system $E_1\xi\zeta$ with its origin at point E_1 in order to produce the profile of the central wheel teeth in the ξ and ζ coordinates on the plane P_1. Point E_2 is where the axis $E_1\xi$ passes through (see Fig. 16.8). Using the following formulas, we changed the coordinates X_{EN}, Y_{EN}, and Z_{EN} to ξ and ζ:

$$\xi_{E_N} = \frac{\left(E_1E_2\right)^2 + v_1^2 + v_2^2}{2\left(E_1E_2\right)}, \zeta_{E_N} = \sqrt{v_1^2 + \xi_{E_N}^2}. \tag{16.46}$$

in which

$$(E_1E_2)^2 = (X_{E_2} - X_{E_1})^2 + (Y_{E_2} - Y_{E_1})^2 + (Z_{E_2} - Z_{E_1})^2. \tag{16.47}$$

$$v_1^2 = (E_1E_N)^2 = (X_{E_N} - X_{E_1})^2 + (Y_{E_N} - Y_{E_1})^2 + (Z_{E_N} - Z_{E_1})^2. \tag{16.48}$$

$$v_2^2 = (E_2E_N)^2 = (X_{E_N} - X_{E_2})^2 + (Y_{E_N} - Y_{E_2})^2 + (Z_{E_N} - Z_{E_2})^2. \tag{16.49}$$

The coordinates of the locations along the curve are described by the equations in (16.46) and when projected onto plane P_1, they produce the profile of the central wheel flanks. By altering the precession angle from $\psi = 0°$ to $\psi = 2\pi Z_2/Z_1$, this is shown in parametric form.

We then aimed to comprehend the special features of the interaction between the central wheel teeth and the profile represented by the ξ and ζ coordinates, as well as the contact between the precessional satellite teeth and the profile described by a circular arc. In order to do this, we created the motion trajectory in plane P_1, starting at point G, which is the radius of the circular arcs, and moving from 0 to $2\pi Z_2/Z_1$ during the course of the precession cycle.

In order to do this, we assigned the coordinates of the intersection points with plane P_1 for the contour line that crosses the origin G and the precession center O to X_{GN}, Y_{GN}, and Z_{GN}. By replacing X_E, Y_E, and Z_E with X_G, Y_G, and Z_G, the coordinates X_{GN}, Y_{GN}, and Z_{GN} were produced. Consequently,

$$X_{G_N} = \frac{X_G\Delta_1}{\Delta_3}, Y_{G_N} = \frac{Y_G X_{G_N}}{X_G}, Z_{G_N} = \frac{Z_G X_{G_N}}{X_G}. \tag{16.50}$$

in which

$$\begin{aligned} \Delta_1 &= (X_{E_2} - X_{E_1})(Y_{E_1}n_Z - Z_{E_1}n_Y) + (Y_{E_2} - Y_{E_1})(Z_{E_1}n_X - X_{E_1}n_Z) \\ &\quad + (Z_{E_2} - Z_{E_1})(X_{E_1}n_Y - Y_{E_1}n_X), \\ \Delta_3 &= (X_{E_2} - X_{E_1})(Y_G n_Z - Z_G n_Y) + (Y_{E_2} - Y_{E_1})(Z_G n_X - X_G n_Z) \\ &\quad + (Z_{E_2} - Z_{E_1})(X_G n_Y - Y_G n_X). \end{aligned} \tag{16.51}$$

We used the following equations to transform the coordinates X_{GN}, Y_{GN}, and Z_{GN} into Cartesian coordinates ξ_{G_N} and ζ_{G_N} in order to generate the motion trajectory of the origin G of the circular arcs in 2D:

$$\xi_{G_N} = \frac{(E_1E_2)^2 + S_1^2 + S_2^2}{2(E_1E_2)}, \zeta_{G_N} = \sqrt{S_1^2 + \xi_{G_N}^2}. \tag{16.52}$$

in which

$$\left(E_1E_2\right)^2 = \left(X_{E_2} - X_{E_1}\right)^2 + \left(Y_{E_2} - Y_{E_1}\right)^2 + \left(Z_{E_2} - Z_{E_1}\right)^2, \tag{16.53}$$

$$S_1^2 = \left(X_{G_N} - X_{E_1}\right)^2 + \left(Y_{G_N} - Y_{E_1}\right)^2 + \left(Z_{G_N} - Z_{E_1}\right)^2, \tag{16.54}$$

$$S_2^2 = \left(X_{G_N} - X_{E_2}\right)^2 + \left(Y_{G_N} - Y_{E_2}\right)^2 + \left(Z_{G_N} - Z_{E_2}\right)^2. \tag{16.55}$$

The motion trajectory of the circular arcs' origin G is projected onto plane P_1 by the function $\zeta_{G_N} = f\left(\xi_{G_N}\right)$ (16.52) and profile of the central wheel teeth made on plane P_1 is described by the function $\zeta_{E_N} = f\left(\xi_{E_N}\right)$ (16.46).

Remark 1 *Two important elements that affect the tooth contact's load-bearing capacity and improve the mechanical efficiency of the A^D-toothed precessional gear are variations in the relative sliding velocity and the radii of curvature of the interacting profiles.*

The CAD/CAM/CAE modeling programs SolidWorks and CATIA V5 were used to develop the profile of the center wheel teeth in the normal section.

16.3.3 The Profile of the Teeth of the Central Wheels Determined by the Winding of the Generator Contour of the Tool

We accept the usual condition that LEM is a curve of the shape of the circle arc (see Fig. 16.9), which arbitrarily prescribes the teeth profile of the crowns of the satellite with spherospatial motion with a fixed point. Curve E_1CE_2 (see Fig. 16.9a) or $E_1E_NE_2$ (see Fig. 16.9b) represents the wrap of the families of the circle arcs LEM with the radius r with the origin G located on the trajectory of its motion $E_1'C'E_2'$ (see Fig. 16.9a) or $E_1'B'C'F'E_2'$ (see Fig. 16.9b) in the immobile coordinate system OXYZ.

The wraps of the positioning of the satellite teeth profiles in a circle arc within a precession cycle $0 \le \psi \le 2\pi Z_2/Z_1$ represent the profile of the central wheel teeth (see Fig. 16.10a) for the gearing with $Z_1 = 29$ and (see Fig. 16.10b) for the gearing $Z_1 = 24$.

Generating the flank profile of the central gear teeth is designed to reproduce the interaction of conjugate teeth in the real precessional transmission, respecting the basic principle of the fundamental law of gearing $\omega_1/\omega_2 = \text{const}$.

The kinematics of the device for generating the flank profile of the teeth of the central wheels is based on the reproduction of the kinematic relationship between

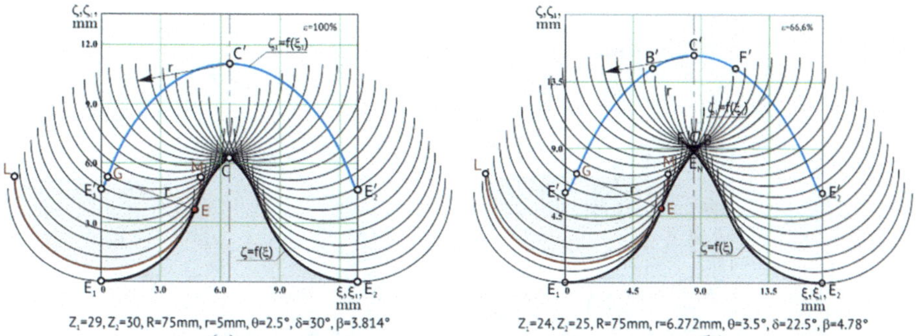

Fig. 16.9 The profile of the central wheel teeth is described by the wrapping of a family of circular arcs in the gearing, with frontal reference multiplicities of ε = 100% (**a**) and ε = 66.6% (**b**)

the "main shaft of the machine tool" and the "machine tool table" (see Fig. 16.10a) identical to the kinematics between the crankshaft and the axis of its own rotation of the satellite wheel in the actual precessional transmission.

In compliance with these conditions, the principle spatial diagram of the process of generating teeth by spatial rolling and the device for its realization was developed.

The device is equipped with a kinematic link mechanism between the precessional tool and the central wheel blank, which imposes on the satellite carrier an additional octoidal movement described with the coordinates X_C, Y_C, Z_C (see chap. 1.8.1 [22]).

To expand the range of tooth flank profiles possible to generate, the kinematic link mechanism between the tool and the blank is placed in the equatorial plane of the sphere with radius R (see Fig. 16.10a).

In this case, the condition of the fundamental law of engagement is ensured $\omega_1/\omega_2 = \text{const}$ for an infinity of profiles designed based on the parametric configuration $[Z_g - \theta, \pm 1]$.

Figure 16.10a, b shows the principle diagram for positioning the circular arcs of the generating contour of the tool during a precession cycle $0 \le \psi \le 2\pi Z_2/Z_1$, the winding which defines the shape of the flank profiles of the teeth of the central wheel, and in Fig. 16.10c its unfolding which represents the interaction of the generating contour of the tool with the central wheel in manufacturing during a precession cycle $0 \le \psi \le 2\pi Z_2/Z_1$.

Figure 16.10d shows the general views of the device and the technological equipment for the generation of tooth flanks with a precessional truncated cone tool according to the process G^{con}_{CX-CV}.

The device and method of generating tooth flank profiles is protected by the invention SU 1663857 A1 1988.

The development of the procedure for generating flank profiles of the teeth precedes an analysis of the compatibility of the shapes of the flank profiles of the teeth

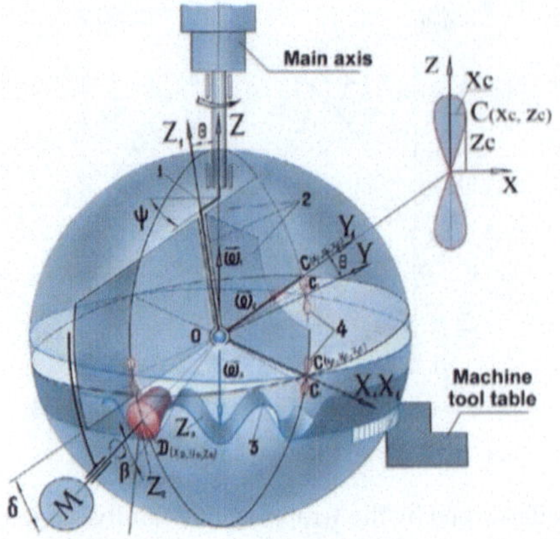

(a)

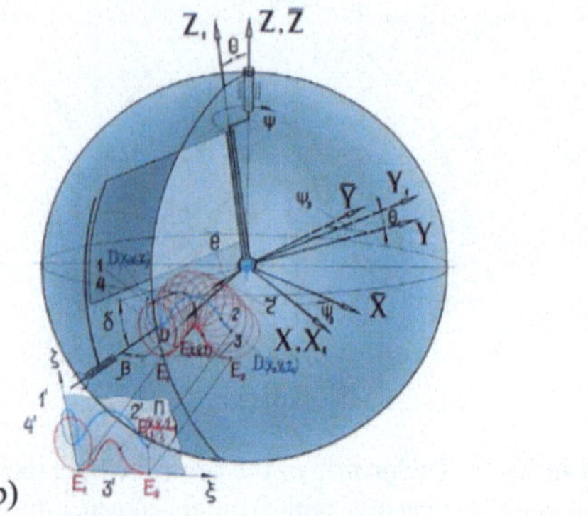

(b)

(a) The main spatial scheme of the teeth generating process by spatial tumbling-rolling (with precessional tool) (b) Determining the wrap of the precessional tool surfaces family

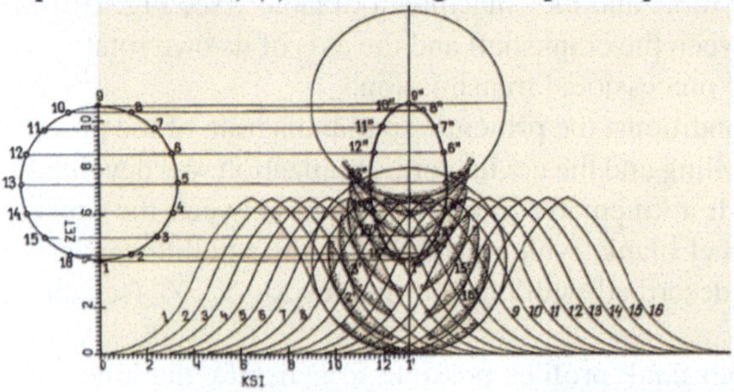

(c) Profilogram of the teeth generation with convex/concave profile with precessional tool

(d) Device and technological equipment for teeth generating with a precessional tool in the shape of a cone trunk

Fig. 16.10 The process of generating flank profiles of teeth in gears A^{D}_{CX-CV} with precessional tool G^{con}_{CX-CV}. (**a**) The main spatial scheme of the teeth generating process by spatial tumbling-rolling (with precessional tool) (**b**) Determining the wrap of the precessional tool surfaces family. (**c**) Profilogram of the teeth generation with convex/concave profile with precessional tool. (**d**) Device and technological equipment for teeth generating with a precessional tool in the shape of a cone trunk described by the winding of the generating contours of the tool and the vector of the speed of the contact point of the gear teeth.

For the analytical description of the profile of the central wheel teeth E_1CE_2 (see Fig. 16.9a) or $E_1E_NE_2$ (see Fig. 16.9b), we identify the equation of the wrap of the family of circle arcs LEM using the equations of the surface of the flanks of the satellite teeth in the circle arc and their relative motion parameters.

To simplify the identification of the wrap, we transfer the coordinates of the origin of the radius of the circle arc G to the mobile coordinate system $O\bar{X}\bar{Y}\bar{Z}$ connected to the central wheel according to the relations:

$$\begin{aligned} \bar{X}_G &= X_G \cos\psi_3 + Y_G \sin\psi_3, \\ \bar{Y}_G &= -X_G \sin\psi_3 + Y_G \cos\psi_3, \\ \bar{Z}_G &= Z_G. \end{aligned} \tag{16.56}$$

where $\bar{X}_G, \bar{Y}_G, \bar{Z}_G$ are the coordinates of the origin of the radius of curvature of the circle arc G in the mobile coordinate system $O\bar{X}\bar{Y}\bar{Z}$, $\psi_3 = \psi/i_{mt}$—the machined wheel rotation angle, i_{mt}—the transmission ratio between the rotation of the "Main shaft" and the "Machine tool table".

The unfolding of the motion trajectory of the origin G of the radius of the circle arc r on the sphere with radius R is represented by Eq. (16.56).

We then represent the working surface of the satellite teeth flanks with circle arc profile in conical shape with the extensions of the generators intersected at the precession center O (in the normal section) in the mobile coordinate system $O\bar{X}\bar{Y}\bar{Z}$ using the condition in differential geometry:

$$\mathbf{r}\mathbf{e} = r\cos\beta, \text{or } X\bar{X}_G + Y\bar{Y}_G + Z\bar{Z}_G = Rr\cos\beta \tag{16.57}$$

where **e** is the unitary vector along the axis of the teeth cone with a circle arc profile, β—the tip angle of the conical surface of the teeth with a circle arc profile.

By concurrently solving the equations describing the family wrap of the surfaces of the satellite teeth flanks with circle arc profile, the equation of the circle arc family wrap LEM (see Fig. 16.9) on the sphere with radius R is found:

$$\begin{aligned} \Phi(X,Y,Z,\psi) &= X\bar{X}_G + Y\bar{Y}_G + Z\bar{Z}_G - Rr\cos\beta = 0, \\ \frac{d\Phi}{d\psi}&(X,Y,Z,\psi) = 0. \end{aligned} \tag{16.58}$$

and the equation of the surface of the sphere:

$$X^2 + Y^2 + Z^2 - R^2 = 0. \tag{16.59}$$

To do this, from the Eqs. (16.58) and (16.59) we write:

$$\frac{d\Phi}{d\psi} = X\frac{\partial \bar{X}_G}{\partial \psi} + Y\frac{\partial \bar{Y}_G}{\partial \psi} + Z\frac{\partial \bar{Z}_G}{\partial \psi},$$
$$\frac{\partial \bar{X}_G}{\partial \psi} = \frac{\partial X_G}{\partial \psi}\cos\psi_3 - \frac{X_G}{i_{mt}}\sin\psi_3 + \frac{\partial Y_G}{\partial \psi}\sin\psi_3 + \frac{Y_G}{i_{mt}}\cos\psi_3,$$
$$\frac{\partial \bar{Y}_G}{\partial \psi} = -\frac{\partial X_G}{\partial \psi}\sin\psi_3 - \frac{X_G}{i_{mt}}\cos\psi_3 + \frac{\partial Y_G}{\partial \psi}\cos\psi_3 - \frac{Y_G}{i_{mt}}\sin\psi_3,$$
$$\frac{\partial \bar{Z}_G}{\partial \psi} = \frac{\partial Z_G}{\partial \psi}. \tag{16.60}$$

$$\frac{\partial X_G}{\partial \psi} = -R\cos\delta(1-\cos\theta)\cos^2\psi - R\sin\delta\sin\theta\cos\psi,$$
$$\frac{\partial Y_G}{\partial \psi} = -R\cos\delta(1-\cos\theta)\sin^2\psi - R\sin\delta\sin\theta\sin\psi,$$
$$\frac{\partial Z_G}{\partial \psi} = -R\cos\delta\sin\theta\sin\psi.$$

After substituting (16.60) in (16.58) and (16.59), we obtain the wrap equations on the sphere:

$$X_w = \frac{-(ab+de) \pm \sqrt{(ab+de)^2 + (1+a^2+d^2)(R^2-b^2-e^2)}}{1+a^2+d^2},$$
$$Y_w = aX_w + b, \tag{16.61}$$
$$Z_w = dX_w + e.$$

in which

$$a = \left(X_G\frac{\partial \bar{Z}_G}{\partial \psi} - Z_G\frac{\partial \bar{X}_G}{\partial \psi}\right) \Big/ \left(Z_G\frac{\partial \bar{Y}_G}{\partial \psi} - Y_G\frac{\partial \bar{Z}_G}{\partial \psi}\right),$$
$$b = \left(-R^2\cos\beta\frac{\partial \bar{Z}_G}{\partial \psi}\right) \Big/ \left(Z_G\frac{\partial \bar{Y}_G}{\partial \psi} - Y_G\frac{\partial \bar{Z}_G}{\partial \psi}\right),$$
$$d = -\frac{X_G - aY_G}{\bar{Z}_G}$$
$$e = \frac{R^2\cos\beta - bY_G}{\bar{Z}_G}$$

The Eq. (16.61) represent the wrap of the family of the circle arcs on the sphere and the profile of the central wheel teeth respectively. In order to identify the profile of the central wheel teeth, in the normal section we project the wrap from the sphere on a plane P perpendicular to two directions passing through two similar and neighboring points on the profile of the teeth on the sphere, for example through E_1 and E_2 corresponding to Fig. 16.8. Figure 16.10b shows the conceptual diagram regarding the description of the profile of the central wheel teeth by wrapping the LEM circle arcs family of the profiles of the satellite wheel teeth in the mobile coordinate system $O\overline{X}\overline{Y}\overline{Z}$.

The equations of the wrap of the circle arcs family (16.62), projected in 2D on the plane P_1 in the coordinates ξ and ζ, are obtained after a number of modifications using the spherical trigonometry principles. The profile of the central wheel teeth in the normal section is represented by the following formulas, which are expressed by the function $\zeta = f(\xi)$, which wraps the profile of the satellite teeth in a circle arc:

$$\xi = \frac{(E_1E_2)^2 + (X - X_1)^2 + (Y - Y_1)^2 + (Z - Z_1)^2 - (X - X_2)^2 - (Y - Y_2)^2 - (Z - Z_2)^2}{2(E_1E_2)} \tag{16.62}$$

$$\zeta = \sqrt{(X - X_1)^2 + (Y - Y_1)^2 + (Z - Z_1)^2 + \xi^2}.$$

A CAD/CAM/CAE modeling system is used to produce the profile of the central wheel teeth in the normal section.

The concept of creating the wrap of the family of circular arcs, as described by Eq. (16.62), with the origins of their radii positioned along the curve defined by Eq. (16.56), serves as the kinematic model for the processes $G_{r.s}$ of generation through the spatial rolling of the central wheel teeth. This approach replicates the geometry and kinematics of the interaction between the teeth in an actual precessional transmission.

Figure 16.11 shows the principle schemes of the general processes of the flank profiles of the teeth with precession tools in the shape of a truncated cone (a) and in the shape of a peripherally profiled disc (b).

The geometric compatibility of the flank profile of the teeth of the central wheels obtained by wrapping the generating contour of the tool on the one hand and of the profile described by the velocity vector of the contact point of the conjugate teeth on the other hand is appreciated by the equations (1.52–1.54) presented in chapter 1.7 [22].

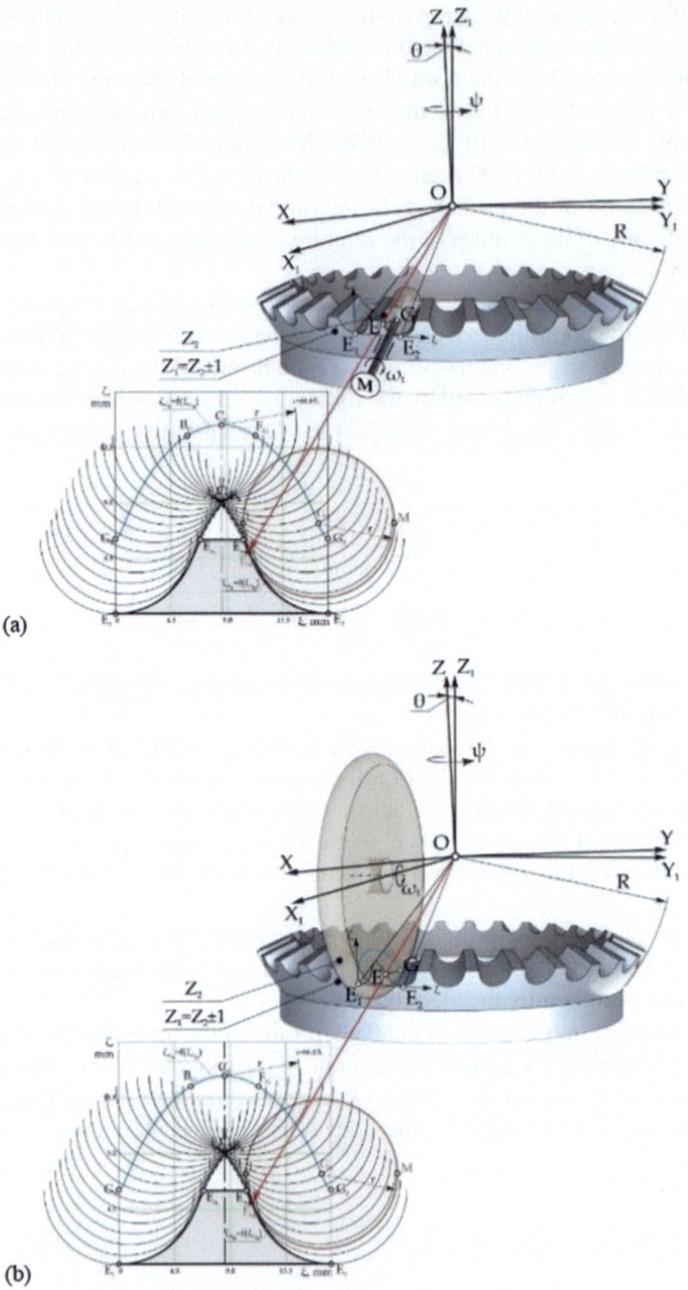

Fig. 16.11 The two methods for creating teeth are: (**a**) using a tool shaped like a “cone trunk” for spatial tumbling and rolling (Patent SU 1663857 A1, 1988) and (**b**) using a tool shaped like a “peripherally profiled disk” (Patent MD 3532 C2, 2008)

16.4 Modification of the Shape of the Teeth in the Multipair Convex-Concave Gearing

Figure 16.12a shows the profilogram of the unmodified gear teeth A^{D}_{CX-CV} with $Z_1 = 24$, $Z_2 = 25$, $\delta_{(1-2)} = 22.5°$, $\beta_2 = 4.78°$ and $\theta = 3.5°$, which successively conjugate at the contact points $k_0 \ldots k_i$ depending on the positioning angle ψ of the crankshaft. The contact points $k_0 \ldots k_i$ are transposed on the active teeth profile of the central wheel and represent, as a whole, the contacts of the same pair of teeth at different positioning angles ψ of the crankshaft within a precessional half-cycle of the planet wheel $0 \leq \psi \leq \pi Z_2/Z_1$.

The tooth profiles of the ring gear teeth and the central gear teeth are displayed in Fig. 16.12b. They are adjusted based on the number of pairs of teeth in the

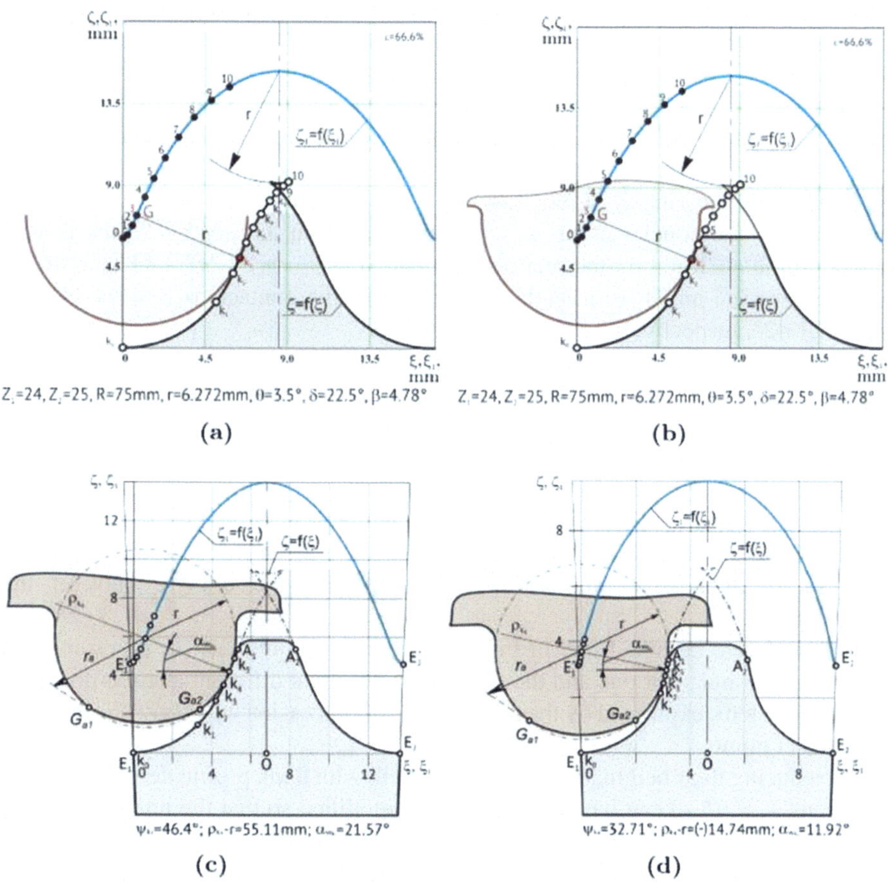

Fig. 16.12 Change in teeth shape depending on contact dispersion k_i: contact points k_i plotted on the unmodified (**a**) and modified (**b**) profile of the central wheel teeth; (**c**), (**d**) dispersion of contact points of the same pair of teeth as a function of ψ

meshing region A^{D}_{CX-CV} and the dispersion of the contact points k_i. By calculating the contact pressure resistance of the tooth flanks and then analyzing the relative sliding velocity with friction between the flanks of the conjugate tooth pairs, the number of concurrently conjugate tooth pairs is established.

Figure 16.12c, d show the profilograms of the conjugate teeth in A^{D}_{CX-CV} gears with $Z_1 = 32$, $Z_2 = 33$, $\delta_{(1-2)} = 15°$, $\beta_2 = 3.5°$ and $\theta = 3.5°$ (see Fig. 16.12a) and respectively $Z_1 = 45$, $Z_2 = 46$, $\delta_{(1-2)} = 22.5°$, $\beta_2 = 2.4°$ and $\theta = 2.5°$ (see Fig. 16.12b), where the tips of the teeth of the satellite wheel are cut off by an arc r so that they do not contact the teeth of the central wheel at point k_0.

Therefore, the geometry modification of the teeth of the satellite wheel with a circular arc profile of radius r (see Fig. 16.12c, d) is achieved by cutting their tips after an arc of radius $r_a > r$ supported symmetrically at two points G_{a1} and G_{a2} on the circle of radius r, so that the semi-arc between the support points G_{a1} and G_{a2} is smaller than the half arc $k_0 \ldots k_1$ with the fulcrum in contact k_1 of the arcuate profile of the satellite wheel teeth and the convex-concave profile of the teeth of the central wheel.

Figure 16.12c, d show the modified shape profilograms of the teeth of the mating gears with contact point dispersion $k_1 \ldots k_5$, characterized by angular coordinates ψ_{ki} and pressure angles α_w between the flanks with different values.

Thus, in the gearing $Z_1 = 32$, $Z_2 = 33$, $\delta_{(1-2)} = 15°$, $\beta_2 = 3.5°$ and $\theta = 3.5°$ (see Fig. 16.12c), the contact points $k_1 \ldots k_5$ are located on the portion of the central wheel profile enclosed by the precession angle $\psi = \psi_{k5}\text{-}\psi_{k1} = 58° - 11.6° = 46.4°$, and the pressure angle from teeth engagement in the contacts $\alpha_{wk1} = 44.44°$ and $\alpha_{wk5} = 21.42°$, respectively.

In the case of the gearing $Z_1 = 45$, $Z_2 = 46$, $\delta_{(1-2)} = 22.5°$, $\beta_2 = 2.4°$ and $\theta = 2.5°$ (Fig. 16.12d), the contact points $k_1 \ldots k_5$ have coordinates between $\psi_{k5} - \psi_{k1} = 32.71°$, and the pressure angle for the conjugate teeth pairs in k_1 and k_5 is $\alpha_{wk1} = 19.14°$ and $\alpha_{wk5} = 12.77°$, respectively.

The small α_w pressure angles between the flanks result in smaller forces in the gearing and, respectively, lower energy losses in the precessional gearing, while the small dispersion of teeth contact location points $k_1 \ldots k_5$ in the gearing is found to increase the flank roll rate within the bounds of the θ-angle knot of the gear-satellite.

It should also be noted that in Fig. 16.12 contact points $k_1 \ldots k_5$ represents the sequence of contacts of one and the same pair of teeth at different angular positions of the crankshaft, expressed by the angular step of precession $\psi_k = 2\pi Z_2/Z_1^2$ between the contact points.

By reducing their height, the central gear teeth with flank profile described by the equation $\zeta_{EN} = f(\xi_{EN})$ can have their geometry modified so that the number of pairs of teeth conjugated simultaneously in the meshing field is equal to or greater than the number found by calculating the teeth's resistance to contact pressure.

By profiling the tips of the satellite wheel's gear teeth along a circular arc of radius $r_a > r$, symmetrically supported at two points G_{a1} and G_{a2} on the circle of radius r, the geometry modification of the gear teeth with a circular arc profile of radius r is accomplished. This way, the semi-arc between the support points G_{a1} and G_{a2} is smaller than the semi-arc k_0k_1 with the support point at the common contact point k_1 of the satellite wheel teeth's arcuate profile and the convex-concave profile of the central wheel teeth.

The teeth tips profiling of the combined gears in the processional gearing A^{D}_{CX-CV} with the teeth number $Z_1 = Z_2 - 1$ is summarized in the following functional and construction-technological aspects:

- reduction of dynamic loads and exclusion of the phenomenon of "head to foot" tooth meshing in the vicinity of the contact point k_0, where the flank pressure angle $\alpha_{wk0} = 90°$;
- ensuring optimal lubrication conditions for active and passive tooth flanks when sliding with skidding.

Figure 16.13 shows the evolution of the positioning of the arcuate profiles of the multipair conjugate sun gear teeth with the convex-concave profiles of the central gear teeth for one and the same position of the crankshaft ψ in A^{D}_{CX-CV} gears with: (Fig. 16.13a) $Z_1 = 32$, $Z_2 = 33$, $R = 19.5$ mm, $r_{sm} = 1.19$ mm, $\theta = 3.5°$, $\delta_{(1-2)} = 15°$; (Fig. 16.13b) $Z_1 = 45$, $Z_2 = 46$, $R = 21$ mm, $r_{sm} = 0.98$ mm, $\theta = 2.5°$, $\delta_{(1-2)} = 22.5°$.

The arcuate teeth of the planetary gear conjugate multiparously with the teeth of the central gear with convex-concave profiles in the contacts k_i, moving successively from one tooth to another when the crankshaft is continuously rotated with an angular velocity $\omega_1 = f(\psi)$. In this scenario, the crankshaft rotates with an angle equal to the angular pitch $\psi_k = 360\ \pi\ Z_2/Z_1^2$ between two contacts of neighboring pairs of teeth.

We draw the following conclusions from the examination of teeth's interaction with spherospatial motion (see Fig. 16.13):

1. Depending on the configuration parameters $[Z_g - \theta, \pm 1]$, the profile shape of the central gear teeth has a variable curvature that continuously increases from the root to the tip of the teeth. This is limited by the fundamental law of gearing's requirement that the constant permanent ratio of motion transformation $\omega_1/\omega_2 = \text{const}$ be maintained for any value of the precession angle ψ.
2. Depending on the precession angle ψ, the contact points of the conjugate flanks of one and the same pair of teeth define the shape of the profile of the central gear teeth, which have a variable, continually increasing curvature from the root to the tip of the teeth.
3. The parametric configuration $[Z_g - \theta, \pm 1]$ and the size of teeth height shortening by cutting define the number of pairs of teeth that concurrently conjugate at the same gearing phase, which is determined by the precession angle ψ.
4. In precessional meshing, the teeth contact on active flank portions, topologically dimensioned with the precession angle within the limits $0 \leq \psi \leq 180 \cdot Z_2/Z_1$ and simultaneously conjugate in multipair contacts k_i with dispersion determined by the relation $\psi_{ki} = i \cdot 360 \cdot Z_2/Z_1^2$, which, in aggregate, define the pressure angles α_{wki} between the flanks of the simultaneously conjugated tooth pairs.

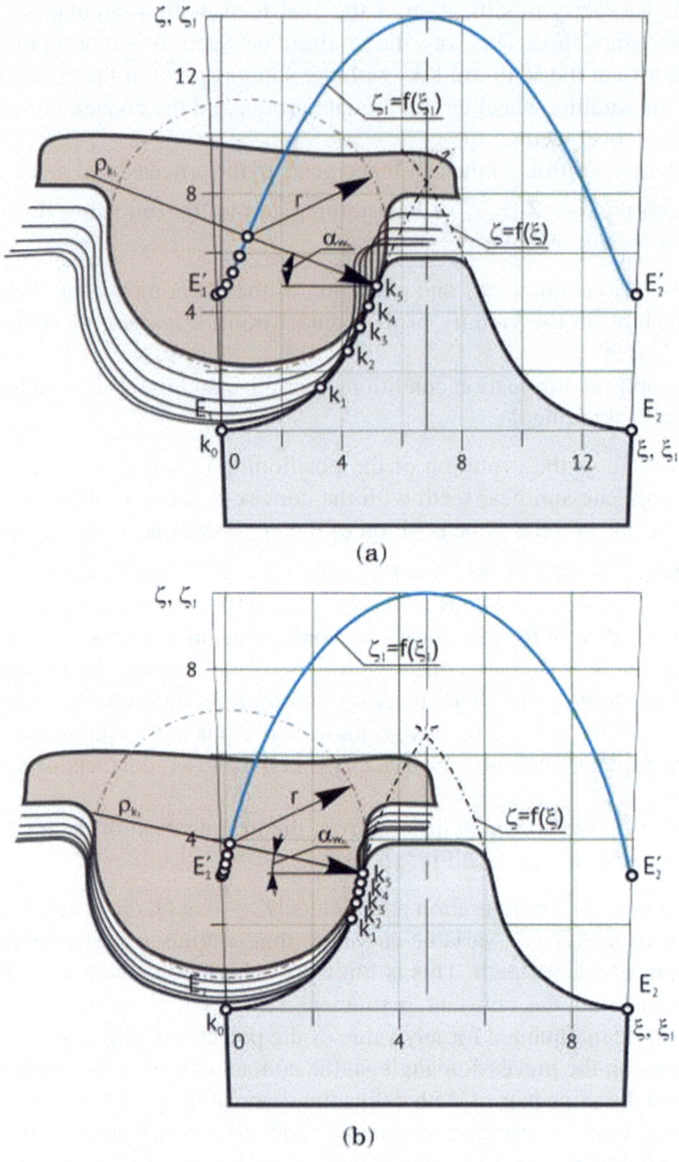

Fig. 16.13 The sequence positioning of multipair conjunctival teeth profiles in contacts $k_0 \ldots k_5$ at one and the same angular coordinate ψ of the crankshaft in the gears A^{D}_{CX-CV} : (**a**) $Z_1 = 32$, $Z_2 = 33$, R = 75 mm, r = 4, 58 mm, θ = 3.5°, $\delta_{(1-2)} = 15°$, $\beta_2 = 3.5°$; (**b**) $Z_1 = 45$, $Z_2 = 46$, R = 75 mm, r = 3.14 mm, θ = 2.5°, $\delta_{(1-2)} = 22.5°$, $\beta_2 = 2.4°$

5. The precession angle ψ is generalising parameter of the profile shape of the central gear teeth and the convex-concave contact geometry. For these reasons, mechanical transmissions with spherospatial interaction of mating wheels are called "precessional transmissions".

Analysis of the details of teeth conjugation in precessional A^{D}_{CX-CV} and Wildhaber-Novikov angulation reveals that: precessional A^{D}_{CX-CV} angulation is similar to Wildhaber-Novikov angulation in that the teeth conjugate in contacts with convex-concave geometries with the same (comparable) small difference in the radii of curvature of the conjugated flanks. On the other hand, precessional angulation is unique in that it involves simultaneous multipair conjugation of teeth under load, which makes up to five times more pair of teeth.

Since both gears have teeth contacts with convex-concave geometries with the same (comparable) small difference in flank curvatures, we can say that precessional gear transmissions with gearing A^{D}_{CX-CV} perform better than Wildhaber-Novikov gears in terms of deformable contact mechanics and multipair contact conditions. In Wildhaber-Novikov gears, only one pair of teeth mesh is present in frontal gearing, whereas in precessional gears, up to three to five pairs of teeth mesh.

According to what has been mentioned in the approaches to the development of precessional gear transmissions with gearing A^{D}_{CX-CV} and in the comparative analyses of contact lift and multipair contact conditions, the following is to be considered—*in precessional transmissions with A^{D}_{CX-CV} gearing and in Wildhaber-Novikov transmissions, the teeth conjugate in contacts with convex-concave geometries with the same (comparable) small difference in flank curvatures, which from the point of view of deformable solid mechanics defines the bearing capacity of the teeth contact, but at the same time, in precessional transmissions, the number of simultaneously conjugated teeth pairs under load is up to 3–5 times higher than in Wildhaber-Novikov transmissions.*

Based on the study of the equations $\xi = f(\xi)$ and $\xi_1 = f(\xi_1)$ it was found that the parameters of the configuration $[Z_g - \theta, \pm 1]$ including the angle of the conical axoid δ (see Fig. 16.14), the nutation angle θ, the angle of the conicity of the satellite-wheel teeth β, the number of conjugated teeth Z and their correlation ± 1 influence differently vectorially and numerically on the geometry of the contact and the kinematics of the contact point of the tooth flanks in gearing.

Figure 16.14a shows the axial section of the precessional transmission with gear A^{D}_{CX-CV} in which the correlation of the tooth numbers of the conjugated wheels $Z_1 = Z_2 - 1$, $Z_4 = Z_3 - 1$ and of the crowns of the satellite wheel $Z_2 > Z_3$ or $Z_2 < Z_3$, and the angles of the conical axoids $\delta_{(1-2)} = 22.5°$ and $\delta_{(3-4)} = 22.5°$.

According to the examination of tooth interactions in the precessional gearing A^{D}_{CX-CV} displaying spherospatial motion around a fixed point, three important aspects affect its load-bearing capability:

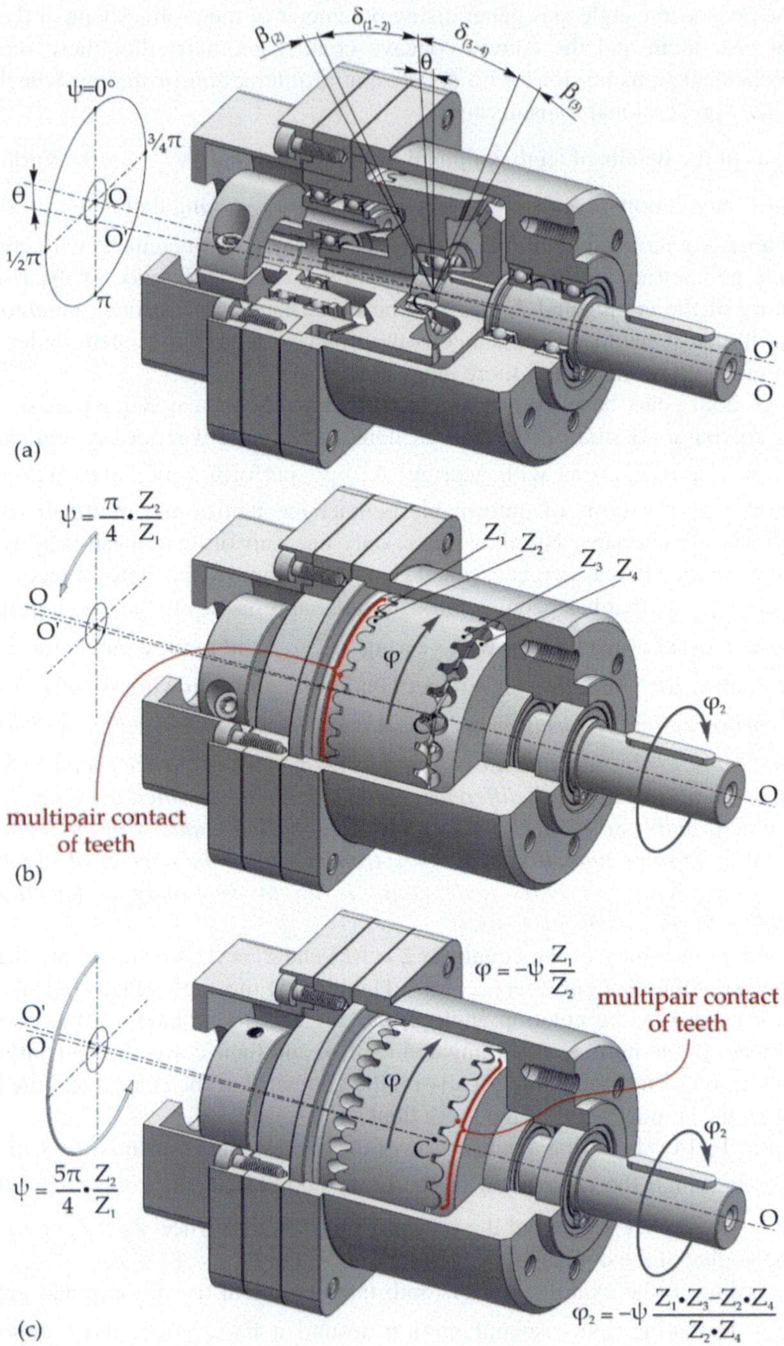

Fig. 16.14 The engagement of the teeth ($Z_1 - Z_2$) and ($Z_3 - Z_4$) in gears A^{D}_{CX-CV} follows the relationships $Z_1 = Z_2 - 1$ and $Z_4 = Z_3 - 1$, depending on the precession angle ψ, the conical the conical axoid angles $\delta_{(1-2)} = \delta_{(3-4)} = 22.5°$: angular positioning of the driving shaft $\psi = 0°$ (**a**); $\psi = \frac{\pi}{4} \cdot \frac{Z_2}{Z_1}$ (**b**) and $\psi = \frac{5\pi}{4} \cdot \frac{Z_2}{Z_1}$ (**c**)

1. According to the mechanics of solid deformable bodies, the geometry of the convex-concave contact between teeth results in a load-bearing capacity that is inversely related to the difference in curvatures of the tooth flanks. In other words, a smaller difference in tooth flank curvatures leads to a higher contact load-bearing capacity, assuming the same operating conditions, material, lubricant, and lubrication parameters.
2. The variability in the profile of the central gear teeth based on the parametric configuration $[Z_g - \theta, \pm 1]$ allows for multiple contacts during one precession cycle $0 \leq \psi \leq 2\pi Z_2/Z_1$, using the angular pitch $\psi = 2\pi Z_2 / Z_1^2$ as the distance between neighboring contacts [35] (p. 9).
3. In accordance with the circumstances listed on p. 2, for the identical crankshaft coordinate, such as $\psi = \frac{\pi}{4} \cdot \frac{Z_2}{Z_1}$ (see Fig. 16.14b) or $\psi = \frac{5\pi}{4} \cdot \frac{Z_2}{Z_1}$ (see Fig. 16.14c), there are simultaneously eight pairs of teeth in the gearing. The profiles of these pairs are modified according to the calculated resistance to contact pressure. There are four pairs that make contact on the teeth's active profile and four pairs on the passive profile.

Figure 16.14 illustrates the changes in the tooth engagement $(Z_1 - Z_2)$ and $(Z_3 - Z_4)$ in the gears A^D_{CX-CV}. It includes the relationships of the tooth counts, where $Z_1 = Z_2 - 1$, $Z_4 = Z_3 - 1$, and the angles of the conical axoids are $\delta_{(1-2)} = \delta_{(3-4)} = 22.5°$, which are influenced by the precession angle ψ.

The teeth are changed by lowering their height in order to reduce frictional sliding between the flanks in the precessional transmission employing gear A^D_{CX-CV}. This modification enables three or four pairs of teeth to engage simultaneously at contacts $k_0, \ldots, k_4$ (see Figs. 16.15, 16.16 and 16.17) on the gear's active side.

Eight pairs of teeth will be present in the gear region, four of which will be passive (not transferring load) if a gear is made to have four pairs of teeth involved in load transmission through contacts $k_0, \ldots, k_4$. In this case, all eight pairs of teeth will rotate at an angle with respect to their neighboring pairs when the drive shaft rotates at an angle $\psi = 360 Z_2 / Z_1^2$. As seen in Figs. 16.16 and 16.17, each pair will go through the gearing phase of the previous pair of teeth.

Figure 16.14a illustrates the arrangement of the tooth pairs that are engaged simultaneously in the gears A^D_{CX-CV} at a precession angle of $\psi = 0°$. As the crankshaft rotates to an angle of $\psi = \frac{\pi}{4} \cdot \frac{Z_2}{Z_1}$, the placement of the tooth pairs simultaneously $(Z_1 - Z_2)$ and $(Z_3 - Z_4)$ is shown in Fig. 16.14b. Additionally, at an angle of $\psi = \frac{5\pi}{4} \cdot \frac{Z_2}{Z_1}$, the arrangement of the simultaneously engaged pairs $(Z_1 - Z_2)$ and $(Z_3 - Z_4)$, is shown in Fig. 16.14c.

By examining the variations in relative friction speeds between the tooth flanks, the energy losses in the precessional transmission using the toothed gear A^D_{CX-CV} in reducer mode are evaluated. Furthermore, the differences in their contact shape are used to assess the bearing capacity.

Examples of tooth profiles that have been modified in accordance with the distribution of contact points k_i of the flanks that are simultaneously engaged in the gears A^{D}_{CX-CV} are shown in Fig. 16.15. $Z_1 = Z_2 - 1$ (see Fig. 16.15a–h) and $Z_1 = Z_2 + 1$ (see Fig. 16.15i, j) are examples of this association. The configuration parameters $[Z_g - \theta, \pm 1]$ are randomly and variable (computer-generated based on Eqs. (16.28) and (16.33)) [30].

The functions $\zeta_{E_N} = f\left(\xi_{E_N}\right)$ and $\zeta_{G_N} = f\left(\xi_{G_N}\right)$ depict, respectively, the trajectory of the center of the circular arc G of the satellite wheel tooth profile and the central wheel tooth profile. With the precession angle ranging from $\psi = 0°$ to $\psi = 2\pi Z_2/Z_1$, they are depicted in the parametric equation of point E in Cartesian coordinates [34, 35].

With a transmission ratio ranging from ±20 to ±3600, the profiles depicted in Fig. 16.15a–h are recommended for the construction of precessional transmissions functioning in reducer mode. The profiles in Fig. 16.15i, j, on the other hand, are designed for precessional transmissions with a transmission ratio ranging from ±7.3 to ±30.3. They operate in multiplier mode if $\delta_{1(4)-2(3)} > 0°$ or in reducer mode if the conical axoid angle $\delta_{1(4)-2(3)} = 0°$.

Each pair of teeth that are engaged simultaneously is represented by ψ, which is used to map the contact points $k_0 \ldots k_5$ (see Fig. 16.15) on the sides of the teeth in the gear area onto the profile of the central wheel teeth.

16.5 Geometry and Kinematics of Tooth Contact with Sliding and Pure Rolling

The interaction of the teeth occurs in a spherospatial motion in precessional gearings A^{D}_{CX-CV} with convex-concave profiles because of this. Relative rolling ω_θ around the contact points ki and relative sliding V_{sl} at the contacts of concurrently engaged tooth pairs k_i are involved in this (see Figs. 16.16 and 16.17).

In relation to the tooth profiles on the central wheel, the relative rolling ω_θ of the satellite wheel crown occurs around their contact points k_i. This interaction mostly depends on the angular pitch between neighboring contact sites $k_0 - k_1$, $k_1 - k_2$, $k_2 - k_3$, and so on, as well as the nutation angle θ of the spherospatial motion.

Vector-conjugated teeth's flank profiles' relative rolling ω_θ is determined by their angular rolling direction ω_θ, which may or may not coincide with the linear movement of the flanks' contact points k_i in the paired teeth that are conjugated simultaneously, along with relative sliding V_{sl}. It has been observed that the contact point k_i of the flanks of these teeth that are conjugated concurrently advances from the root to the tips along the active profile of the teeth on the central wheels (see Fig. 16.16).

The angular pitch of the contact point can be used to characterize the angular coordinates of the contact points on the flank profiles of the simultaneously

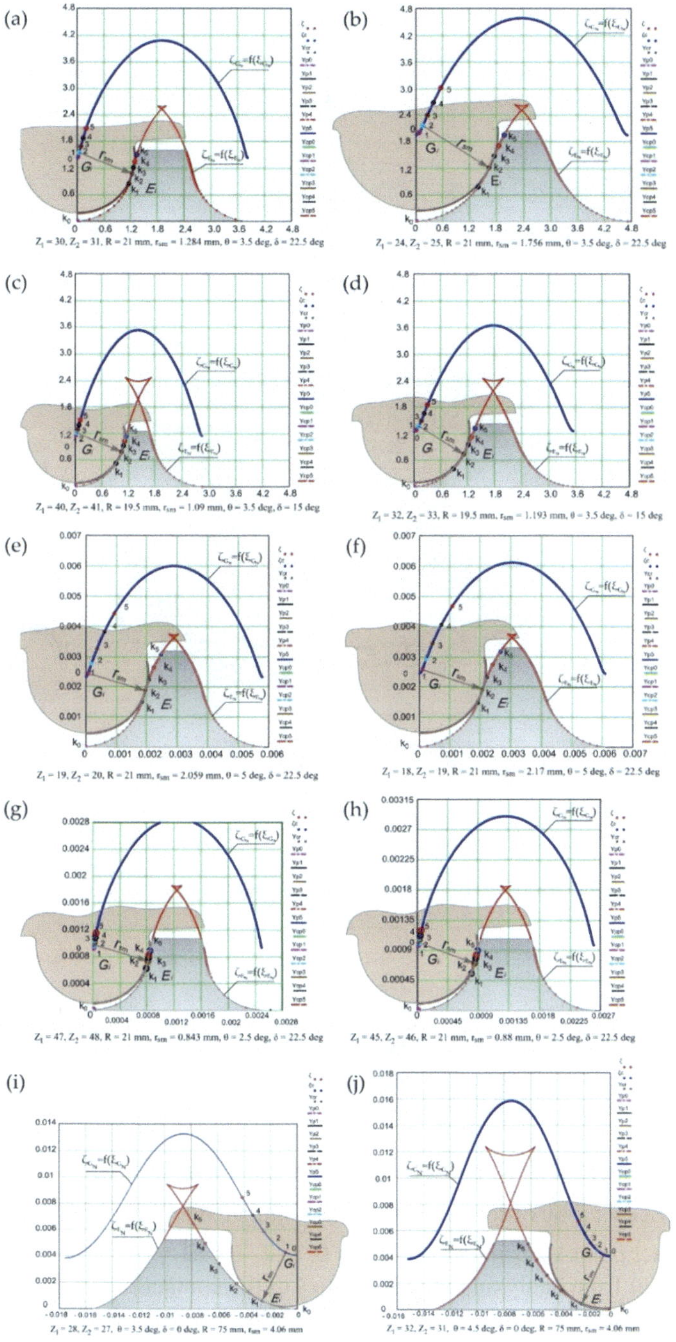

Fig. 16.15 Flank profiles adjusted based on the distribution of contact points k_i in the precessional gearing A^{D}_{CX-CV} of the teeth that are engaged simultaneously for $Z_{1(4)} = Z_{2(3)} - 1$ (**a–h**) and for $Z_{1(4)} = Z_{2(3)} + 1$ (**i, j**)

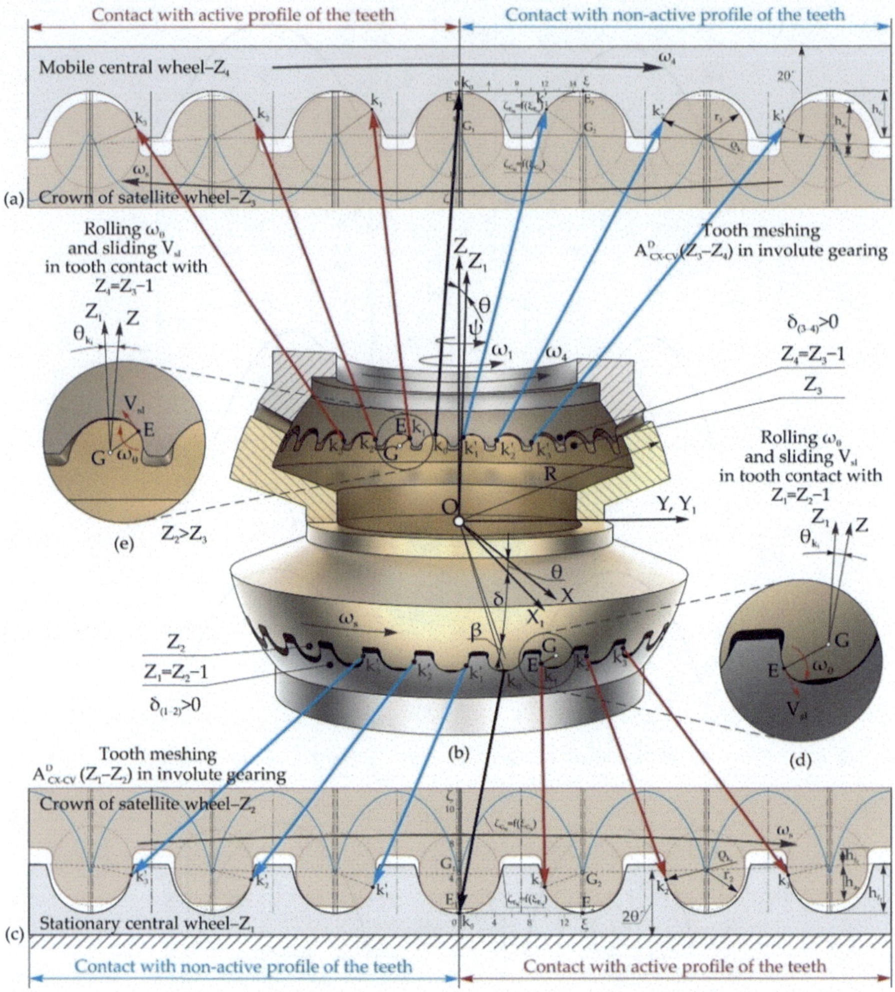

Fig. 16.16 In the gearing A^{D}_{CX-CV} (**a, c**), the relative sliding V_{sl} and rolling ω_0 of the flank profiles (**d, e**) at the contact points k_i of the conjugated teeth take place with $Z_1 = Z_2 - 1$, $Z_4 = Z_3 - 1$, $\delta_{(1-2)} > 0°$, $\delta_{(3-4)} > 0°$, and $Z_2 > Z_3$ or $Z_2 < Z_3$ (**b**)

conjugated teeth, according to an analysis of Figs. 16.16 and 16.17. In the equation $\psi_{k_i} = i \cdot 2\pi \cdot Z_2 / Z_1^2$, the order number of the conjugated tooth pairs, starting with the pair at contact point k_0, is indicated by i = 0, 1, 2, 3…

It is evident from analyzing the kinematics of the contact points k_i of the concurrently conjugated teeth in the gears A^{D}_{CX-CV} in Fig. 16.16d, e that the spherospatial interaction of the flank profiles causes the conjugated flanks ω_0 to roll angularly around the contact points E. This movement can happen in the opposite direction of the relative sliding vector between the flanks V_{sl} or in the same direction. The profile

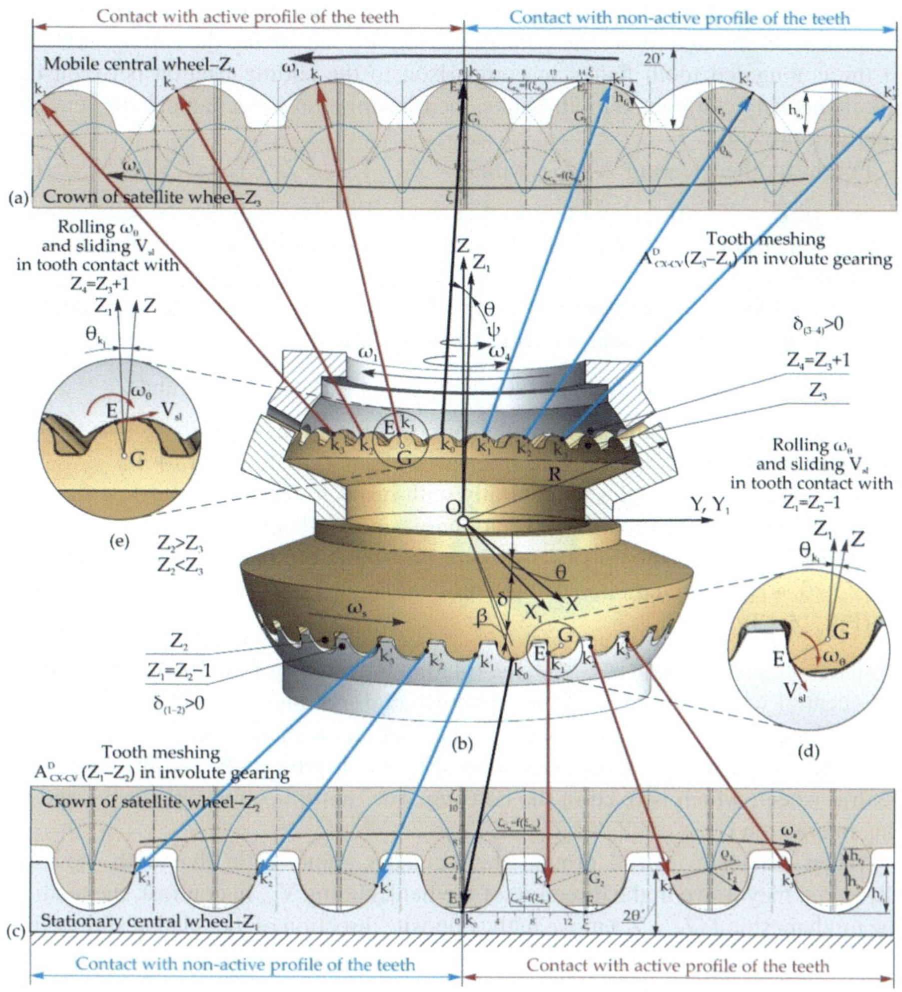

Fig. 16.17 In the gearing A^{D}_{CX-CV} (**a**, **c**), the relative sliding V_{sl} and rolling ω_0 of the flank profiles (**d**, **e**) at the contact points k_i of the conjugated teeth take place with $Z_1 = Z_2 - 1$, $Z_4 = Z_3 + 1$, $\delta_{(1-2)} > 0°$, $\delta_{(3-4)} > 0°$, and $Z_2 > Z_3$ or $Z_2 < Z_3$ (**b**)

rolling and the energy loss between the conjugated flanks are thus either increased or decreased as a result of this interaction.

As the precession angle increases within the range of $0 \le \psi \le 2\pi Z_2 / Z_1^2$, the teeth interact in the precessional transmission involving toothed gear A^{D}_{CX-CV} where the tooth count ratios of the conjugated wheels are $Z_1 = Z_2 - 1$, $Z_4 = Z_3 - 1$, with $\delta_{(1-2)} > 0°$ and $\delta_{(3-4)} > 0°$ and $Z_2 > Z_3$ or $Z_2 < Z_3$. As they travel from the roots to the tips of the teeth, frictional forces and relative sliding V_{sl} are present at the contact locations k_i. The flank profiles move in the same direction and simultaneously show reciprocal rolling ω_0 around the contact sites k_i. By increasing the rolling angle ω_0 of the conjugated tooth flanks in comparison to the sliding friction between the simultaneously conjugated tooth flanks, the transmission's mechanical efficiency is finally improved.

For the toothed gearing A^{D}_{CX-CV} with $Z_1 = Z_2 - 1, Z_4 = Z_3 + 1, \delta_{(1-2)} > 0°, \delta_{(3-4)} > 0°$, $Z_2 > Z_3$ or $Z_2 < Z_3$, operating in multiplier mode, Fig. 16.17 shows the relative rolling ω_0 and sliding V_{sl} of the tooth flank profiles at the contact points $k_0, k_1, k_2, k_3 \ldots$ and $\acute{k}_0, \acute{k}_1, \acute{k}_2, \acute{k}_3 \ldots$.

These particular circumstances allow the 2K–H precessional transmission to function efficiently in a multiplier (or differential) mode:

- With the conical axoid angle satisfying $\delta_{(1-2)} > 0°$, $\delta_{(3-4)} > 0°$ and $Z_2 > Z_3$ or $Z_2 < Z_3$, the relationship between the number of teeth in one of the gears A^{D}_{CX-CV} can be written as $Z_1 = Z_2 + 1$, $Z_4 = Z_3 - 1$ or $Z_1 = Z_2 - 1$, $Z_4 = Z_3 + 1$;
- The satellite wheel slides tangentially with respect to the central gears, floating axially between the moving and stationary central wheels;
- In one of the engagements $(Z_1 - Z_2)$ or $(Z_3 - Z_4)$, the parametric configuration $[Z_g - \theta, +1]$ must guarantee that the pressure angle between the teeth is $\alpha_\omega > 45°$ (Fig. 16.17).

With a tooth profile angle $\alpha_\omega > 45°$, the rotating movement of the shaft and the movable central wheel is converted into spherospatial motion of the satellite wheel by the inclined wedge. Thus, the teeth Z_2 and Z_3 interact with the teeth Z_1 and Z_4 of the movable central wheel, which is connected to the driving shaft, and the stationary central wheel, which is fixed in the housing. The multiplication ratio at the end of this interaction is $i = -Z_2Z_4/(Z_1Z_3 - Z_2Z_4)$.

When the teeth ω_0 roll around their contact points k_i in the tooth meshing $(Z_1 - Z_2)$, they move in the direction of the flank sliding V_{sl}. In contrast, the teeth in the tooth meshing $(Z_3 - Z_4)$ move in the opposite direction of the relative sliding due to the rolling of the teeth around their contact points k_i, which also causes friction between the flanks.

Figures 16.16 and 16.17 show the sliding with friction at the teeth's ki contacts and the concurrent rolling of the flanks of the engaged teeth, where the following claims are accurate:

Remark 2 *The teeth in precessional gearing* A^{D}_{CX-CV} *move spherospatially, showing relative rolling* ω_0 *around the flank profiles' contact locations* k_i *and relative sliding* V_{sl} *at those contact points.*

Remark 3 *The difference in linear velocities* $\left(V_{sl} = V_{E_1} - V_{E_2}\right)$ *at the contact points of teeth* Z_1 *and* Z_2 *(or* Z_3 *and* Z_4*) affects the relative sliding* V_{sl}*, which is quantitatively related to the parameters of the* $[Z_g - \theta, \pm 1]$ *configuration.*

Remark 4 *The flank profiles at the contact locations k_i are tangentially aligned with the relative velocity vector V_{sl} between the engaged tooth flanks. The teeth move from the root to the tip as the precession angle $0 \leq \psi \leq 2\pi \cdot Z_2 / Z_1^2$ rises.*

Remark 5 *The angular position of the contact points k_i on the conjugated flank profiles and the nutation angle θ affect the relative rolling ω_0 of the flank profiles of concurrently engaged teeth around the rolling centers E (contact points).*

Remark 6 *The relative rolling ω_0 increases in relation to the relative sliding when the migration direction of the rolling center E (contact points) of the flank profiles aligns with the vector of relative sliding V_{sl}. This reduces energy losses during the meshing of the teeth, and vice versa.*

16.6 The Contact Point E_1's Linear Velocity on the Tooth Profile of the Central Wheel

With relative sliding V_{sl} between the flanks, conjugated teeth interact in precessional gearing A^{D}_{CX-CV}.

The configuration of the gear's geometric characteristics, denoted by $[Z_g - \theta, \pm 1]$, affects the size of the tooth contact and the kinematics at the contact point. With the tooth correlation being either $Z_1 = Z_2 - 1$ or $Z_1 = Z_2 + 1$, these parameters are Z_1, Z_2, δ, θ, and β.

Using the vector equation, the position vector of contact point E_1 on the conjugated teeth, which is situated on the central wheel tooth profile (see Fig. 16.18), is located:

$$\mathbf{r}_{E_1} = \mathbf{r}_{G_2} + \mathbf{G_2E_1} = \mathbf{r}_{G_2} + \mathbf{r}, \tag{16.63}$$

where $\mathbf{G_2E_1} = \mathbf{r} = R\sin\beta \dfrac{\mathbf{V}_{G_2} \times \mathbf{r}_{G_2}}{\left|\mathbf{V}_{G_2} \times \mathbf{r}_{G_2}\right|}$ and $r_{G_2} = r_G\cos\beta$. In this case, r is the position vector of point E_1 with respect to G_2, and its magnitude is equal to the radius of curvature of the arc profile of the satellite teeth. The collinear vectors $\mathbf{r}_G$ and $\mathbf{r}_{G_2}$ are oriented along OG and originate from the radius of curvature of the teeth in the satellite's arc on the sphere (see Fig. 16.18).

The coordinates X_{G_2}, Y_{G_2}, and Z_{G_2} can be used to describe the position vector for the origin of the radius of curvature r_{G_2}:

$$\mathbf{r}_{G_2} = X_{G_2}\mathbf{i} + Y_{G_2}\mathbf{j} + Z_{G_2}\mathbf{k}, \tag{16.64}$$

The velocity vector $\mathbf{V}_{G_2}$ of point G_2 is given by $\mathbf{V}_{G_2} = \dot{\mathbf{r}}_{G_2} = \dot{X}_{G_2}\mathbf{i} + \dot{Y}_{G_2}\mathbf{j} + \dot{Z}_{G_2}\mathbf{k}$.

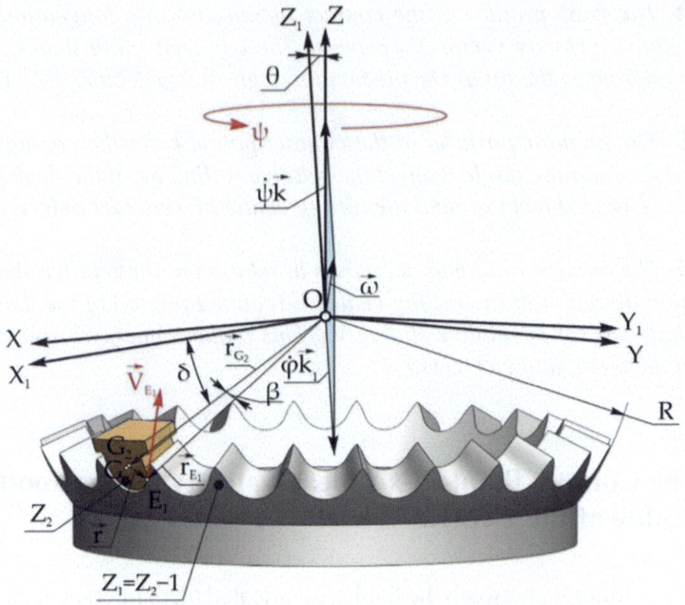

Fig. 16.18 The velocity vector position V_{E_1} at the contact point on the central wheel tooth profile

Here, $\dot{X}_{G_2}$, $\dot{Y}_{G_2}$, and $\dot{Z}_{G_2}$ are point G_2's velocity components along the fixed coordinate system's axes OXYZ.

The axis OG passing through point G_2 in its normal section at contact point E_1 is taken into consideration in order to determine the position vector of contact point E_1 (see Fig. 16.18). The motion trajectory in the fixed coordinate system OXYZ is identified by the coordinates X_{G_2}, Y_{G_2}, and Z_{G_2} as a function of ψ:

$$\begin{aligned} X_{G_2} &= R\cos\beta\cos\delta(-\cos\psi\sin\varphi + \sin\psi\cos\varphi\cos\theta) - R\cos\beta\sin\delta\sin\psi\sin\theta, \\ Y_{G_2} &= -R\cos\beta\cos\delta(\sin\psi\sin\varphi + \cos\psi\cos\varphi\cos\theta) + R\cos\beta\sin\delta\cos\psi\sin\theta, \\ Z_{G_2} &= -R\cos\beta\cos\delta\cos\varphi\sin\theta - R\cos\beta\sin\delta\cos\theta. \end{aligned} \tag{16.65}$$

where β is the satellite tooth profile's tip angle.

The velocity vector V_{G_2} is projected onto the OX, OY, and OZ axes as follows:

$$V_{G_2X} = R\omega\cos\beta\cos\delta\left[\left(1-\frac{Z_1}{Z_2}\cos\theta\right)\sin\psi\sin\left(\psi\frac{Z_1}{Z_2}\right)\right.$$
$$\left.+\left(\cos\theta-\frac{Z_1}{Z_2}\right)\cos\psi\cos\left(\psi\frac{Z_1}{Z_2}\right)\right]-R\omega\cos\beta\sin\delta\cos\psi\sin\theta,$$
$$V_{G_2Y} = -R\omega\cos\beta\cos\delta\left[\left(1-\frac{Z_1}{Z_2}\cos\theta\right)\cos\psi\sin\left(\psi\frac{Z_1}{Z_2}\right)\right.$$
$$\left.-\left(\cos\theta+\frac{Z_1}{Z_2}\right)\sin\psi\cos\left(\psi\frac{Z_1}{Z_2}\right)\right]-R\omega\cos\beta\sin\delta\sin\psi\sin\theta, \tag{16.66}$$
$$V_{G_2Z} = R\omega\cos\beta\frac{Z_1}{Z_2}\cos\delta\sin\left(\psi\frac{Z_1}{Z_2}\right)\sin\theta.$$

where Z_1 is the number of teeth on the central wheel; Z_2 is the number of teeth on the satellite; and $\omega=\dot{\psi}$ represents the angular velocity of the crankshaft. With a rotation frequency of n = 3000 rpm, is $\omega = 2\pi n\ s^{-1}$.

We can represent the vector product $\mathbf{V}_{G_2}\times\mathbf{r}_{G_2}$ in Eq. (16.63) as **a**:

$$\mathbf{a}\equiv\mathbf{V}_{G_2}\times\mathbf{r}_{G_2}=a_X\mathbf{i}+a_Y\mathbf{j}+a_Z\mathbf{k}, \tag{16.67}$$

in which

$$\begin{aligned} a_X &= V_{G_2Y}Z_{G_2}-Y_{G_2}V_{G_2Z},\\ a_Y &= V_{G_2Z}X_{G_2}-Z_{G_2}V_{G_2X},\\ a_Z &= V_{G_2X}Y_{G_2}-X_{G_2}V_{G_2Y}. \end{aligned} \tag{16.68}$$

Using the notation in (16.68), we derive the vector **r**:

$$\mathbf{r}=R\sin\beta\frac{\mathbf{a}}{|\mathbf{a}|},\ \text{where}\ |\mathbf{a}|=\sqrt{a_X^2+a_Y^2+a_Z^2}. \tag{16.69}$$

The time derivative of the vector $\dot{\mathbf{r}}$ was determined using the following relation:

$$\dot{\mathbf{r}}=R\sin\beta\frac{\dot{\mathbf{a}}|\mathbf{a}|-\mathbf{a}\dfrac{\mathbf{a}\cdot\dot{\mathbf{a}}}{|\mathbf{a}|}}{|\mathbf{a}|^2}=\frac{R\sin\beta\left[\dot{\mathbf{a}}|\mathbf{a}|^2-\mathbf{a}(\mathbf{a}\cdot\dot{\mathbf{a}})\right]}{|\mathbf{a}|^3}, \tag{16.70}$$

where $\dot{\mathbf{a}}=\left(\dot{a}_X,\dot{a}_Y,\dot{a}_Z\right)$ was introduced:

$$\begin{aligned}\dot{a}_X &= a_{G_2Y}Z_{G_2} - Y_{G_2}a_{G_2Z},\\ \dot{a}_Y &= a_{G_2Z}X_{G_2} - Z_{G_2}a_{G_2X},\\ \dot{a}_Z &= a_{G_2X}Y_{G_2} - X_{G_2}a_{G_2Y}.\end{aligned} \tag{16.71}$$

The acceleration of point G_2 is projected onto the OX, OY, and OZ axes using the following expressions:

$$\begin{aligned}a_{G_2X} &= R\omega^2\cos\beta\cos\delta\left\{\left[\cos\psi\sin\left(\psi\frac{Z_1}{Z_2}\right)+\frac{Z_1}{Z_2}\sin\psi\cos\left(\psi\frac{Z_1}{Z_2}\right)\right]\left(1-\frac{Z_1}{Z_2}\cos\theta\right)\right.\\ &+\left.\left[\sin\psi\cos\left(\psi\frac{Z_1}{Z_2}\right)+\frac{Z_1}{Z_2}\cos\psi\sin\left(\psi\frac{Z_1}{Z_2}\right)\right]\left(\frac{Z_1}{Z_2}-\cos\theta\right)\right\}+R\omega^2\cos\beta\sin\delta\sin\psi\sin\theta,\\ a_{G_2Y} &= -R\omega^2\cos\beta\cos\delta\left\{\left[\frac{Z_1}{Z_2}\cos\psi\cos\left(\psi\frac{Z_1}{Z_2}\right)-\sin\psi\sin\left(\psi\frac{Z_1}{Z_2}\right)\right]\left(1-\frac{Z_1}{Z_2}\cos\theta\right)\right.\\ &+\left.\left[\cos\psi\cos\left(\psi\frac{Z_1}{Z_2}\right)-\frac{Z_1}{Z_2}\sin\psi\sin\left(\psi\frac{Z_1}{Z_2}\right)\right]\left(\frac{Z_1}{Z_2}-\cos\theta\right)\right\}+R\omega^2\cos\beta\sin\delta\cos\psi\sin\theta,\\ a_{G_2Z} &= R\omega^2\cos\beta\left(\frac{Z_1}{Z_2}\right)^2\cos\delta\cos\left(\psi\frac{Z_1}{Z_2}\right)\sin\theta.\end{aligned} \tag{16.72}$$

The linear velocity of the contact point on the conjugated teeth, which is a component of the central wheel tooth profile E_1, can be found by differentiating equation (16.63) in relation to time. The result is:

$$\begin{aligned}\dot{\mathbf{r}}_{E_1} &= \mathbf{V}_{E_1} = \dot{\mathbf{r}}_{G_2} + \mathbf{r} = \dot{X}_{G_2}\mathbf{i}+\dot{Y}_{G_2}\mathbf{j}+\dot{Z}_{G_2}\mathbf{k}\\ &+\frac{R\sin\beta}{\left(\sqrt{a_X^2+a_Y^2+a_Z^2}\right)^3}\left[\left(a_X^2+a_Y^2+a_Z^2\right)\left(\dot{a}_X\mathbf{i}+\dot{a}_Y\mathbf{j}+\dot{a}_Z\mathbf{k}\right)\right.\\ &\left.-\left(a_X\dot{a}_X+a_Y\dot{a}_Y+a_Z\dot{a}_Z\right)\left(a_X\mathbf{i}+a_Y\mathbf{j}+a_Z\mathbf{k}\right)\right],\end{aligned} \tag{16.73}$$

The following formulae provide the projections of the vector V_{E_1} on the OX, OY, and OZ axes:

$$\begin{aligned}V_{E_1X} &= V_{G_2X}+\frac{R\sin\beta}{\left(\sqrt{a_X^2+a_Y^2+a_Z^2}\right)^3}\left[\left(a_X^2+a_Y^2+a_Z^2\right)\dot{a}_X-\left(a_X\dot{a}_X+a_Y\dot{a}_Y+a_Z\dot{a}_Z\right)a_X\right],\\ V_{E_1Y} &= V_{G_2Y}+\frac{R\sin\beta}{\left(\sqrt{a_X^2+a_Y^2+a_Z^2}\right)^3}\left[\left(a_X^2+a_Y^2+a_Z^2\right)\dot{a}_Y-\left(a_X\dot{a}_X+a_Y\dot{a}_Y+a_Z\dot{a}_Z\right)a_Y\right],\\ V_{E_1Z} &= V_{G_2Z}+\frac{R\sin\beta}{\left(\sqrt{a_X^2+a_Y^2+a_Z^2}\right)^3}\left[\left(a_X^2+a_Y^2+a_Z^2\right)\dot{a}_Z-\left(a_X\dot{a}_X+a_Y\dot{a}_Y+a_Z\dot{a}_Z\right)a_Z\right].\end{aligned} \tag{16.74}$$

The contact point E_1's velocity magnitude is computed using

$$V_{E_1} = \sqrt{V_{E_1X}^2 + V_{E_1Y}^2 + V_{E_1Z}^2}. \tag{16.75}$$

16.7 The Contact Point E_2's Linear Velocity on the Tooth Profile of the Satellite Wheel

It is possible to use analytical equations to express the spherical motion of the satellite wheel, which is securely fixed to the mobile coordinate system $OX_1Y_1Z_1$, in respect to the stationary system OXYZ.

The velocity vector for the contact point E_2, which is connected to the satellite wheel's tooth profile, is positioned in accordance with the following condition, as shown in the vector diagram in Fig. 16.19:

$$\mathbf{r}_{E_1} = \mathbf{OE}_1 = \mathbf{OE}_2 = \mathbf{r}_{E_2}, \tag{16.76}$$

in which

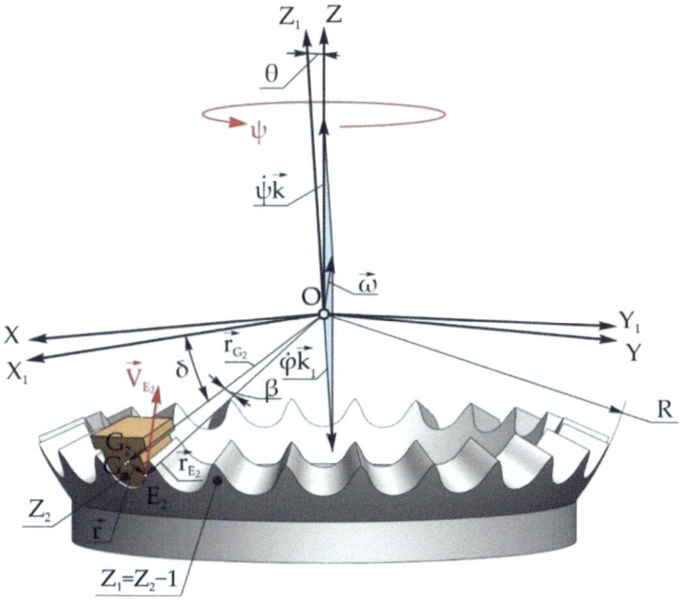

Fig. 16.19 The velocity vector position V_{E_2} at the contact point on the satellite wheel tooth profile

$$\mathbf{OE}_1 = X_{E_1}\mathbf{i} + Y_{E_1}\mathbf{j} + Z_{E_1}\mathbf{k}, \mathbf{OE}_2 = X_{E_2}\mathbf{i}_1 + Y_{E_2}\mathbf{j}_1 + Z_{E_2}\mathbf{k}_1, \quad (16.77)$$

We used the unit vectors $\mathbf{i}_1$, $\mathbf{j}_1$, and $\mathbf{k}_1$ to represent the satellite wheel's location in order to determine the coordinates of the contact point E_2. These vectors depend on Euler's angles ψ, θ, and φ in addition to the unit vectors **i**, **j**, and **k**.

$$\begin{aligned} X_{1E_2} &= \left(X_{E_1}\mathbf{i} + Y_{E_1}\mathbf{j} + Z_{E_1}\mathbf{k}\right)\cdot\mathbf{i}_1, \\ Y_{1E_2} &= \left(X_{E_1}\mathbf{i} + Y_{E_1}\mathbf{j} + Z_{E_1}\mathbf{k}\right)\cdot\mathbf{j}_1, \\ Z_{1E_2} &= \left(X_{E_1}\mathbf{i} + Y_{E_1}\mathbf{j} + Z_{E_1}\mathbf{k}\right)\cdot\mathbf{k}_1. \end{aligned} \quad (16.78)$$

The following formulas were used to get the coordinates of the contact point E_2 on the satellite wheel tooth profile in the mobile coordinate system $OX_1Y_1Z_1$:

$$\begin{aligned} X_{1E_2} &= X_{E_1}\left[\cos\psi\cos\left(\psi\frac{Z_1}{Z_2}\right) + \sin\psi\sin\left(\psi\frac{Z_1}{Z_2}\right)\cos\theta\right] \\ &+ Y_{E_1}\left(\sin\psi\cos\left(\psi\frac{Z_1}{Z_2}\right) - \cos\psi\sin\left(\psi\frac{Z_1}{Z_2}\right)\cos\theta\right) - Z_{E_1}\sin\left(\psi\frac{Z_1}{Z_2}\right)\sin\theta, \\ Y_{1E_2} &= X_{E_1}\left[\cos\psi\sin\left(\psi\frac{Z_1}{Z_2}\right) - \sin\psi\cos\left(\psi\frac{Z_1}{Z_2}\right)\cos\theta\right] \\ &+ Y_{E_1}\left[\sin\psi\sin\left(\psi\frac{Z_1}{Z_2}\right) + \cos\psi\cos\left(\psi\frac{Z_1}{Z_2}\right)\cos\theta\right] + Z_{E_1}\cos\left(\psi\frac{Z_1}{Z_2}\right)\sin\theta, \\ Z_{1E_2} &= X_{E_1}\sin\psi\sin\theta - Y_{E_1}\cos\psi\sin\theta + Z_{E_1}\cos\theta. \end{aligned} \quad (16.79)$$

By differentiating the coordinates X_{1E_2}, Y_{1E_2}, and Z_{1E_2} with respect to time—which were influenced by the coordinates X_{E_1}, Y_{E_1}, Z_{E_1}, and Euler's angles—the projections of the velocity vector at point E_2 on the satellite wheel's tooth profile were obtained.

Consequently, we obtain the velocity projection formulas of point E_2 along the coordinate axes OX_1, OY_1, and OZ_1.

$$
\begin{aligned}
V_{E_2X_1} &= V_{E_1X}\left[\cos\psi\cos\left(\psi\frac{Z_1}{Z_2}\right)+\sin\psi\sin\left(\psi\frac{Z_1}{Z_2}\right)\cos\theta\right]\\
&+V_{E_1Y}\left(\sin\psi\cos\left(\psi\frac{Z_1}{Z_2}\right)-\cos\psi\sin\left(\psi\frac{Z_1}{Z_2}\right)\cos\theta\right)-V_{E_1Z}\sin\left(\psi\frac{Z_1}{Z_2}\right)\sin\theta\\
&+\omega\left\{X_{E_1}\left[-\left(1-\frac{Z_1}{Z_2}\cos\theta\right)\sin\psi\cos\left(\psi\frac{Z_1}{Z_2}\right)+\left(-\frac{Z_1}{Z_2}+\cos\theta\right)\cos\psi\sin\left(\psi\frac{Z_1}{Z_2}\right)\right]\right.\\
&+Y_{E_1}\left[\left(1-\frac{Z_1}{Z_2}\cos\theta\right)\cos\psi\cos\left(\psi\frac{Z_1}{Z_2}\right)+\left(-\frac{Z_1}{Z_2}+\cos\theta\right)\sin\psi\sin\left(\psi\frac{Z_1}{Z_2}\right)\right]\\
&\left.-Z_{E_1}\frac{Z_1}{Z_2}\cos\left(\psi\frac{Z_1}{Z_2}\right)\sin\theta\right\},\\
V_{E_2Y_1} &= V_{E_1X}\left(\cos\psi\sin\left(\psi\frac{Z_1}{Z_2}\right)-\sin\psi\cos\left(\psi\frac{Z_1}{Z_2}\right)\cos\theta\right)\\
&+V_{E_1Y}\left(\sin\psi\sin\left(\psi\frac{Z_1}{Z_2}\right)+\cos\psi\cos\left(\psi\frac{Z_1}{Z_2}\right)\cos\theta\right)+V_{E_1Z}\cos\left(\psi\frac{Z_1}{Z_2}\right)\sin\theta\\
&+\omega\left\{X_{E_1}\left[-\left(1-\frac{Z_1}{Z_2}\cos\theta\right)\sin\psi\sin\left(\psi\frac{Z_1}{Z_2}\right)-\left(-\frac{Z_1}{Z_2}+\cos\theta\right)\cos\psi\cos\left(\psi\frac{Z_1}{Z_2}\right)\right]\right.\\
&+Y_{E_1}\left[\left(1-\frac{Z_1}{Z_2}\cos\theta\right)\cos\psi\sin\left(\psi\frac{Z_1}{Z_2}\right)-\left(-\frac{Z_1}{Z_2}+\cos\theta\right)\sin\psi\cos\left(\psi\frac{Z_1}{Z_2}\right)\right]\\
&\left.-Z_{E_1}\frac{Z_1}{Z_2}\sin\left(\psi\frac{Z_1}{Z_2}\right)\sin\theta\right\},\\
V_{E_2Z_1} &= V_{E_1X}\sin\psi\sin\theta-V_{E_1Y}\cos\psi\sin\theta+V_{E_1Z}\cos\theta\\
&+\omega\left(X_{E_1}\cos\psi\sin\theta+Y_{E_1}\sin\psi\sin\theta\right).
\end{aligned}
\tag{16.80}
$$

We calculated the velocity projections at contact point E_2 using relation (16.80), and then used the following equation to find the velocity magnitude at the contact point E_2 on the satellite tooth profile:

$$
V_{E_2}=\sqrt{V_{E_2X_1}^2+V_{E_2Y_1}^2+V_{E_2Z_1}^2},
\tag{16.81}
$$

As a function of the crankshaft's positioning angle ψ, the difference in linear velocities of the contact points on the profiles of the central wheel tooth V_{E_1} and the satellite wheel tooth V_{E_2} is used to calculate the relative sliding between the tooth flanks at contact point E.

$$
V_{sl(\psi)}=V_{E_1(\psi)}-V_{E_2(\psi)}.
\tag{16.82}
$$

16.8 Relative Sliding Between Teeth with Modified Flanks

Based on the angular coordinate of the contact point ψ_{k_i} for each pair of teeth that are engaged concurrently, the relative sliding with friction between the flanks of the engaged teeth in the precessional gearing A^{D}_{CX-CV} varies throughout the course of a precessional cycle. This coordinate is then obtained from relation (16.82) and is affected by the crankshaft's precession angle ψ.

The change in the relative sliding velocity V_{sl} as a function of the precession angle ψ is shown in Fig. 16.20a for the precessional gearing A^{D}_{CX-CV} with a conical axoid angle $\delta = 22.5°$ and a tooth correlation of $Z_1 = Z_2 - 1$. The correlation $Z_1 = Z_2 + 1$ with $\delta = 0°$ is shown in Fig. 16.20b, whereas the correlation $Z_1 = Z_2 + 1$ with $\delta = 22.5°$ is shown in Fig. 16.20c.

Importantly, for pairs of teeth that are engaged concurrently, the reference frontal multiplicity ε_a can reach 100% in the toothed precessional gearing A^{D}_{CX-CV}.

The gearing must have a certain number of concurrently engaged pairs of teeth for a given crankshaft position identified by the precession angle ψ, separated by the angular pitch $\psi = 360Z_2 / Z_1^2$.

The contact points where the profiles of the central wheels' teeth meet the top profiles of the satellite wheel's teeth exhibit the least amount of relative sliding, according to the analysis of the variations in relative velocities between the flanks of the engaged teeth displayed in Fig. 16.20a–c.

For these reasons, it is recommended to change the teeth by lowering their height by trimming the tips of the center wheel teeth in order to decrease energy losses in the toothed gearings A^{D}_{CX-CV} [33, 34]. The teeth's resistance to contact pressure, σ_H, can be calculated to determine how many pairs of teeth can be kept in engagement by shortening them in this way.

On the active profiles of the teeth, for example, four pairs of teeth with an angular pitch $\psi_{k_i} = 360Z_2 / Z_1^2$ engage at a crankshaft precession angle of $\psi = 62.4°$ in the gearing with the correlation $Z_1 = Z_2 - 1$ and $\delta = 22.5°$. At a crankshaft precession angle of $\psi = 55.2°$, four pairs of teeth engage in gearings with $Z_1 = Z_2 + 1$ and $\delta = 0°$ or $\delta = 22.5°$.

Specifically, 4041 steel central wheels paired with PEEK plastic satellite wheels and vice versa, with PEEK plastic central wheels and 4041 steel satellite wheels, were used to design and build precessional transmissions in order to investigate the energy losses in gearings brought on by the relative sliding between the tooth surfaces V_{sl}.

Figure 16.21a shows how the precession angle ψ for the gearing with $Z_1 = Z_2 - 1$ and $\delta_{(1-2)} = 22.5°$ (where $Z_1 = 24$, $Z_2 = 25$, $\theta = 3.5°$, $r = 6.27$ mm, $R = 75$ mm) affects the velocity magnitudes V_{E_1} and V_{E_2} at the contact points E_1 and E_2 on the tooth profiles of the satellite wheel and the central wheel, respectively. Additionally, the sliding velocity V_{sl} with friction at point E is also shown.

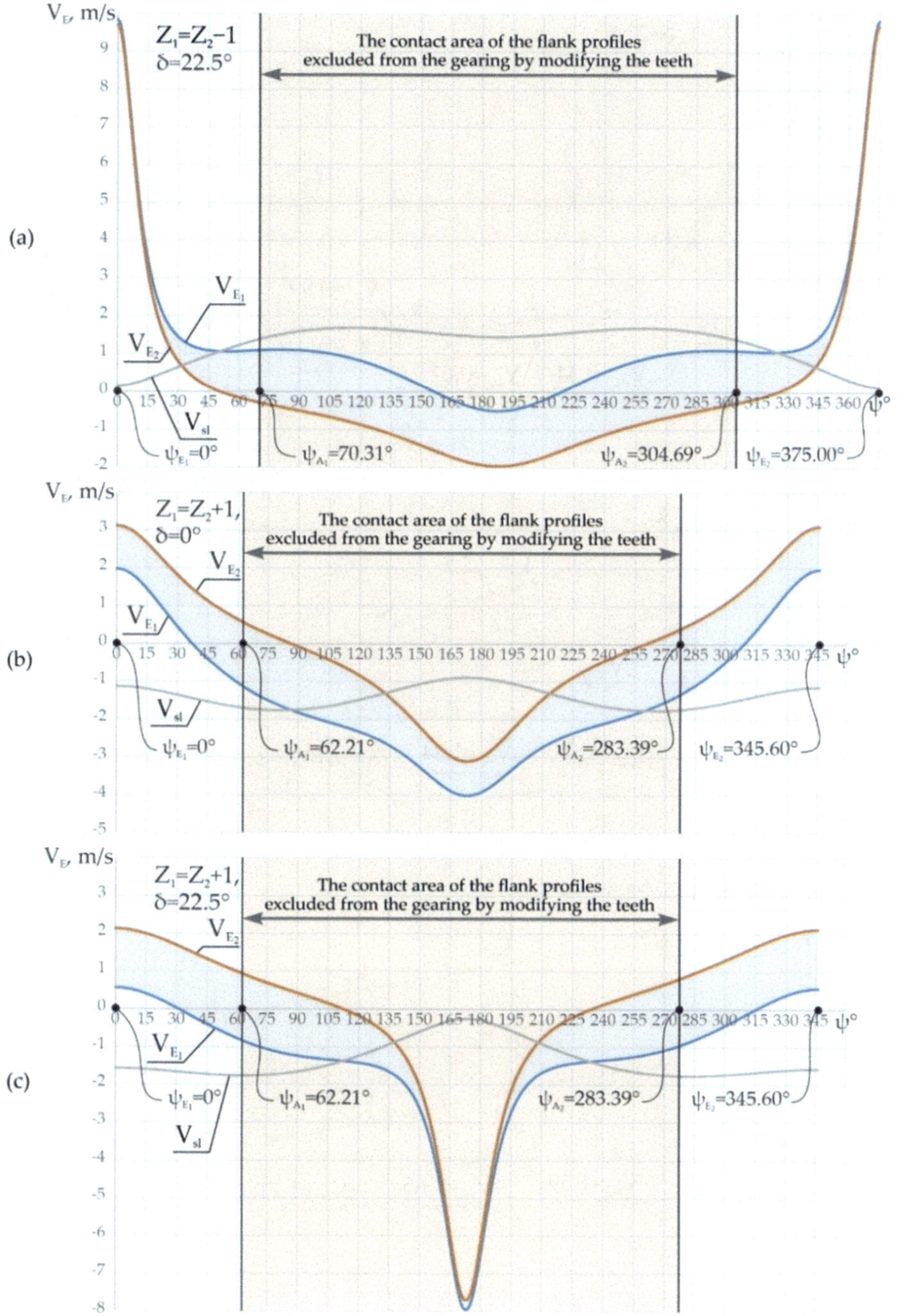

Fig. 16.20 The linear velocities at contact points V_{E_1}, V_{E_2}, and the relative sliding velocity V_{sl} between the teeth in the precessional gearing A^{D}_{CX-CV} analyzed under the following conditions: for $Z_1 = Z_2 - 1$, with $Z_1 = 24$ and $\delta = 22.5°$ (**a**); for $Z_1 = Z_2 + 1$, with $Z_1 = 25$ and $\delta = 0°$ (**b**); and for $Z_1 = Z_2 + 1$, with $Z_1 = 25$ and $\delta = 22.5°$ (**c**)

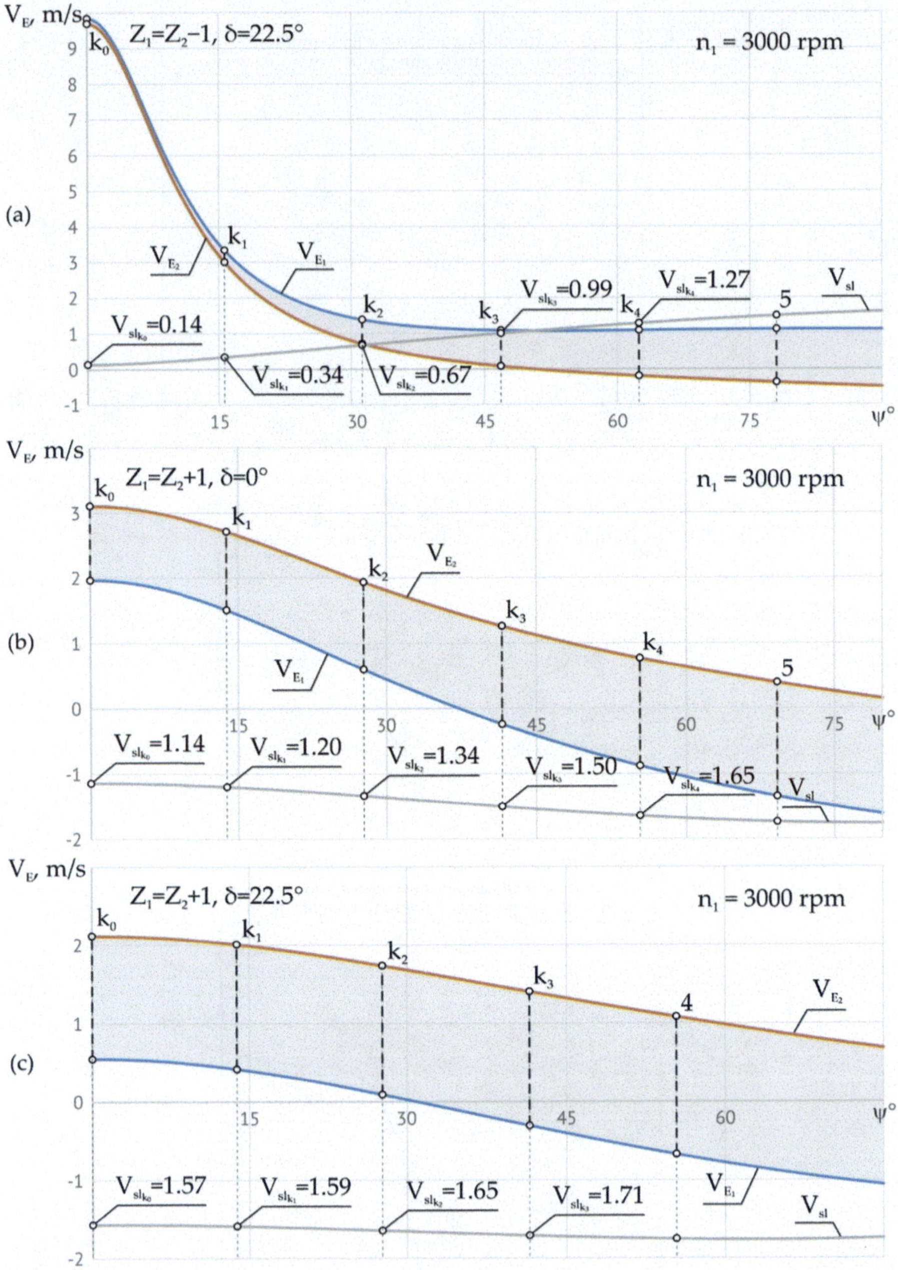

Fig. 16.21 The linear velocities V_{E_1}, V_{E_2}, and the sliding velocity V_{sl} in the contact kinematics vary depending on ψ under the following conditions: for $Z_1 = Z_2 - 1$, with $Z_1 = 24$, and $\delta = 22.5°$ (**a**); $Z_1 = Z_2 + 1$, with $Z_1 = 25$, and $\delta = 0°$ (**b**); and $Z_1 = Z_2 + 1$, with $Z_1 = 25$, and $\delta = 22.5°$ (**c**)

As shown in Fig. 16.21a, we may calculate the linear velocities V_{E_1} and V_{E_2}, as well as the sliding velocity V_{sl}, when the crankshaft rotates at a speed of $n_1 = 3000$ rpm. While the load-bearing capacity of the convex-concave contact is dependent on their size and the variation in their radii of curvature, the mechanical efficiency of the gear is a reflection of the energy losses resulting from sliding friction between the engaged flanks.

The impact of several parametric configurations $[Z_g - \theta, \pm 1]$ on the geometry and kinematics of the contact was therefore examined. The only differences between these configurations were the conical axoid angle $\delta \geq 0°$ and the equation $Z_1 = Z_2 \pm 1$.

According to the analysis of the contact kinematics of the gear teeth in the gear systems [35] depicted in Fig. 16.21a–c, the relative sliding velocity V_{sl} at contact points k_0, k_1, k_2, and k_3 is comparatively low for the parametric configuration $[Z_g - \theta, -1]$, where the number of teeth is $Z_1 = Z_2 - 1$ and the conical axoid angle $\delta = 22.5°$ (see Fig. 16.21a). Accordingly, this arrangement might work well for creating precessional gears for use in reducers. On the other hand, the configuration $[Z_g - \theta, +1]$ with a conical axoid angle $\delta > 0°$ (see Fig. 16.21b) is advised for use in multiplier-functioning transmission systems, especially with a conical axoid angle of $\delta = 0°$ in reducers, and for small transmission ratios between −7.3 and −29.3 for $\delta_{(1-2)} = 0°$ with $Z_1 = Z_2 + 1$, and between +8.3 and +30.3 for $\delta_{(3-4)} = 0°$ with $Z_3 = Z_4 + 1$.

16.9 Variation in Relative Sliding Between the Conjugated Teeth's Flanks

Any other gearing $(Z_3 - Z_4)$ with which it conjugates has constant values for the relative sliding between the teeth in gearing $(Z_1 - Z_2)$, and vice versa for the same crank-shaft speed. Accordingly, only the configuration $[Z_g - \theta, \pm 1]$ of the corresponding gear determines the geometry of the contact and the kinematics of the contact point of the conjugated teeth in the gear $(Z_1 - Z_2)$ or $(Z_3 - Z_4)$.

Figure 16.22 shows how the configuration parameters $[Z_g - \theta, \pm 1]$ affect the relative sliding speed obtained from precessional transmissions with transmission ratios of i = −124, i = −164, i = +15, and i = −14. Additionally, for the more complex comparison study, the relative sliding in the gears A^{D}_{CX-CV} with the ratios i = −360, i = −1080, and i = −2115 are included.

The analysis of Fig. 16.22 shows that the relative sliding V_{sl} depends on the relationship between the geared wheels' tooth count and $Z_{1(4)} = Z_{2(3)} - 1$ or $Z_{1(4)} = Z_{2(3)} + 1$. The curves V_{sl_2}, V_{sl_3}, and $V_{sl_5} \ldots V_{sl_{14}}$ give the relative sliding velocities for the correlation of the number of teeth $Z_{1(4)} = Z_{2(3)} - 1$. The curves V_{sl_2}, V_{sl_3}, and $V_{sl_5} \ldots V_{sl_{14}}$ representing the respective sliding velocities for the correlation of the number of teeth $Z_{1(4)} = Z_{2(3)} - 1$ extend from the minimum in the pair of teeth conjugated in

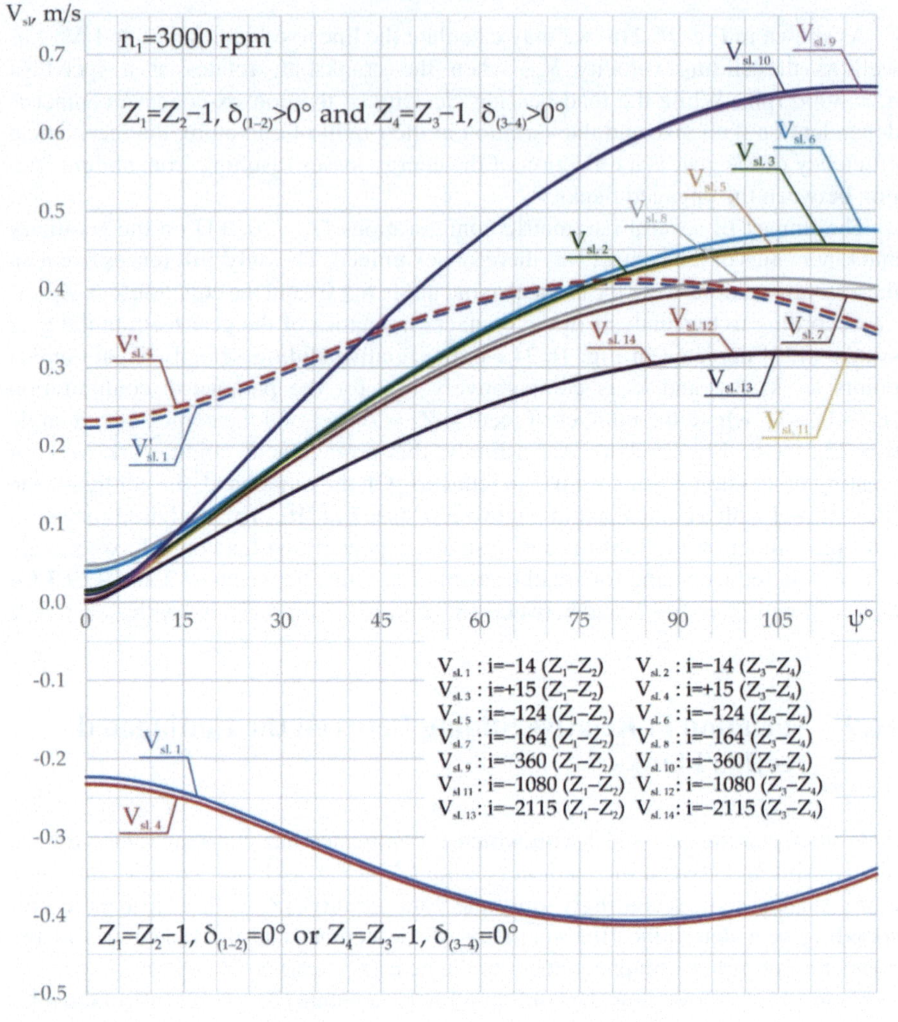

Fig. 16.22 The impact of configuration parameters $[Z_g - \theta, \pm 1]$ on the relative sliding velocity V_{sl} between the teeth's conjugated flanks

contact k_0 (see Fig. 16.22) to the maximum in the pair of teeth conjugated in contact k_4.

The curves V_{sl_1} and V_{sl_4} display the relative sliding velocities for the correlation $Z_{1(4)} = Z_{2(3)} + 1$. They demonstrate that the relative sliding is greater in pairs of teeth with angular coordinates up to $\psi_k = 40°$ than in the case of the correlation $Z_{1(4)} = Z_{2(3)} - 1$.

The sliding velocity modules V_{sl_1} and V_{sl_4} vary according to ψ_k, as shown by the dotted curves. This facilitates the comparison between the sliding velocities in the gears A^{D}_{CX-CV} with the correlation $Z_{1(4)} = Z_{2(3)} - 1$ and the sliding velocities for the correlation $Z_{1(4)} = Z_{2(3)} + 1$.

The relative sliding speeds at the contact points of the profiles in precessional transmissions are lower than those of involute, cycloidal, and Wildhaber-Novicov gears.

16.10 Precession Kinematic Transmissions with Different Gearings Characteristics and Materials Tribocouples

The central wheels, both movable and stationary, along with the teeth of the satellite wheel, were constructed from *4140* carbon steel and 30% *Glass-Filled PEEK (Poly Ether Ether Ketone)*. Depending on the load transmitted and the operational conditions, the toothed wheels can be configured in various transmissions, allowing the paired teeth to create tribological couples of *plastic-plastic*, *steel-plastic*, and *steel-steel* (see Table 16.1). These central wheels and satellite wheels were produced using CNC machinery in collaboration with Gleason Corporation in Rochester, NY, USA, based on technical documentation created by the authors at TUM.

In Fig. 16.23a–l samples of *2K-H* kinematic precessional transmissions with convex-concave gears, with straight teeth from couples of materials with different tribological behavior are shown. The wheels were manufactured by machining steel *4140* and solid *Glass-Filled PEEK* plastic on CNC machine tools. The configuration parameters $[Z_g - \theta, \pm 1]$ of the manufactured transmissions and their functional characteristics are included in Table 16.1.

The samples presented in the Fig. 16.23 differ from each other by the values of the angles of the conical axoids $\delta_{1(4)\text{-}2(3)}$ the correlation of the number of teeth $Z_{1(4)} = Z_{2(3)} \pm 1$ and the couple of the conjugated wheel materials, thus representing distinctive features: (a), (b), (c)—metal-metal, $\delta_{1(4)\text{-}2(3)} = 15°$ and $Z_{1(4)} = Z_{2(3)} - 1$; (d)—metal-plastic, $\delta_{1(4)\text{-}2(3)} = 22.5°$, $Z_{1(4)} = Z_{2(3)} - 1$; (e), (f)—metal-plastic, $\delta_{(1\text{-}2)} = 22.5°$, $\delta_{(3\text{-}4)} = 0°$, $Z_1 = Z_2 - 1$, $Z_3 = Z_4 + 1$; (g), (h), (i)—metal-plastic, $\delta_{1(4)\text{-}2(3)} = 15°$, $Z_{1(4)} = Z_{2(3)} - 1$; (j), (k), (l)—metal-plastic, $\delta_{1(4)\text{-}2(3)} = 22.5°$, $Z_{1(4)} = Z_{2(3)} - 1$.

Figure 16.24 shows the 2K-H precessional transmission with the transmission ratio $i = -164$, achieved in gears $A^{D}_{CX\text{-}CV}$ with $Z_1 = 40$; $Z_2 = 41$, $\delta_{(1\text{-}2)} = 15°$, $\beta_2 = 3.2°$, $Z_3 = 33$, $Z_4 = 32$, $\delta_{(3\text{-}4)} = 15°$, $\beta_3 = 3.5°$, $\theta = 3.5°$ and $D_e = 42$ mm.

In the Fig. 16.24a the general view of the transmission is shown, in (b), (c)—the profilograms describe by $\zeta = \xi(f)$ of the conjugated teeth in the gears $(Z_1 - Z_2)$ and $(Z_3 - Z_4)$, respectively, in (d)—the toothed wheels in axial unfolding in various pairs of materials, steel *4140* and plastic *Glass-Filled PEEK*.

Figure 16.25a shows the general view of the *2K-H* precessional transmission, with the transmission ratio $i = +15$. The A^D gearing consists of gears $A^{D}_{CX\text{-}CV}$ with configuration parameters $[Z_g - \theta, \pm 1]$: $Z_1 = 28$, $Z_2 = 29$, $\delta_{(1\text{-}2)} = 22.5°$, $\beta_3 = 3.7°$, $Z_3 = 29$, $Z_4 = 30$, $\delta_{(3\text{-}4)} = 0°$, $\beta_3 = 3.5°$, $\theta = 3.5°$ sand outer diameter $D_e = 42$ *mm*. From the analysis of the profilograms of the teeth presented in the Fig. 16.25b, c it follows

Table 16.1 Parametric characteristics of precessional transmissions with A^{D}_{CX-CV} gearings fabricated on CNC machine tools

Nr.	Transmission ratio, *i*	Nutation angle, θ, deg.	Engagement, A^{D}_{CX-CV}	Number of central wheel teeth, $Z_{1(4)}$	Number of satellite wheel teeth, $Z_{2(3)}$	Angle of the conical axoid, $\delta_{1(4)-2(3)}$, deg.	Taper angle of the teeth of the satellite, $\beta_{2(3)}$, deg.	Radius of curvature of the flank profile of the satellite tooth in average section, $r_{2(3)}$, mm	Average radius of the central wheels, $R_{m1(4)}$, mm
1	−14	3.5	$(Z_1 - Z_2)$	30	29	0	3.5	1.13	18.5
			$(Z_4 - Z_3)$	28	29	22.5	3.7	1.36	21
2	+15	3.5	$(Z_1 - Z_2)$	28	29	22.5	3.7	1.36	21
			$(Z_4 - Z_3)$	30	29	0	3.5	1.13	18.5
3	−45	3.5	$(Z_1 - Z_2)$	44	45	15	2.4	0.88	21
			$(Z_4 - Z_3)$	22	23	27.5	4.7	1.72	21
4	−124	3.5	$(Z_1 - Z_2)$	30	31	22.5	3.5	1.28	21
			$(Z_4 - Z_3)$	24	25	22.5	4.78	1.75	21
5	−164	3.5	$(Z_1 - Z_2)$	40	41	15	3.2	1.12	20
			$(Z_4 - Z_3)$	32	33	15	3.5	1.22	20
6	−360	5	$(Z_1 - Z_2)$	19	20	22.5	5.6	2.05	21
			$(Z_4 - Z_3)$	18	19	22.5	5.9	2.16	21
7	−1080	2.5	$(Z_1 - Z_2)$	47	48	22.5	2.3	0.84	21
			$(Z_4 - Z_3)$	45	46	22.5	2.4	0.88	21
8	−2115	2.5	$(Z_1 - Z_2)$	46	47	22.5	2.35	0.86	21
			$(Z_4 - Z_3)$	45	46	22.5	2.4	0.88	21

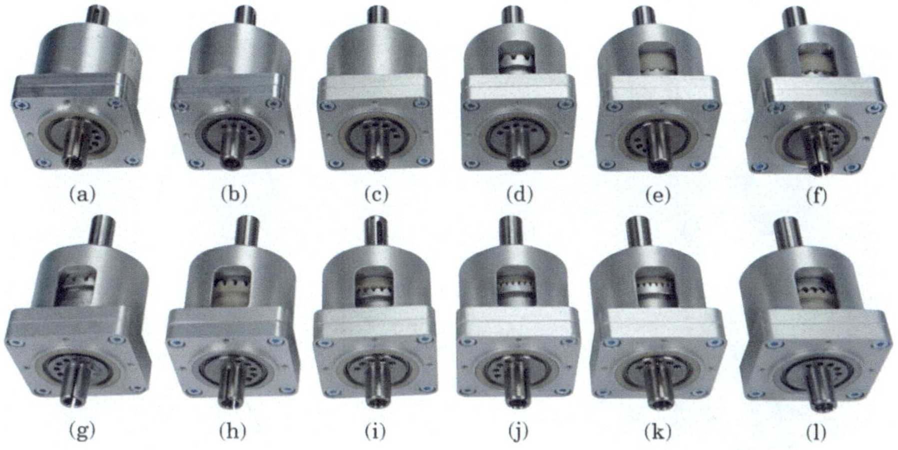

Fig. 16.23 Samples of 2K–H kinematic PT with A^{D}_{CX-CV} convex-concave gears with wheels made of steel *4140* and *Glass-Filled PEEK* plastic with different angles of the conical axoids $\delta_{1(4)\text{-}2(3)}$, correlations of the number of teeth $Z_{1(4)} = Z_{2(3)} \pm 1$ and various pairs of materials of conjugated gears: **(a–c)**—metal-metal, $\delta_{1(4)\text{-}2(3)} = 15°$ and $Z_{1(4)} = Z_{2(3)} - 1$; **(d)**—metal-plastic, $\delta_{1(4)\text{-}2(3)} = 22.5°$, $Z_{1(4)} = Z_{2(3)} - 1$; **(e, f)**—metal-plastic, $\delta_{(1\text{-}2)} = 22.5°$, $\delta_{(3\text{-}4)} = 0°$, $Z_1 = Z_2 - 1$, $Z_3 = Z_4 + 1$; **(g–i)**—metal-plastic, $\delta_{1(4)\text{-}2(3)}$,$=15°$, $Z_{1(4)} = Z_{2(3)} - 1$; **(j–l)**—metal-plastic, $\delta_{1(4)\text{-}2(3)} = 22.5°$, $Z_{1(4)} = Z_{2(3)} - 1$ (samples of manufactured transmissions)

that the correlation of the numbers of teeth $Z_1 = Z_2 - 1$ or $Z_4 = Z_3 + 1$ essentially influences the shape of the profiles of the teeth of the central wheels.

Figure 16.25d shows the central wheels made of plastic Glass-Filled PEEK, and the toothing of the satellite wheel—made of carbon steel 4140.

It should be mentioned that due to the correlation of the numbers of teeth of the mobile central wheel and the satellite wheel $Z_4 = Z_3 + 1$, with the angle of the conical axoid $\delta_{(3\text{-}4)} = 0°$, it is ensured the expansion of the range of transmission ratios to small values, and the direction of rotation of the driving and driven shafts coincides.

The 2K-H precessional transmission with gears A^{D}_{CX-CV} and a transmission ratio of $i = -45$ is shown in general view in Fig. 16.25h. The configuration parameters are $[Z_g - \theta, -1]$: $Z_1 = 44$, $Z_2 = 45$, $\delta_{(1\text{-}2)} = 15°$, $\beta_2 = 2.4°$, $Z_3 = 23$, $Z_4 = 22$, $\delta_{(3\text{-}4)} = 22.5°$, $\beta_3 = 4.7°$, $\theta = 3.5°$, and outer diameter $D_e = 42$ mm.

The profileograms of the teeth $(Z_1 - Z_2)$ and $(Z_3 - Z_4)$ are displayed in Fig. 16.25f, g, respectively, in (e). The satellite wheel and the central wheels are unfolded and made using CNC machine tools using materials from the following couples: steel-plastic, steel-steel, and plastic-plastic. The driving and driven shafts revolve in the opposite direction because the teeth numbers of the conjugated wheels, $Z_{1(4)} = Z_{2(3)} - 1$ and $Z_2 > Z_3$, correlate.

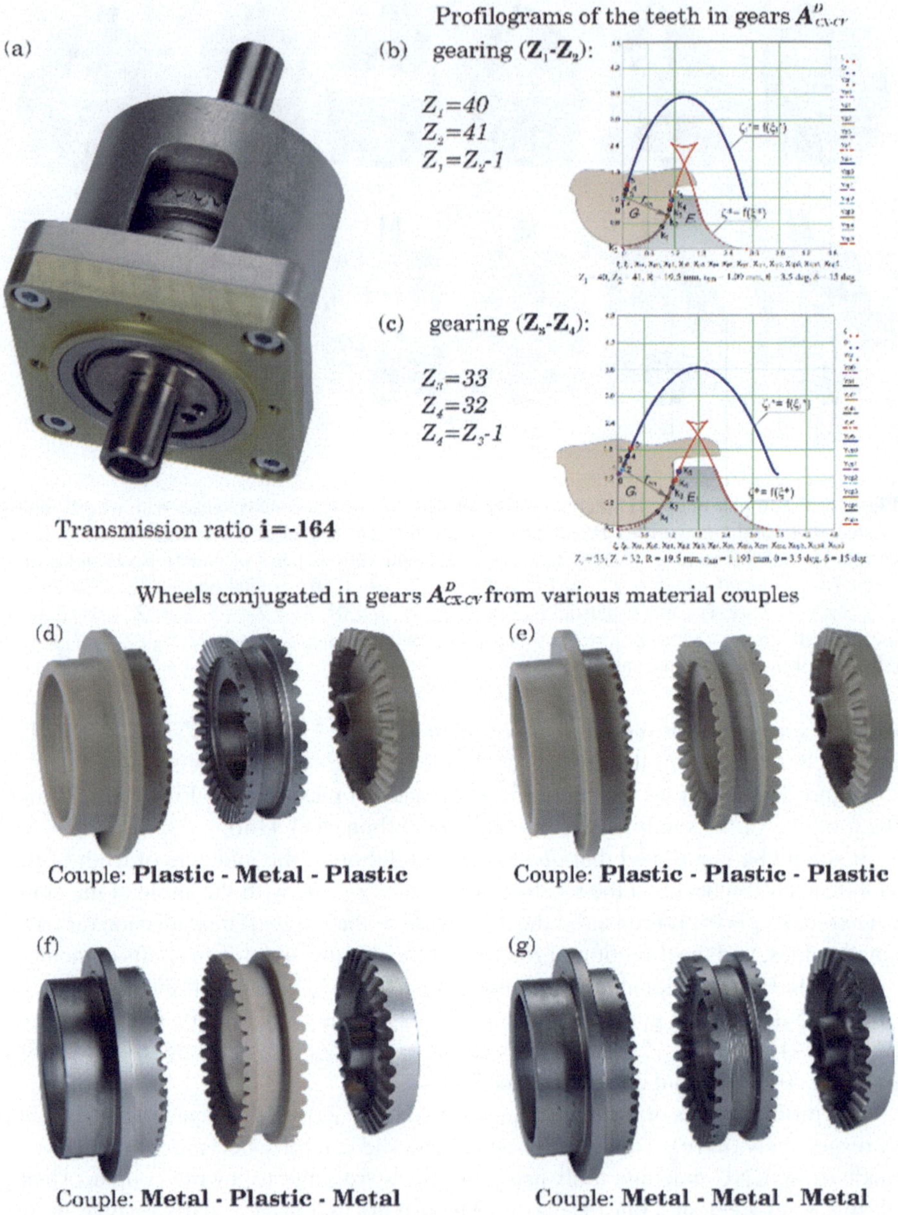

Fig. 16.24 2K–H PT, transmission ratio i = −164 with gears A^{D}_{CX-CV} ($Z_1 = 40$, $Z_2 = 41$, $Z_3 = 33$, $Z_4 = 32$, $\delta_{(1-2)} = \delta_{(3-4)} = 15°$, $\beta_2 = 3.2°$, $\beta_3 = 3.5°$, $\theta = 3.5°$ and $D_e = 42$ mm): (**a**)—general view; (**b**, **c**)—profilograms of teeth ($Z_1 - Z_2$) and ($Z_3 - Z_4$), respectively; (**d**–**g**)—toothed wheels in axial unfolding of paired materials, steel *4140* and plastic *Glass-Filled PEEK*, manufactured on CNC machine tools (physical component display—manufactured)

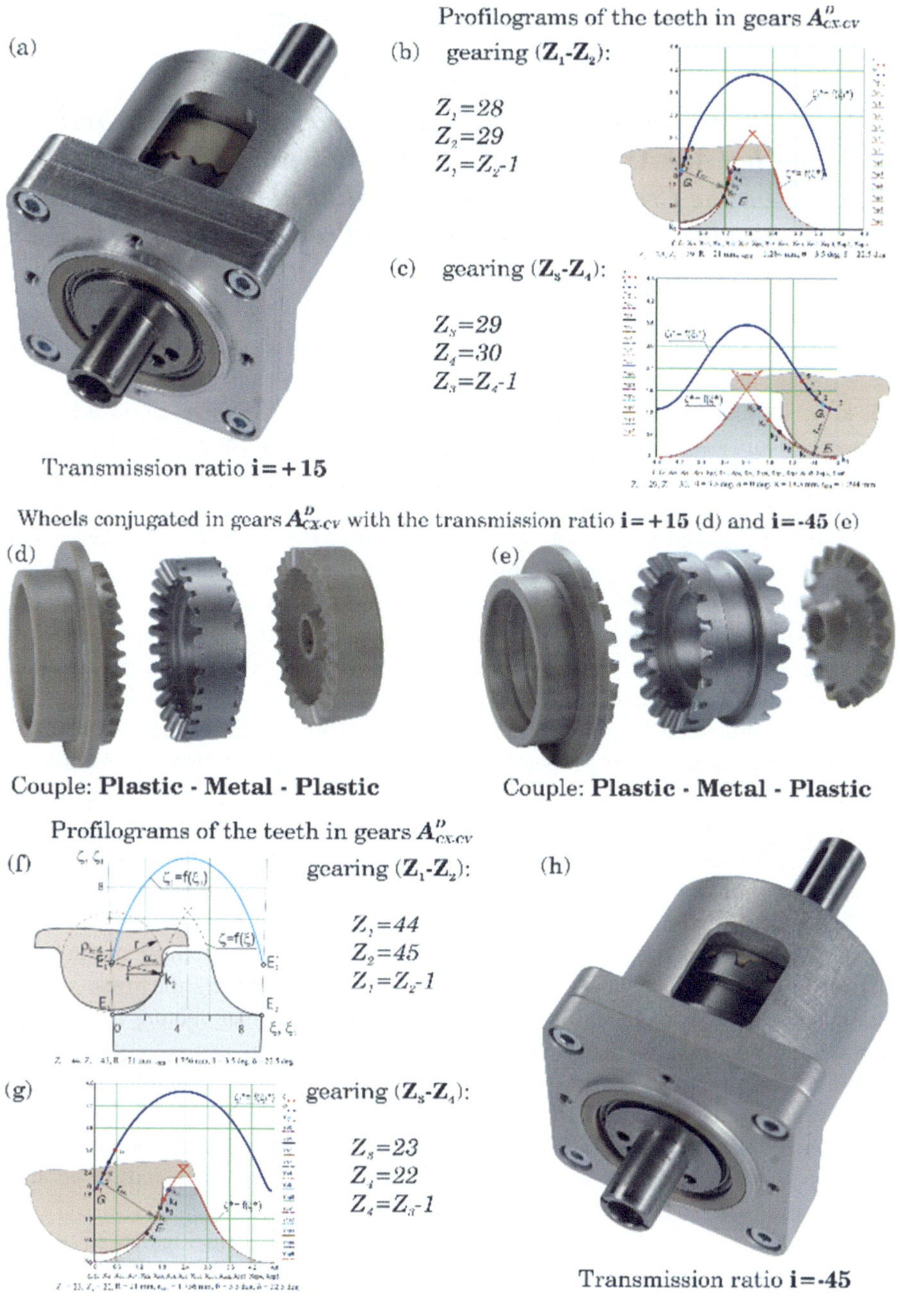

Fig. 16.25 2K–H kinematic precessional transmissions, with the transmission ratio i = +15 (**a**), achieved in the gears A^{D}_{CX-CV} (**b**) and (**c**), and with the transmission ratio i = −45 (**h**), achieved in the gears A^{D}_{CX-CV} (**f**) and (**g**), with toothed wheels (**d**) and (**e**) manufactured on CNC machine tools (physical components exposure—manufactured)

16.11 Low-Power Precessional Transmissions: Constructive and Functional Aspects

16.11.1 Precessional Transmissions Featuring Satellite Wheels that Are Axially Mounted and Able to Float on the Crankshaft, with Axes Aligned at the Precession Center (Wheels Produced Using CNC Machining)

Figure 16.26 illustrates the 2K–H precessional transmission featuring a ratio of $i = -164$ in an axial view, showcasing the conjugated wheels in gearing A^{D}_{CX-CV}. The setup includes straight teeth engaged in multiple convex-concave contact pairs, with specifications: $Z_1 = 40$, $Z_2 = 41$, $Z_3 = 33$, $Z_4 = 32$, $\theta = 3.5°$, $\delta_{(1-2)} = 15°$, $\beta_{(2)} = 3.2°$, $\delta_{(3-4)} = 15°$, and $\beta_{(3)} = 3.5°$.

Satellite wheel 2 is axially mounted and can float within radial-axial bearings 6 and 7 on the inclined section of crankshaft 5, with axes aligned at the precession center and at a nutation angle θ relative to the shared axis of radial bearings 8 and 9. The immobile central wheel 3 is secured within cover 11 and is axially centered with housing 1. The central mobile wheel 4 is attached to the flange of driven shaft 10, while satellite wheel 2, along with crankshaft 5, is also axially floating in radial bearings 8 and 9.

It is recommended for power transmissions that need high transmission ratios and where the forces in the gears are significant to use a precessional transmission with a satellite wheel that is axially floating on the crankshaft, with axes aligned at the precession center and positioned at a nuclear angle θ. The central wheel tooth profiles should have low pressure angles between the conjugated flanks, specifically $\alpha_w < 20°$, to reduce the stresses on the transmission shafts, especially the bearings.

The axially floating precessional node, pressure angles between the flanks of $\alpha_w < 20°$, and the spherical contact of satellite wheel 2's teeth with those of the laterally positioned central wheels 3 and 4 all contribute to reducing the stress on the kinematic couplings. As a result, the precision needed to manufacture the precessional node's component pieces is likewise decreased.

With a transmission ratio of $i = -164$, Fig. 16.26b shows examples of 2K–H precessional transmissions with gears A^{D}_{CX-CV} that have straight teeth engaged in numerous convex-concave contacts. With axes aligned at the precession center at a nutation angle θ, satellite wheel 2 is axially floating on crankshaft 5 in this configuration.

The satellite wheel (d) and the center wheels (c) and (e) in Fig. 16.26 are all CNC-machined from 4140 steel at Gleason Corporation in the USA.

The driving and driven shafts rotate in opposing directions when the gearing A^{D}_{CX-CV} with $Z_1 = 40$, $Z_2 = 41$, $Z_3 = 33$, and $Z_4 = 32$ produces a transmission ratio of $i = -164$. In contrast, the transmission ratio becomes $i = +165$, enabling both shafts to rotate in the same direction, if the center wheel configuration is changed to $Z_1 = 32$, $Z_2 = 33$, $Z_3 = 41$, and $Z_4 = 40$.

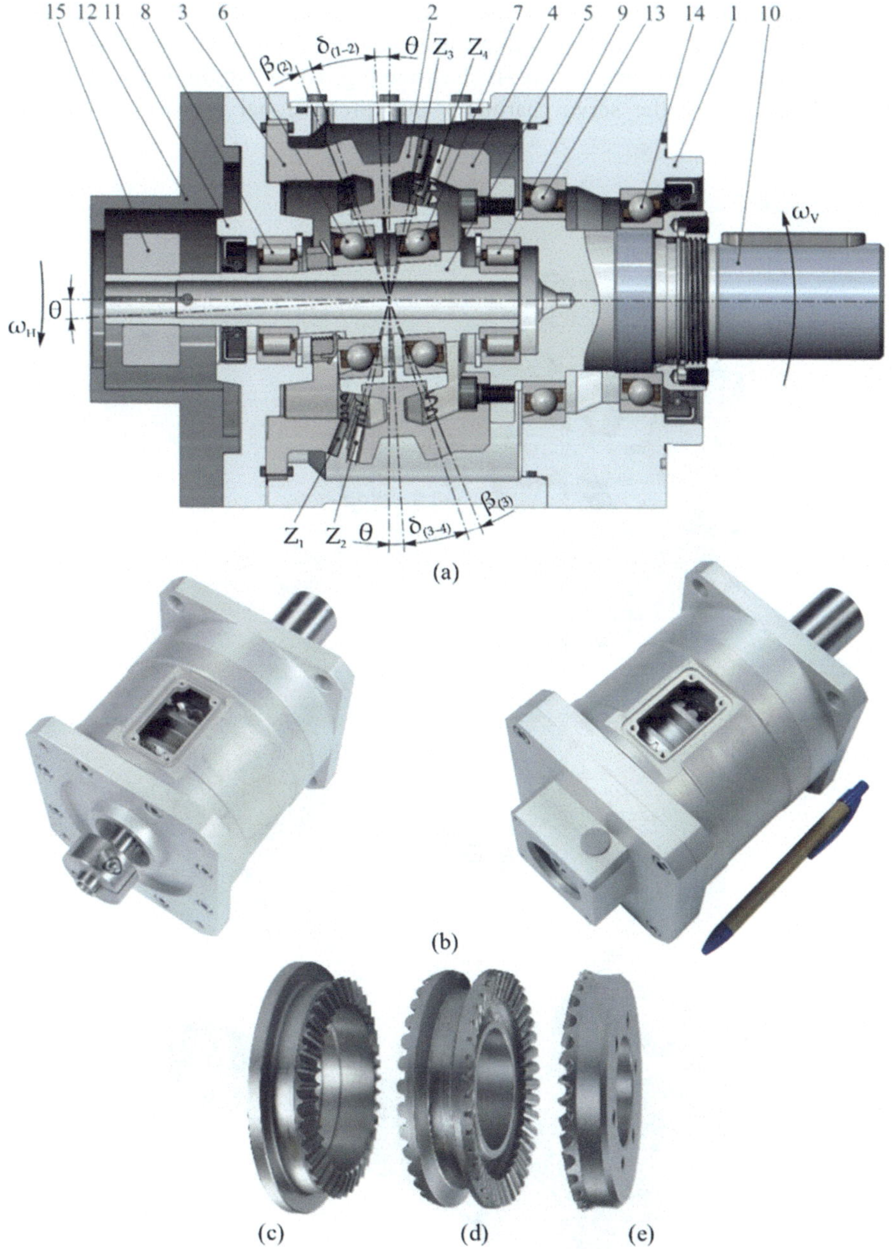

Fig. 16.26 The 2K–H low-power precessional transmission (i = −164) features gears A^{D}_{CX-CV} with straight teeth and a satellite wheel that is axially floating on the crankshaft, with axes aligned at the precession center at a nutation angle θ. The presentation includes a general drawing (**a**), examples of the transmission (**b**), central wheels (**c**, **e**), and satellite wheels (**d**)

A three-dimensional view of the precessional transmission with straight-toothed gearing A^{D}_{CX-CV} and a satellite wheel axially floating on the crankshaft and oriented at the precession center (a) is shown in Fig. 16.27. Additionally, an experimental sample of the servo-gearmotor and precessional transmission (b) is included.

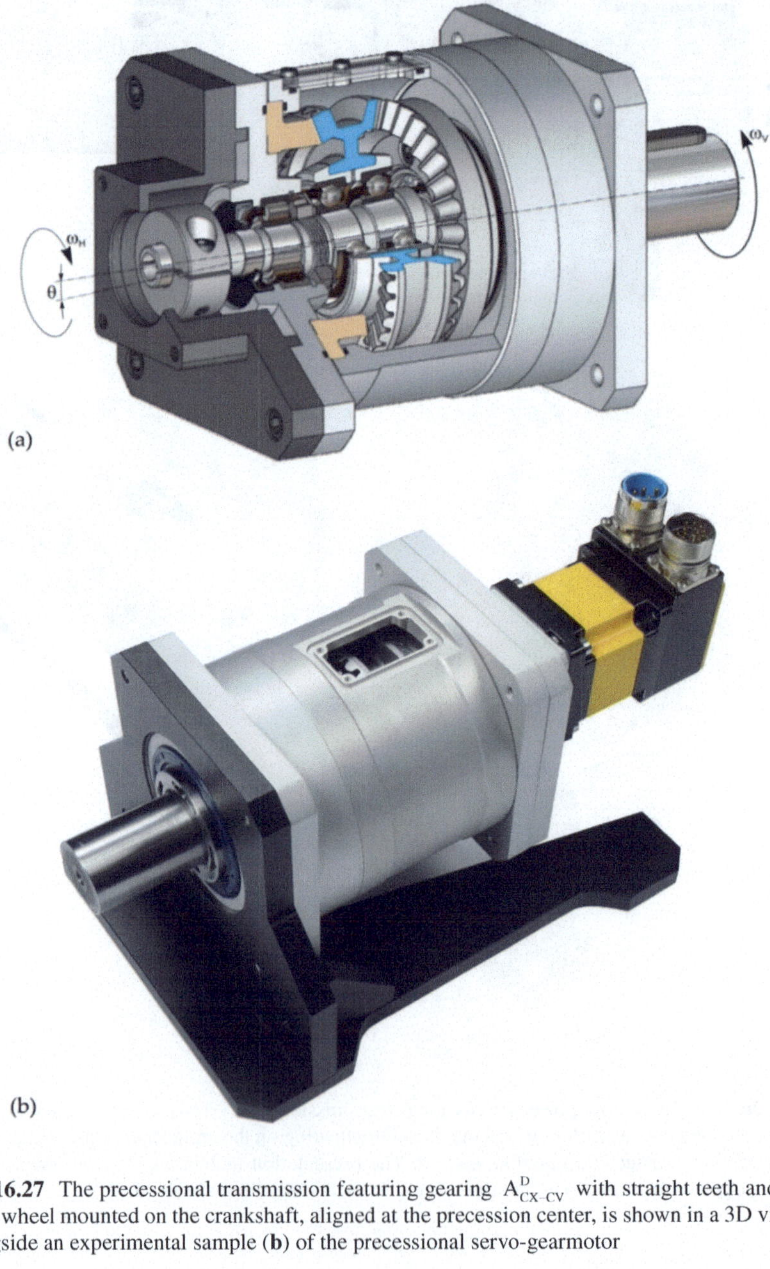

Fig. 16.27 The precessional transmission featuring gearing A^{D}_{CX-CV} with straight teeth and a satellite wheel mounted on the crankshaft, aligned at the precession center, is shown in a 3D view (**a**) alongside an experimental sample (**b**) of the precessional servo-gearmotor

16.11.2 Precessional Transmissions Featuring Circular Arc Teeth, with the Satellite Wheel Axially Floating on the Crankshaft with Concurrent Axes in the Precession Center

The 2K–H precessional transmission illustrates in Fig. 16.28 has a transmission ratio of i = −164 achieved through gearing $A^{D,\beta}_{CX-CV}$ that features circular arc teeth. The parameters include: $\theta = 3.5°$, $\beta_m = 35°$, $Z_1 = 40$, $Z_2 = 41$, $\delta_{(1-2)} = 15°$, $\beta_{(2)} = 3.2°$,

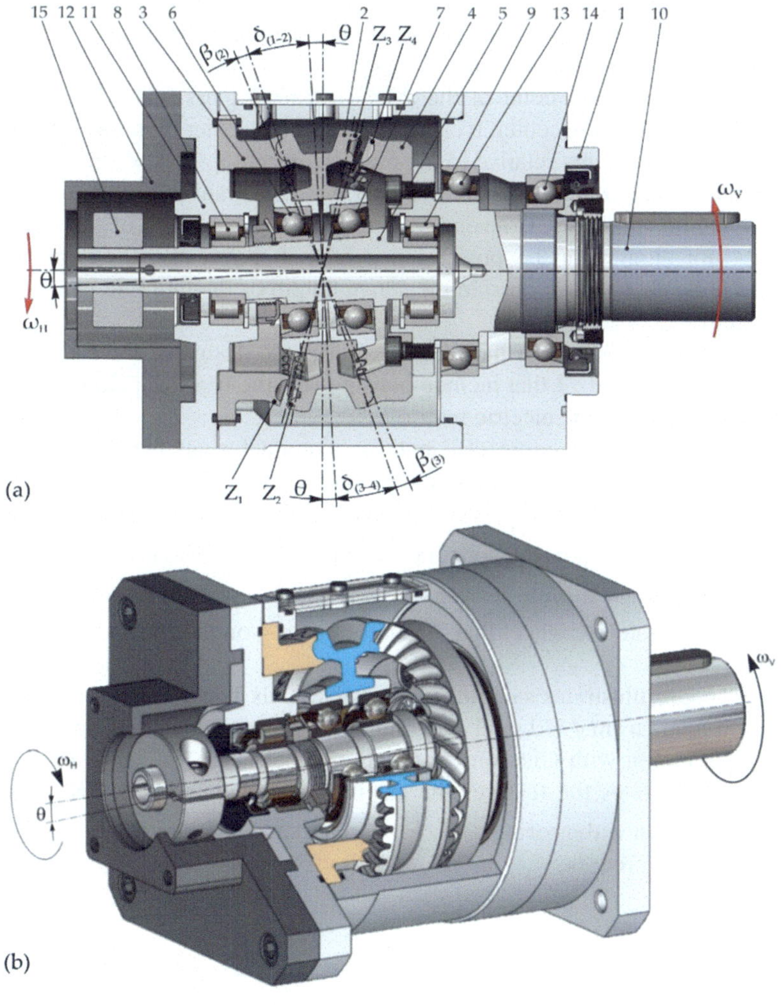

Fig. 16.28 The 2K–H precessional transmission (i = −164) features gearing $A^{D,\beta}_{CX-CV}$ with circular arc teeth and a satellite wheel that is axially floating on the crankshaft, aligned at the precession center at a nutation angle θ. This is illustrated in a general drawing (**a**) and a 3D overview (**b**)

$Z_3 = 33$, $Z_4 = 32$, $\delta_{(3-4)} = 15°$, $\beta_{(3)} = 3.5°$, $R_m = 40$ mm, and $R_c = 40$ mm.

The driving shaft 5, equipped with a crank function, features a section that is inclined at a nutation angle θ relative to the axis of the axial bearings 8 and 9. A satellite wheel 2 is positioned on this inclined section of driving shaft 5, supported by radial-axial bearings 6 and 7. The axes of the bearing pairs 8 and 9, as well as 6 and 7, converge at the precession center O.

As the crankshaft 5 rotates, the satellite wheel 2, which is mounted in the radial-axial bearings 6 and 7, undergoes spherospatial motion around the precession center O at the nutation angle θ. It also moves axially between the stationary central wheel 3 and the movable central wheel 4. The axial positioning rates of both the stationary central wheel 3 and the movable central wheel 4 are limited by deviation fields, ensuring that the intersection point of the axes of bearing pairs 6 and 7 and 8 and 9 coincides with the precession center.

When the intersection point of the axes of bearing pairs 6 and 7 and 8 and 9 aligns with the precession center, it causes a shift in the center of mass of the precessional node's components relative to the precession center. This results in dynamic imbalance and additional stress on the bearings and crankshafts, as outlined by Euler's dynamic equations. Conversely, if the intersection point of the crankshaft axes does not align with the precession center, it may cause the contact point between the flanks to move outside the active length of the multiple pairs of interlocking teeth.

Another notable feature of the transmission is the design of the casing, which comprises bodies 1 and 11 that include housings for the bearings, along with cover 12 for the assembly of the electric motor.

This casing design is more efficient for manufacturing, facilitating modular assembly and disassembly of components, along with the ability to adjust clearance or constraints in gearing $A^{D,\beta}_{CX-CV}$ in accordance with the design specifications.

Figure 16.29a illustrates the constructed sample of the 2K–H (i = −164) precessional transmission featuring gearings $A^{D,\beta}_{CX-CV}$ with circular arc tooth profiles. It also illustrates a satellite wheel that is axially floating on the crankshaft, with concurrent axes.

Figure 16.29b illustrates samples of central wheels 1 and 3, as well as satellite wheel 2, all made from *4140* steel using numerically controlled multi-axis machining in collaboration with Gleason Corporation, Rochester, NY, USA. Meanwhile, Fig. 16.29c illustrates the features the original components of the 2K–H precessional transmission with gearings $A^{D,\beta}_{CX-CV}$ and a transmission ratio of i = −164 during the unfolding process.

(a)

(b)

Toothed wheels of the precessional gearing:
1–immobile central wheel; 2–satellite–wheel; 3–mobile central wheel

(c)

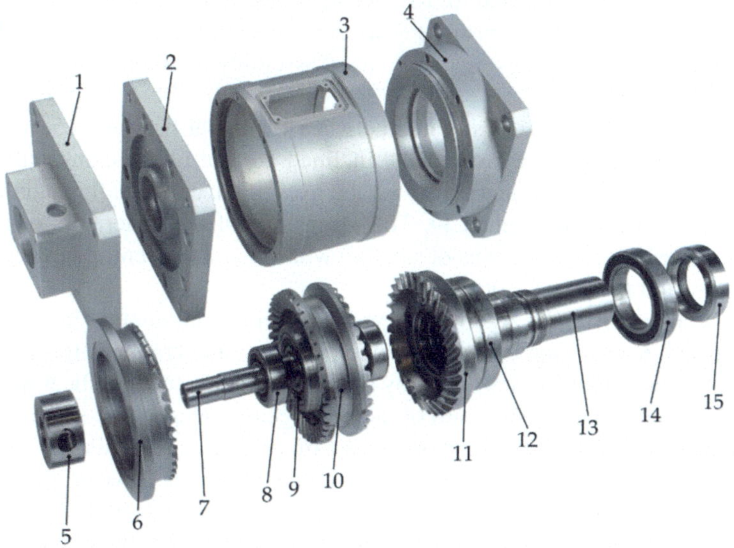

Precessional transmission components in the process of unfolding: 1-intermediate cover;
2, 4–covers; 3–body; 5–shaft clamp; 6–immobile central wheel; 7–crankshaft;
8, 9, 12, 14–bearings; 10–satellite-wheel; 11–mobile central wheel; 13–driven shaft; 15–nut.

Fig. 16.29 A 2K–H low-power precessional transmission (i = −164) features gearing $A^{D,\beta}_{CX-CV}$ that includes circular arc teeth. The satellite wheel is axially mounted and floats on the crankshaft, aligning with the precession center (**a**). Additionally, there are samples of the central wheels and satellite wheel (**b**) that have curved teeth, crafted from 4140 steel using CNC machinery. The transmission components are also shown in the unfolding process (**c**)

16.11.3 Precessional Transmissions Featuring Circular Arc Teeth, with the Satellite Wheel Axially Mounted and Floating on the Crankshaft, with Planar Axial Cams Inclined at an Angle Due to Nutation

Figure 16.30 illustrates the precessional transmission with gearings A^{D}_{CX-CV} or $A^{D,\beta}_{CX-CV}$ with satellite wheel 2 driven by inclined side cams at the nutation angle θ through axial bearings 14 and 15.

Precessional transmissions comprises a housing 1 containing satellite wheel 2 with two bevel gear crowns with Z_2, Z_3, a driving shaft with a crank 5, a driven shaft 6, and two central bevel gears 3, 4. One gear, 3, is fixed and rigidly connected to the housing 1, while 4 is mobile and linked to the driven shaft 6.

The teeth of the central wheels engage in multipair convex-concave meshing with the teeth of the satellite wheel 2, which have profiles in the form of circular arcs with a radius *r*. The teeth of the central wheels 3 and 4 have curvilinear profiles with continuously increasing curvature towards their tips [22, 25–27]. The fixed central wheel 3 is mounted at the junction of the housing 1 and the cover 7 with a preload on its flange, being radially centered simultaneously on the inner ribs provided in both the housing 1 and the cover 7. The mobile central wheel 4 is installed on the flange of the driven shaft 6, which is mounted in the bearings 18 in the housing 1 through the auxiliary bushing 8, assembled with the possibility of displacement in the axial direction using the threaded assembly between the cover 7 and the auxiliary bushing 8.

The satellite wheel 2 is positioned between two axial bearings 15, mounted in planes perpendicular to the axis of the inclined portion of the crankshaft 5 at the angle θ, the axis of which intersects with the axes of the crankshaft 5 and the driven shaft 6 at the precession center O. The extensions of the generators of the tooth profiles of the bevel gear crowns of satellite wheel 2 and the central bevel gears 3, 4 also intersect at the precession center O.

The satellite wheel 2 is additionally mounted with axial floating in a radial bearing 16, which is positioned on the inclined portion of the crank shaft 5 at the angle θ. Also, the crankshaft 5 is supported by a coaxially mounted radial bearing 17 at the end of the driven shaft 6. The teeth of the bevel gear crowns of the satellite wheel 2 are designed with a flank profile in the shape of a circular arc, while the teeth of the central bevel gears 3 and 4 are designed with a curvilinear flank profile with continuously increasing curvature from the root to the tip of the teeth [22, 23].

The profile shapes depend on the angles θ, δ, the radius of the circular arc r, the number and the ratio of the numbers of teeth in the gears $(Z_1 - Z_2)$ and $(Z_3 - Z_4)$.

The parameter with the configuration $[\underline{Z}_g - \theta, \pm 1]$ influences the shape of the tooth profile, determines the degree of frontal overlap ε_f, expressed by the number of pairs of teeth simultaneously engaged, including the value of the pressure angle α_w between the conjugated flanks and the sliding speed with friction between the flanks V_{al}.

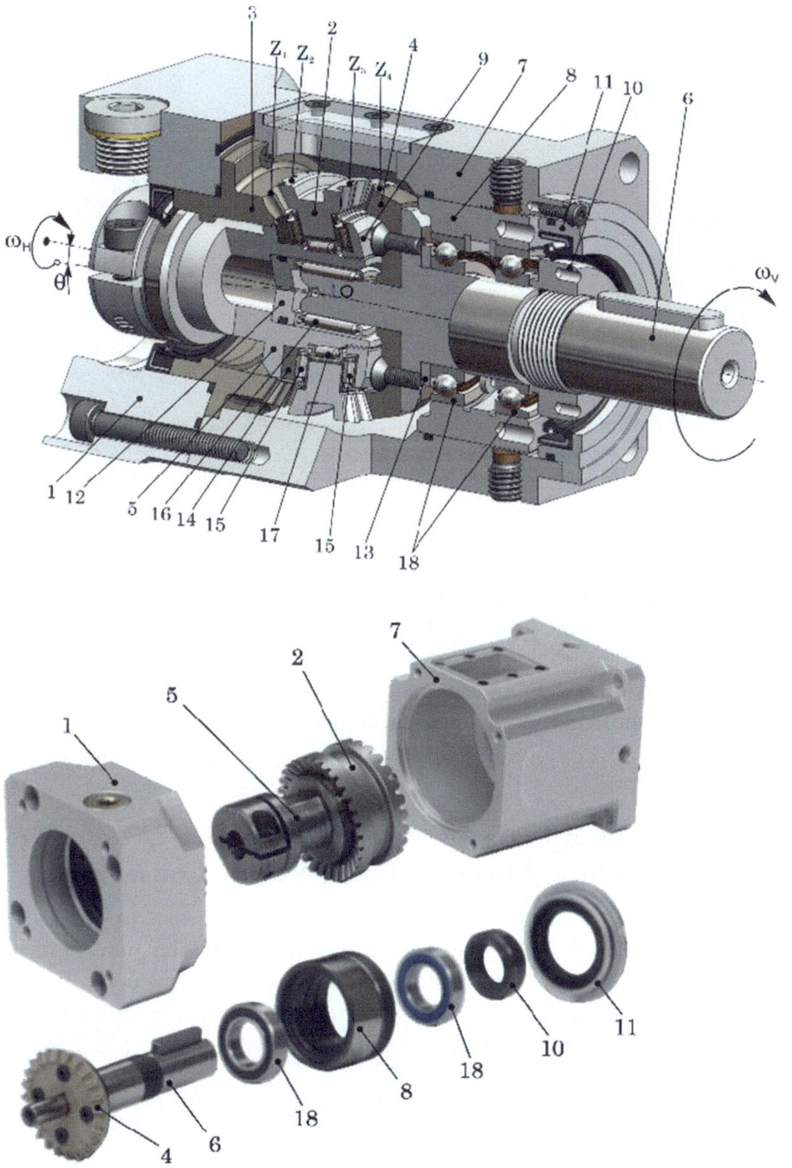

Fig. 16.30 Model of the precessional reducer with the driven satellite-wheel with axial cams

By mounting the satellite wheel 2 between two axial bearings 15 at an angle of nutation θ in relation to the common axis of the driven shaft 6, and the central bevel gears 3 and 4, it ensures the transformation of the rotational motion of the crank-shaft 5 into spherospatial motion of the satellite wheel 2 with axial floating between the central bevel gears 3 and 4. This leads to reduced precision requirements in the

(a) (b) (c)

Fig. 16.31 Physical samples of the precessional transmissions with satellite-wheels driven by cams inclined at the nutation angle θ: (**a**) $i = -13.3$, $Z_1 = 25$, $Z_2 = 24$, $Z_3 = 32$, $Z_4 = 31$, $\theta = 3.5^{\circ}$, $\delta_{(1-2)} = 0^{\circ}$, $\delta_{(3-4)} = 22.5^{\circ}$; (**b**) $i = -124$, $Z_1 = 30$, $Z_2 = 31$, $Z_3 = 25$, $Z_4 = 24$, $\theta = 3.5^{\circ}$, $\delta_{1(3)\text{-}2(4)} = 22.5^{\circ}$; (**c**) $i = +15$, $Z_1 = 28$, $Z_2 = 29$, $Z_3 = 29$, $Z_4 = 30$, $\theta = 3.5^{\circ}$, $\delta_{(1-2)} = 22.5^{\circ}$, $\delta_{(3-4)} = 0^{\circ}$ and experimental models of precessional servomotreducers

execution of the precessional transmission components and a more uniform distribution of the load along the length of the meshed teeth.

Figure 16.30 shows samples of precessional reducers with satellite wheels driven with inclined cams under the nutation angle θ, and in Fig. 16.31 presents the construction of the manufactured precessional transmission.

16.12 Precessional Transmissions with Gears A^{D}_{CX-CV} with Satellite Wheels Coupled Axially Consecutively

Precessional transmissions with consecutively coupled satellites may be of interest in the development of actuation mechanisms for special-purpose equipment with transmission ratios in the order of millions. The functional characteristics of these precessional transmissions are further examined only under the kinematic aspect [22, 23, 26, 27].

These transmissions can be with two satellite wheels coupled axially consecutively with the kinematic structure 4K-2H, which ensure transmission ratios $\pm 55 \leq i \leq \pm 13{\cdot}10^6$ and with three satellite wheels with the kinematic structure 6K-2H, which ensure transmission ratios $\pm 389 \leq i \leq \pm 4.2{\cdot}10^{10}$.

In the Fig. 16.32 is presented: (a) the general view of the sample of the *6K-3H* precessional transmission with gears A^{D}_{CX-CV}, consisting of three consecutive axially coupled satellite wheels the teeth of which kinematically form the gears $(Z_1 - Z_2)$, $(Z_3 - Z_4)$, $(Z_5 - Z_6)$, $(Z_7 - Z_8)$, $(Z_9 - Z_{10})$ and $(Z_{11} - Z_{12})$, with the correlation of the teeth numbers of the conjugated wheels $Z_{1(4)} = Z_{2(3)} - 1$, $Z_{5(8)} = Z_{6(7)} - 1$ and $Z_{9(12)} = Z_{10(11)} - 1$.; (b, c)—parts of the transmission in development; (d)—sample of the experimental precessional gearmotor.

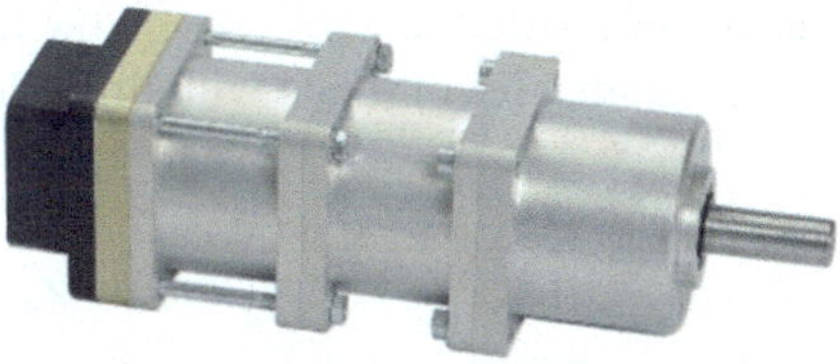

General view of the physical sample

(a)

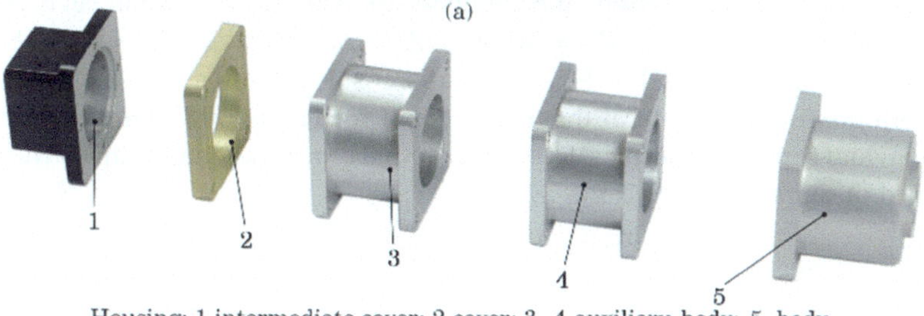

Housing: 1-intermediate cover; 2-cover; 3, 4-auxiliary body; 5- body

(b)

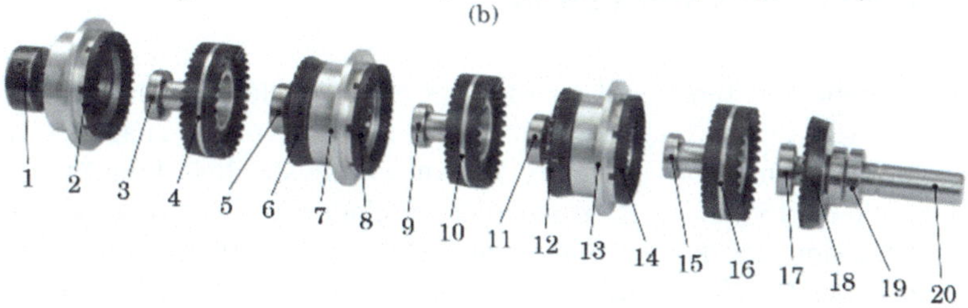

Precessional node with plastic injected wheels: 1-crank; 2-immobile central wheel; 3, 9, 15-bearing; 4, 10, 16-satellite-wheel with semi-axle; 5, 11-spherical bearing; 6, 12-mobile central wheel; 7, 13- immobile central wheel; 8, 14-driven shaft; 17-spherical bearing; 18-mobile central wheel; 19-bearings; 20-driven shaft

(c)

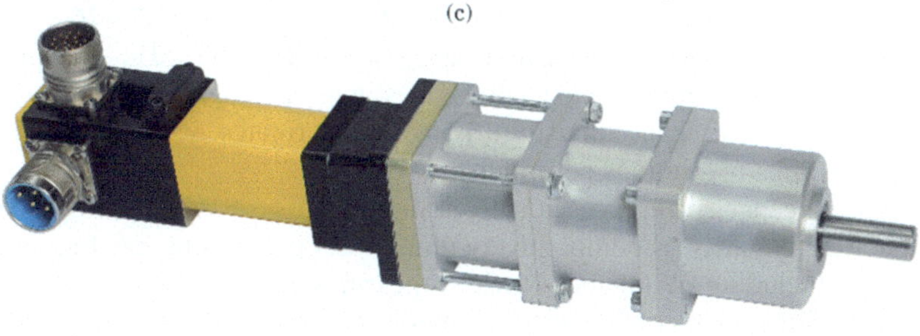

Sample of the experimental precesional servo-gearmotor (manufactured)

(d)

Fig. 16.32 6K-3H type precessional transmission with gears A^{D}_{CX-CV}, transmission ratio $i = -4.41 \cdot 10^6$

As a result of the gearing of the satellite wheel 3 with the immobile 9 and mobile 10 central wheels, the driven shaft 11 will rotate with the reduction

$$i = \frac{Z_2 Z_4 Z_6 Z_8 Z_{10} Z_{12}}{(Z_1 Z_3 - Z_2 Z_4)(Z_5 Z_7 - Z_6 Z_8)(Z_9 Z_{11} - Z_{10} Z_{12})}. \quad (16.83)$$

If, in the 6K-3H transmission, the numbers of teeth of the conjugated wheels $Z_1 = Z_5 = Z_9 = 59$, $Z_2 = Z_6 = Z_{10} = 60$, $Z_3 = Z_7 = Z_{11} = 59$, $Z_4 = Z_8 = Z_{12} = 58$, the transmission ratio $i = -4.21 \times 10^{10}$. This ratio ensures that the driven shaft rotates one revolution in 58.5 years, if the driving shaft rotates $n_1 = 1370\ \text{min}^{-1}$. The construction and operating principle of the 6K-3H precessional transmission are described in section 8.9.2 [22, 26, 27].

16.13 Conclusions and Findings

1. The kinematic and functional possibilities of precessional transmissions with gearings A^{D}_{CX-CV}, $A^{D,\beta}_{CX-CV}$ are considered unique without analogies among mechanical transmissions currently known worldwide. It is found that if the number of teeth $15 \le Z_{1,2,3,4} \le 60$ and their correlation $Z_1 = Z_2 - 1$ and $Z_4 = Z_3 - 1$, the transmission ratio i varies in the range $\pm 20 \le i \le \pm 3600$ (see Fig. 16.5a), for the correlation $Z_1 = Z_2 - 1$ and $Z_4 = Z_3 + 1$ $+8.3 \le i \le +30.3$ (see Fig. 16.5b), and for the correlation $Z_1 = Z_2 + 1$ and $Z_4 = Z_3 - 1$ $-7.3 \le i \le 29.3$ (see Fig. 16.5c). It is also found that in the precessional transmission of the 6K-3H type shown in Fig. 16.32 with three satellite wheels axially coupled in a kinematic chain with the numbers of teeth $Z_1 = Z_5 = Z_9 = 59$, $Z_2 = Z_6 = Z_{10} = 60$, $Z_3 = Z_7 = Z_{11} = 59$ and $Z_4 = Z_8 = Z_{12} = 58$, the transmission ratio $i = -4.21 \times 10^{10}$. This transmission ratio ensures that the driven shaft rotates with a single revolution during 58.5 years, if the continuously driving shaft rotates with the frequency $n_1 = 1370\ \text{min}^{-1}$ revolutions per minute.
2. The bearing capacity of precessional gearings A^{D}_{CX-CV} or $A^{D,\beta}_{CX-CV}$, in the case of parametric similarity, is competitive with the best performing existing involute gearings, including Wildhaber-Novicov gearings due to the convex-concave geometry of the contact of the teeth with minimal difference in flank curves and, at the same time, due to the degree of frontal and longitudinal coverage of the teeth in higher gearing $1.5 \le \varepsilon \le 4$.
3. The mechanical efficiency of precessional gearings A^{D}_{CX-CV} or $A^{D,\beta}_{CX-CV}$ is higher by 5…7% compared to planetary or harmonic transmissions with the same transmission ratio i, due to the mutual rolling of the flank profiles of the teeth according to the nutation angle θ and the reduction of the relative sliding with friction between the flanks of multiple pair conjugated teeth in gearings, which leads to the reduction of the normal forces in the tooth contacts and, respectively, to the

reduction of the frictional forces. At the same time, gearings with gearing angles $\alpha < 20°$ can be created with the reduction of axial forces in the tooth contacts.
4. Considering the advantages stated in p. 1, 2, 3 and respecting the parametric similarity, precessional transmissions with gearings A^{D}_{CX-CV} or $A^{D,\beta}_{CX-CV}$ are characterized by high bearing capacity, extended kinematic possibilities, constructive simplicity, small dimensions, low consumption of material and manufacturing costs.
5. The manufacture of toothed crowns of wheels of conical gearings A^{D}_{CX-CV} and $A^{D,\beta}_{CX-CV}$ is accessible by generation with spatial rolling of the precessional tool, in the shape of a *cone trunk* or a *peripherally profiled disc*, by pressure molding from plastics, by sintering from metal powders, by additive technologies with 3D printing and by generation with rolling in the Gleason system on numerically controlled multi-axis machine tools.

16.14 Recommendations

1. To decrease the relative sliding velocity V_{sl} during contact, the tooth shape is modified by shortening their height through trimming the tips. This adjustment ensures that only a sufficient number of tooth pairs remain in convex-concave contact. The necessary quantity was established using the contact pressure resistance calculation σ_H at points $k_0 - k_3$ or $k_1 - k_3$ (k_4).
2. It is recommended that precessional transmissions working in a reduction mode employ the toothed precessional gear A^D, which is defined by the relationships $Z_1 = Z_2 - 1$ and $Z_4 = Z_3 - 1$, as well as an angle of the conical axoid $\delta > 0°$. This suggestion is predicated on the convex-concave contact's shape, which reduces friction between the mating flanks due to a small difference in radii of curvature and a low relative sliding velocity.
3. Transmissions in multiplier mode are advised to utilize the precessional gear A^D, which is defined by the following characteristics: an angle of the conical axoid $\delta \geq 0°$, a tooth profile angle of the central wheel more than $\alpha_w > 45°$, and the connection $Z_4 = Z_3 + 1$ (or $Z_1 = Z_2 + 1$). This is because of advantageous kinetostatics, which enable the satellite wheel's spherospatial motion to be converted from the central wheel's rotating motion, with a precession cycle frequency of $n_s = Z_4$.
4. The functional characteristics of precessional transmissions with spherospatial motion and a fixed point can be improved in two ways:

   - By utilizing gears that engage multiple pairs of teeth simultaneously;
   - By adopting gearings that feature convex-concave tooth flank contact geometries with minor variances in flank curvatures..

5. The precession angle ψ and the crankshaft's angular speed were used to analyze the link between relative sliding in convex-concave tooth contacts and the rolling of profiles around the contact sites.
6. Various technical solutions were suggested to implement mechanisms that convert the crankshaft's rotary motion into the spherospatial motion of the satellite wheel. This approach facilitates the spherospatial interaction of the teeth and diminishes the rotational movement.
7. In accordance with the fundamental law of engagement, technical solutions were put up for the creation of technologies that provide convex-concave flank profiles, encouraging multiple tooth interactions.

16.15 Applications

Geared precessional transmissions are recommended for the development of drive mechanisms for machines, devices and technological equipment, with general and special purposes for applications in the fields of:

- Machine building;
- Automotive industry;
- Robotics;
- Precision mechanical systems;
- Chemical and atomic industry for transmission of motion and load in tight spaces;
- The cosmic technique of flying and tracking the movement of celestial bodies.

16.16 Patents

The profile of the central gear teeth varies, exhibiting a steadily increasing curvature from the base to the tip. The shape of this variable profile for the teeth on both the stationary and movable center wheels is determined by the configuration parameters [$Z_g - \theta, \pm 1$] is safeguarded by patents in the United States and Germany [34, 35].

16.17 Notations Used in the Paper

PT	Precessional Transmission;
K-H-V	precessional planetary transmission with a central wheel (K), a crankshaft (H) and a satellite link to the driven shaft (V);
2K-H	precessional planetary transmission with two central wheels (2 K) and a crankshaft (H);

4K-2H	precessional planetary transmission with two central wheels (4 K) and a crankshaft (2H);
6K-3H	precessional planetary transmission with two central wheels (6 K) and a crankshaft (3H);
PTB	Precessional Transmission with Bolts;
TPT	Toothed Precessional Transmission;
A^{B}_{CX-CV}	precessional gearing with bolts with convex-concave contact;
A^{D}_{CX-CV}	precessional toothed gearing with convex-concave contact;
$A^{D,\beta}_{CX-CV}$	precessional toothed gearing with convex-concave contact of inclined teeth;
ECFMC POB	Extraction Complex for Ferro-Manganese Concretions from the Planetary Ocean Bottom;
A^{B}	precessional gear with bolts;
A^{D}	precessional gear with straight teeth;
$A^{D,\beta}$	precessional gear with inclined teeth;
A^{B}_{CX-R}	precessional gearing with bolts with convex-rectilinear contact;
A^{D}_{CX-CX}	precessional toothed gearing with convex-convex contact;
$[Z_g - \theta, -1]$	parametric configuration of the precessional gear with the correlation of the numbers of teeth of the conjugated wheels $Z_{1(4)} = Z_{2(3)} - 1$ and reducer operating mode;
$[Z_g - \theta, +1]$	parametric configuration of the precessional gear with the correlation $Z_{1(4)} = Z_{2(3)} + 1$ and multiplier operating mode (of reducer in case $\delta_{1(4)-2(3)} = 0°$);
$G^{con}_{r.s}$	generation of tooth flanks by spatial tumbling-rolling with a cone trunk tool shape;
$G^{cil}_{m.ax}$	generation of straight teeth with cylindrical tool on multi-axis CNC machine tools;
$G^{cil,\beta}_{m.ax}$	generation of inclined teeth with cylindrical tool on multi-axis CNC machine tools;
K_{CX-CV}	convex-concave contact of straight teeth;
K^{β}_{CX-CV}	convex-concave contact of inclined teeth;
V_{E_1}	linear velocity of the contact point E_1 on the teeth profile of the central wheel;
V_{E_2}	linear velocity of the contact point E_2 on the teeth profile of the satellite wheel;
V_{al}	relative sliding in the contacts of the tooth flank profiles;
ω_{θ}	relative rolling in the contacts of the tooth flank profiles.

References

1. Braren LK (1932) Gear transmission. U.S. Patent 1,694,031, 12 July 1932
2. Sensinger IW (2010) Unified approach to cycloid drive profile, stress and efficiency optimization. J Mech Des 132:024503
3. Vorobyov YV (1984) Analytical basis for assessing the durability of work surfaces in rolling with slippage. Mashinovedenie 4:67–76. (In Russian)
4. Dion JL, Pawelski Z, Chianca V, Zdziennicki Z, Peyret N, Uszpolewicz G, Ormezowski J, Mitukiewicz G, Lelasseux X (2020) Theoretical and experimental study for an improved cycloid drive model. J Appl Mech 87:011002. https://doi.org/10.1115/1.4044456
5. Musser CW (1959) Strain wave gearing. U.S. Patent 2,906,143, 29 September 1959
6. Ivanov MN, Shuvalov SA, Finogenov VA (1968) Experimental determination of the number of simultaneously engaging teeth and the values of their deformation in wave transmissions. Mashinostroenie 9:37–40. (In Russian)
7. Radford NA, Strawser P, Hambuchen K, Mehling JS, Verdeyen WK, Donnan AS, Holley J, Sanchez J, Nguyen V, Bridgwater L et al (2015) Valkyrie: NASA's first bipedal humanoid robot. J Field Robot 32:397–419
8. Ginsburg EG (1974) Harmonic gear drives. AD-783578. Foreign Technology Division Wright-Patterson Air Force Base, Dayton
9. Wildhaber E (1926) Helical gearing. U.S. Patent 1,601,750, 5 October 1926
10. Wildhaber E (1966) Gear tooth shape. U.S. Patent 3,251,236, 17 May 1966
11. Dyson A, Evans HP, Snidle RW (1989) Wildhaber-Novicov circular arc gear: some properties of relevance to their design. Proc R Soc Lond A Math Phys Sci 425:341–363
12. Stadtfeld HI (2014) Gleason bevel gear technology—the sciences of gear engineering and modern manufacturing methods for angular transmissions. Gleason Publication, New York
13. Radzevich SP (2018) An examination of high-conformal gearing. Gear Solutions, Pelham, pp 31–39
14. Kedrowski K, Slimak P (1993) Wobbling gear drivetrain for cordless screwdriver honors theses. Western Michigan University, Kalamazoo
15. Lemanski A, Monahan TJ (2004) Non-traction pericyclic CVTs, vol 40. SAE International, Warrendale, p 5
16. Nelson CA, Cipra R (2005) Similarity and equivalence of nutating mechanisms to bevel epicyclic gear trains for modeling and analysis. J Mech Des 127:269–277
17. Jasem MA, Krauinsh PY (2019) Kinematical analysis of the nutation speed reducer. IOP Conf Ser Mater Sci Eng 510:7
18. Jasem MA, Krauinsh PY (2019) Features of the engagement of the nutation speed reducer. J Adv Electr Devices MAT J 4:27–32
19. Gracia PL, Crispel S, Saerens E, Cerstraen T, Lefeber D (2020) Compact gearboxes for modern robotics a review. Front Robot AI 7:103
20. Barbagelata A, Ellero S, Lando R (2016) Planetary gearbox. European Patent No. EP2975296A3, 31 August 2016
21. Barbagelata A, Corsini R (2000) Riduttore Ad Ingranaggi Conici Basculanti. Italian Patent No. IT SV20000049A1, 19 October 2000
22. Bostan IA (1991) Precessional transmissions with multiple pair gear. Stiinta, Chisinau, 356p. (In Russian)
23. Bostan V (2014) Mathematical models in engineering. Bons Offices, Chişinău, 470p, ISBN 978, 9975-80-831-6. (In Romanian)
24. Bostan I, Babaian I (1988) Manufacturing procedure of the modified teeth of precessional gears. Patent S.U. 1663857 A1, 05 January 1988. (In Russian)
25. Bostan I, Mazuru S, Scaticailov S (2003) Precessional gear tooth machining procedure. Patent MD 2120 C2, 31 March 2003
26. Bostan I, Mazuru S, Olevschi A (2008) Precessional gear tooth machining procedure. MD 3532 C2, 31 March 2008

27. Bostan V, Bostan I, Vaculenco M (2021) Precessional gear transmissions. No WO/2021/137682, 08 July 2021
28. Bostan I, Bostan V, Vaculenco M (2022) Deutsches Patent-und Markenamt; Zahnradubertragung Mit Prazessions Bewegung; DE 212020000799 U1, 03 November 2022
29. Bostan V, Bostan I, Vaculenco M (2024) Precessional gear transmission. U.S. 11,913,523 B2, 27 February 2024
30. Bostan I (2012) Planetary precessional transmissions: synthesis and generation technologies. In: Power transmissions, Proceedings of the 4th International Conference, Sinaia, Romania, 20–30 June 2012. Springer, Berlin/Heidelberg, pp 21–42
31. Bostan IA (1989) Creation of planetary-precessional gears with multi-pair gearing. Doctoral Thesis, Bauman Moscow State Technical University, Moscow, Russia, vol 1, 511p, vol 2 (App.), 236p. (In Russian)
32. Bostan IA (1989) Precessional gear transmission. Patent SU 1455094 A1, 30 January 1989. (In Russian)
33. Tronin I (1991) The wonderful world of reduction gears created by Ion Bostan. Partn J 5:20+46
34. Gromyko PN, Zholobov AA, Statsenko AA, Khatetovskiy SN, Puskov OM (2000) Planetary presession transmissions. Kinematic, power and technological aspects of their creation. Belarusian State Polytechnical Academy, Belarus, 252p. (In Russian)
35. Skoybeda AT (2001) Kinematic-cylindrical precessional gearboxes. Belarusian State Polytechnical Academy, Minsk, p 199p. (In Russian)
36. Skoybeda AT, Gromyko PN, Khatetovskiy SN (2001) Regarding to the problem of shaping the gear tooth profile of planetary precession gear wheels. NASB, Minsk, pp 69–73. (In Russian)

ARW Photos

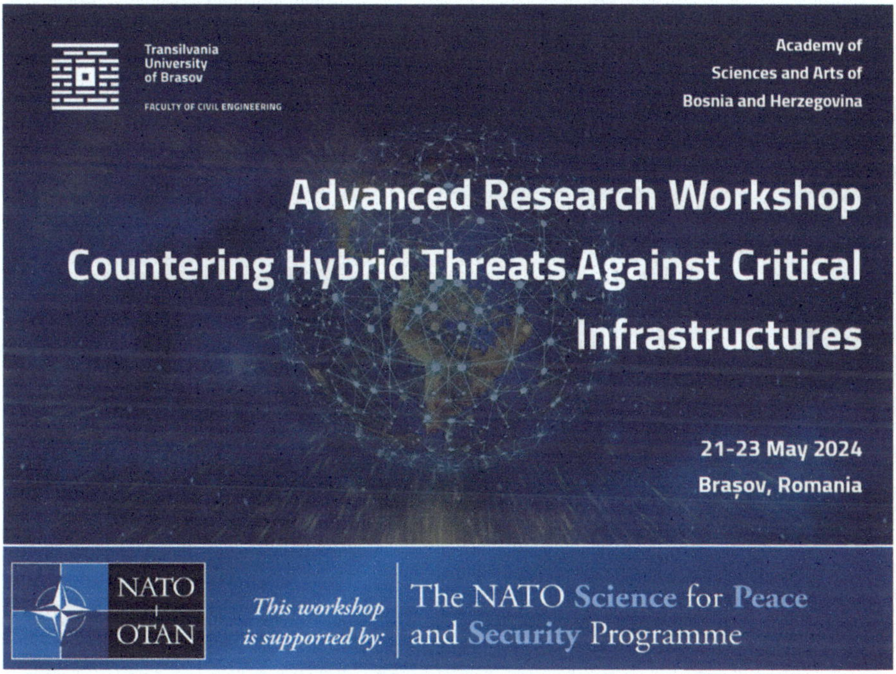

D. Radu et al. (eds.), *Countering Hybrid Threats Against Critical Infrastructures*, NATO Science for Peace and Security Series B: Physics and Biophysics, https://doi.org/10.1007/978-94-024-2304-4

WG 2 – Technology Innovations to Counter Threat Vectors
WG 3 – Risk assessment
NATO
OTAN
The NATO Science for Peace and Security Programme
Advanced Research
Workshop

ARW Format
Working Groups
WG 1 – Hybrid threats, risk assessment and vulnerabilities
WG 2 – Technology innovations to Counter Threat Vectors
WG 3 – Risk assessment
NATO
OTAN
The NATO Science for Peace and Security Programme
Advanced Research Workshop Countering Hybrid Threats Against Critical Infrastructures

Advanced Research
Workshop
Countering Hybrid
Threats Against Critical
Infrastructures

Advanced Research Workshop
Countering Hybrid Threats Against Critical
Infrastructures
The NATO Science for Peace
and Security Programme

Advanced Research Workshop
Countering Hybrid Threats Against Critical
Infrastructures

Author Index

D. Radu et al. (eds.), *Countering Hybrid Threats Against Critical Infrastructures*, NATO Science for Peace and Security Series B: Physics and Biophysics, https://doi.org/10.1007/978-94-024-2304-4

Keyword Index

D. Radu et al. (eds.), *Countering Hybrid Threats Against Critical Infrastructures*, NATO Science for Peace and Security Series B: Physics and Biophysics, https://doi.org/10.1007/978-94-024-2304-4

The manufacturer's authorised representative in the EU is Springer
Nature Customer Service Centre GmbH, Europaplatz 3, 69115 Heidelberg,
Germany. If you have any concerns regarding our products, please
contact ProductSafety@springernature.com

Printed and bound by CPI Group (UK) Ltd, Croydon, CR0 4YY
30/09/2026
02225568-0013